Astronomically Speaking

A Dictionary of Quotations on Astronomy and Physics

About the Compilers

Carl C. Gaither was born in 1944 in San Antonio, Texas. He has conducted research work for the Texas Department of Corrections and for the Louisiana Department of Corrections. Additionally he worked as an Operations Research Analyst for ten years. He received his undergraduate degree (Psychology) from the University of Hawaii and has graduate degrees from McNeese State University (Psychology), North East Louisiana University (Criminal Justice), and the University of Southwestern Louisiana (Mathematical Statistics).

Alma E. Cavazos-Gaither was born in 1955 in San Juan, Texas. She has worked in quality control, material control, and as a bilingual data collector. She is a Petty Officer First Class in the United States Navy Reserve. She received her associate degree (Telecommunications) from Central Texas College and her BA (Spanish) from Mary Hardin-Baylor University.

Together they selected and arranged quotations for the books *Statistically Speaking: A Dictionary of Quotations* (Institute of Physics Publishing, 1996), *Physically Speaking: A Dictionary of Quotations on Physics and Astronomy* (Institute of Physics Publishing, 1997), *Mathematically Speaking: A Dictionary of Quotations* (Institute of Physics Publishing, 1998), *Practically Speaking: A Dictionary of Quotations on Engineering, Technology, and Architecture* (Institute of Physics Publishing, 1998), *Medically Speaking: A Dictionary of Quotations on Dentistry, Medicine and Nursing* (Institute of Physics Publishing, 1999), *Scientifically Speaking: A Dictionary of Quotations* (Institute of Physics Publishing, 2000), *Naturally Speaking: A Dictionary of Quotations on Biology, Botany, Nature, and Zoology* (Institute of Physics Publishing, 2001) and *Chemically Speaking: A Dictionary of Quotations* (Institute of Physics Publishing, 2002).

About the Illustrator

Andrew Slocombe was born in Bristol in 1955. He spent four years of his life at Art College where he attained his Honours Degree (Graphic Design). Since then he has tried to see the funny side to everything and considers that seeing the funny side to science has tested him to the full! He would like to thank Carl and Alma for the challenge!

Astronomically Speaking
A Dictionary of Quotations on Astronomy and Physics

Selected and Arranged by

Carl C. Gaither

and

Alma E. Cavazos-Gaither

Illustrated by Andrew Slocombe

Published in 2003 by
Taylor & Francis Group
270 Madison Avenue
New York, NY 10016

Printed in the United States of America on acid-free paper
10 9 8 7 6 5 4 3 2

International Standard Book Number-10: 0-7503-0868-0 (Softcover)

Library of Congress Cataloging-in-Publication Data

Catalog record is available from the Library of Congress

Taylor & Francis Group
is the Academic Division of T&F Informa plc.

Visit the Taylor & Francis Web site at
http://www.taylorandfrancis.com

We respectfully dedicate this book to the students, faculty and administrators of the Marlin High School, Marlin, Texas

In memory of
Pearl Stevenson Gaither
(August 9, 1917–October 29, 2002)
Wife, Mother, and Friend
You will be missed

Distant scintillating star,
Shall I tell you what you are?
Nay, for I can merely know
What you were some years ago.

For, the rays that reach me here
May have left your photosphere
Ere the fight of Waterloo—
Ere the pterodactyl flew!

Many stars have passed away
Since your aether-shaking ray
On its lengthy journey sped—
So that you, perhaps, are dead!

Smashed in some tremendous war
With another mighty star—
You and all your planets just
Scattered into cosmic dust!

Strange, if you have vanished quite,
That we still behold your light,
Playing for so long a time
Some celestial pantomime!

But, supposing all is well,
What you're made of, can I tell?
Yes, 'twill be an easy task
If my spectroscope I ask.

Minchin, G.M.
Nature
14 April 1898

CONTENTS

No prefatory remarks can so clearly indicate the scope of this book as its Table of Contents, to which the reader is referred. Such a table is a skeleton of the subject matter presented, but it does not describe the flesh which this bare framework carries.

Moulton, Forest Ray
Astronomy
Preface (p. v)

CONTENTS vii

PREFACE xv

ACCURACY 1

AGE 2

ANALOGY 3

ARBITRARY 4

ASTEROID 5

ASTROLOGY 6

ASTRONAUTS 8

ASTRONOMER 10

ASTRONOMICAL 17

ASTRONOMICAL SONGS 19

ASTRONOMY 22

ASTRONOMY MNEMONICS 29

ASTROPHYSICS 32

ATOM 34

ATOMIC POWER 37

ATOMISM 38

AURORA BOREALIS 39

AXIAL TILT 42

AXIOM 43

BEAUTY 44

BIG BANG 48

BLACK HOLE 50

BOOK 52

CELESTIAL MOTION 54

CHANCE 55

CHANGE 56

CHAOS 57

COMET 58

COMMUNICATION 63

COMPULSORY 64

CONCEPT 65

CONSTELLATION 67

- ANDROMEDA 70
- ARCTURUS 70
- ARIES 71
- CANIS MAJOR 71
- CAPRICORNUS 71
- DRACO 71
- LIBRA 72
- LOST PLEIAD 72
- ORION 72
- PLEIADES 73
- SAGITTARIUS 73
- SCORPIO 73
- SOUTHERN CROSS 74
- VIRGO 74

COSMOCHEMISTRY 75

COSMOGONY 76

COSMOLOGY 77

COSMOS 79

CREATION	81
CURIOSITY	82
DARK MATTER	84
DATA	85
DEPLETION	86
DESIGN	87
DETERMINISM	88
DIFFERENTIAL EQUATIONS	89
DIMENSION	91
DISCOVERY	92
DISTANCE	96
DOGMA	97
DUST	98
ECLIPSE	100
ELECTRON	104
ELLIPSE	105
ENERGY	106
EQUATION	109
ERROR	110
ETERNITY	111
EVENT	113
EXPERIMENT	114
EXTRATERRESTRIAL LIFE	115
FACT	124
FORCE	128
FORMULA	129
FUSION	130
FUTURE	131
GALAXY	132
GEOMETRY	134
GOD	136

GRAIN	140
GRAVITATIONAL LENS	141
GRAVITY	142
HEAVENS	144
HYPOTHESIS	148
IDEA	150
IGNORANCE	152
IMAGINATION	153
IMPOSSIBLE	157
INFINITE	158
INSTRUMENT	161
INTERACTION	164
KNOWLEDGE	165
LAWS	169
LEARN	172
LIFE	173
LIGHT	175
LOGIC	177
MAGNETIC	179
MAN	180
MATTER	182
MEASUREMENT	184
MECHANICS	185
MERE	186
METAPHOR	187
METEOR	188
MILKY WAY	192
MIND	195
MODEL	198
MOLECULE	199
MOMENTUM	200

MOON 201
MOON LANDING 206
MOTION 208
MUON 210
NATURE 211
NEUTRINO 217
NIGHT 218
NOTATION 220
NOVAE 221
OBSERVATION 222
OBSERVATORY 228
OBSERVER 229
ORDER 230
OTHER WORLDS 232
PARADOX 233
PARTICLE 234
PAST 236
PATTERNS 237
PHENOMENON 239
PHILOSOPHY 241
PHOTONS 244
PHYSICIST 245
PHYSICS 248
PION 256
PLANET 257
MERCURY 260
VENUS 261
EARTH 261
MARS 264
JUPITER 265
SATURN 265
URANUS 266
NEPTUNE 266
PLUTO 267

POSITRON 268
PROBLEM 269
PROGRESS 271
PROOF 273
QUANTUM 277
QUASAR 280
QUESTION 281
RADIO ASTRONOMY 283
REALITY 285
REASON 288
RED SHIFT 290
RELATIVITY 292
RESEARCH 293
SCATTERING 294
SCIENCE 295
SCIENTIFIC 308
SCIENTIST 310
SENSES 312
SHADOW 313
SIMPLICITY 315
SKY 317
SOLAR SYSTEM 321
SPACE 323
SPACETIME 328
SPACE TRAVEL 330
SPECTRA 334
SPIN 336
SPIRAL ARMS 337
STAR 338
STUDY 354
STUPIDITY 355

SUN 356
SUNSPOT 360
SUPERNOVA 361
SYMMETRY 363
TEACH 365
TELESCOPE 366
THEORY 371
THERMODYNAMICS 376
TIME 378
TIME TRAVEL 382
TRUTH 383
UFO 385
UNCERTAINTY 386
UNDERSTAND 387
UNIVERSE 389
COSMOGENESIS 408
DYING 410
SAINT AUGUSTINE ERA 414
UNKNOWN 416
VACUUM 418
VERNAL EQUINOX 419
VERNIER 420
WAVE 421
WISDOM 422
WORK 423
WORLD 424
WRITING 426
BIBLIOGRAPHY 428
PERMISSIONS 460
SUBJECT BY AUTHOR INDEX 466
AUTHOR BY SUBJECT INDEX 521

PREFACE

The overwhelming majority of authors have very little to say. If we suppose, rather charitably, that in a typical book of fifteen chapters there are only eight passages worthy of quotation, then simple mathematics will convince us that in short order there will be no original quotations left for chapter headings. The implication is obvious...

Armand Blague
How to Write

The wisdom of the wise and the experience of ages may be preserved by quotations.

Disraeli, Isaac
Curiosities of Literature

Somewhere we had read, "*How often we rake in the litter of the printing press whilst a crown of gold and rubies is offered us in vain*". It has been our concern that a great amount of the gold and rubies, the wit and wisdom that is published will, after being read once, end up on the library shelf and be heard of no more. While we in no way claim to have found all of this wit and wisdom we have, at least, captured some of it so that it will not become lost.

People generally buy a dictionary of quotations for one of two reasons—either to help in locating the source for wording of some half-remembered saying or else to better express themselves.

Astronomically Speaking: A Dictionary of Quotations on Astronomy and Physics is the largest compilation of astronomy quotations published to date. There are many books of quotations. A large number of these books have small sections about astronomy and astronomy-related topics. *Astronomically Speaking* is a quotation book that is devoted to the topic of astronomy.

With so many well-written books of quotations on the market is it necessary that another book of quotations is needed? We and our publisher agreed that there was a need since the standard dictionaries of quotations, for whatever cause, are sorely weak in providing entries devoted to quotations on astronomy. *Astronomically Speaking* fills that need.

The understanding of the history, the accomplishments and failures, and the meanings of astronomy requires a knowledge of what has been said by the authoritative and the not so authoritative philosophers, novelists, playwrights, poets, scientists and laymen about astronomy. Because of the multidisciplinary interrelationships that exists it is virtually impossible for an individual to keep abreast of the literature outside of their own particular specialization. With this in mind, *Astronomically Speaking* assumes a particularly important role as a guide to what has been said in the past through the present about astronomy.

Astronomically Speaking was designed as an aid for the general reader who has an interest in astronomical topics as well as for the experienced scientist. The general reader with no knowledge of astronomy who reads *Astronomically Speaking* can form a pretty accurate picture of what astronomy is. Students can use the book to increase their understanding of the complexity and richness that exists within the scientific disciplines. Finally, the experienced scientist will find *Astronomically Speaking* useful as a source of quotes for use in the classroom, in papers and in presentations. We have striven to compile the book so that any reader can easily and quickly access the wit and wisdom that exists and a quick glance through the table of contents will show the variety of topics discussed.

A book of quotations, even as restricted in scope as *Astronomically Speaking* is, can never be complete. Many quotations worthy of entry have, no doubt, been omitted because we did not know of them. However, we have tried to make it fairly comprehensive and have searched far and wide for the material. If you are aware of any quotes that should be included please send them in for the second edition.

Quite a few of the quotations have been used frequently and will be recognized while others have probably not been used before. All of the quotations included in *Astronomically Speaking* were compiled with the hope that they will be found useful. The authority for each quotations has been given with the fullest possible information that we could find so as to help you pinpoint the quotation in its appropriate context or discover more quotations in the original source. When the original source could not be located we indicated where we found the quote. Sometimes, however, we only had the quote and not the source. When this happened we listed the source as unknown and included the quotation anyway so that it would not become lost in time.

How to Use This Book

A quotation for a given subject may be found by looking for that subject in the alphabetical arrangement of the book itself. This arrangement will be approved, we believe, by the reader in making it easier to locate a quotation. To illustrate, if a quotation on ideas is wanted, you will find 23 quotations listed under the heading IDEA. The arrangement of quotations in this book under each subject heading constitutes a collective composition that incorporates the sayings of a range of people.

To find all the quotations pertaining to a subject and the individuals quoted use the SUBJECT BY AUTHOR INDEX. This index will help guide you to the specific statement that is sought. A brief extract of each quotation is included in this index.

It will be admitted that at times there are obvious conveniences in an index under author's names. If you recall the name appearing in the attribution or if you wish to read all of an individual author's contributions that are included in this book then you will want to use the AUTHOR BY SUBJECT INDEX. Here the authors are listed alphabetically along with their quotations. The birth and death dates are provided for the authors whenever we could determine them.

Thanks

It is never superfluous to say thanks where thanks are due. First, we want to thank Jim Revill and Simon Laurenson, of IOP Publishing, who have assisted us so very much with our books. Next, we have been very fortunate to have our books reviewed by some people who gave us page-by-page comments. We cannot begin to express our gratitude for their time and energy. We also thank the following libraries for allowing us to use their resources: The Jesse H. Jones Library and the Moody Memorial Library, Baylor University; the main library of the University of Mary-Hardin Baylor; the main library of the Central Texas College; the Undergraduate Library, the Engineering Library, the Law Library, the Physics–Math–Astronomy Library and the Humanities Research Center all of the University of Texas at Austin. Again, we wish to thank all of the librarians of the Perry- Castañeda Library for putting up with us when we were checking out the hundreds of books.

A great amount of work goes into the preparation of any book. When the book is finished there is then time for the editors and authors to enjoy what they have written. It is our hope that this book stimulated your imagination and interests in matters about astronomy, physics or mathematics and this hope has been eloquently expressed by Helen Hill:

If what we have within our book
Can to the reader pleasure lend,
We have accomplished what we wished,
Our means have gained our end.

In Llewellyn Nathaniel Edwards
A Record of History and Evolution of Early American Bridges (p. xii)

Carl Gaither
Alma Cavazos-Gaither
6 February 2003

ACCURACY

Mitchell, Maria

The training of a girl fits her for delicate work. The touch of her fingers upon the delicate screws of an astronomical instrument might become wonderfully accurate in results; a woman's eyes are trained to nicety of color. The eye that directs a needle in the delicate meshes of embroidery will equally well bisect a star with the spider web of the micrometer. Routine observations, too, dull as they are, are less dull than the endless repetition of the same pattern in crochet-work.

In Phebe Mitchell Kendall
Maria Mitchell: Life, Letters, and Journals
Chapter XI (pp. 237–8)

AGE

Dirac, P.A.M.
Age is of course a fever chill
That every physicist must fear.
He's better dead than living still
When once he's past his thirtieth year.

In Leon Lederman
The God Particle
Chapter 5 (p. 168)

Eliot, George
The young ones have always a claim to the old to help them forward.

Middlemarch
Chapter 56 (p. 537)

Hardy, G. H.
No mathematician should ever allow himself to forget that mathematics, more than any other art or science, is a young man's game.

A Mathematician's Apology
Section 4 (p. 70)

Yudowitch, K.L.
The knowledge that so many important discoveries in physics have been made by young men comes as a surprise to most students—and a pleasant surprise. Students never fail to look with new interest upon work done by a man at very nearly their own age. Physics is revitalized in the minds of the students by the knowledge that it is a field for *young* men—men like themselves.

American Journal of Physics
Young Men in Physics (p. 191)
Volume 15, Number 2, March–April 1947

ANALOGY

Bernstein, Jeremy

It is probably no exaggeration to say that all of theoretical physics proceeds by analogy.

Elementary Particles and their Currents
Philosophical Preface (p. vii)

Campbell, Norman Robert

. . . analogies are not "aids" to the establishment of theories; they are an utterly essential part of theories, without which theories would be completely valueless and unworthy of the name. It is often suggested that the analogy leads to the formulation of the theory, but that once the theory is formulated the analogy has served its purpose and may be removed or forgotten. Such a suggestion is absolutely false and perniciously misleading.

Physics, The Elements
Chapter VI (p. 129)

Heinlein, Robert

Analogy is even slipperier than logic.

Stranger in a Strange Land
Part III
Chapter XXIV (p. 318)

Melville, Herman

O Nature, and O soul of man! How far beyond all utterance are your linked analogies! Not the smallest atom stirs or lives on matter, but has its cunning duplicate in mind.

Moby Dick
Chapter 70

ARBITRARY

Poincaré, Henri

Are the law of acceleration, the rule of the composition of forces only arbitrary conventions? Conventions, yes; arbitrary, no; they would be so if we lost sight of the experiments which led the creators of the science to adopt them, and which, imperfect as they may be, suffice to justify them. It is well that from time to time our attention is carried back to the experimental origin of these conventions.

The Foundations of Science
Science and Hypothesis
Part I
Chapter VI (p. 106)

ASTEROID

Asphaung, Erik

Neither rocks nor planets, they are something of Earth and Heaven.

Scientific American
The Small Planets (p. 55)
Volume 282, Number 5, May 2000

ASTROLOGY

Byron, George Gordon
Ye stars! Which are the poetry of Heaven,
If in your bright leaves we would read the fate
Of men and empires,—'tis to be forgiven,
That in our aspirations to be great,
Our destinies o'erleap their mortal state,
And claim a kindred with you.

The Complete Poetical Works of Byron
Child Harold
Canto iii
Stanza 88

Durant, Will
...astrology antedated—and perhaps will survive—astronomy; simple souls are more interested in telling futures than in telling time.

The Story of Civilization
Part I
Our Oriental Heritage
Introduction
Chapter I (p. 80)

Emerson, Ralph Waldo
Astronomy to the selfish becomes astrology;...

Essays and Lectures
Essays
Second Series
Nature (p. 546)

Johnson, Severance
Astronomy
Is for the mind of gods, astrology

For simpletons.

The Dictator and the Devil

Shakespeare, William

It is the stars,
The stars above us, govern our condition.

King Lear
Act IV, scene iii, L. 34–5

The fault, dear Brutus, is not in the stars, but in ourselves.

Julius Caesar
Act I, scene ii, L. 134

ASTRONAUTS

Armstrong, Neil
That's one small step for man; one giant leap for mankind.

Apollo 11

Apollo 11
Here men from the planet Earth first set foot upon the Moon July 1969, A.D. WE CAME IN PEACE FOR ALL MANKIND.

Plaque left on moon

Conrad, Pete
Whoopee! Man, that may have been a small one for Neil, but it's a long one for me.

Apollo 12

Swigert, Jack
Okay, Houston; we've had a problem.

Apollo 13

Shepherd, Alan
It's been a long way, but we're here.

Apollo 14

Scott, Dave
Man must explore. And this is exploration at its greatest.

Apollo 15

Young, John
There you are: Mysterious and Unknown Descartes. Highland plains. Apollo 16 is gonna change your image. I'm sure glad they got ol' Brer Rabbit, here, back in the briar patch where he belongs.

Apollo 16

Okay, Houston; we've had a problem.
Jack Swigert – (See p. 8)

Cernan, Gene
I'm on the footpad. And, Houston, as I step off at the surface at Taurus-Littrow, I'd like to dedicate the first step of Apollo 17 to all those who made it possible.

Apollo 17

Bob, this is Gene, and I'm on the surface; and, as I take man's last step from the surface, back home for some time to come—but we believe not too long into the future—I'd like to just (say) what I believe history will record. That America's challenge of today has forged man's destiny of tomorrow. And, as we leave the Moon at Taurus-Littrow, we leave as we came and, God willing, as we shall return, with peace and hope for all mankind. Godspeed the crew of Apollo 17.

Apollo 17

ASTRONOMER

Calder, Nigel

When astronomers express dissatisfaction with both the Big Bang and the Steady State concepts of the universe, they are in trouble, because it is hard to imagine radical alternatives.

The Violent Universe
Chapter III (pp. 121–2)

Cunningham, Clifford J.

Today's astronomers live and die by journals and conferences.

Sky and Telescope
The Baron and His Celestial Police (p. 271)
Volume 75, Number 3, March 1988

Donne, John

If then th' Astronomers, whereas they spie
A new-found Starre, their Opticks magnifie,
How brave are those, who with their Engine, can
Bring man to heaven, and heaven againe to man?

In Charles M. Coffin (ed.)
The Complete Poetry and Selected Prose of John Donne
To Mr Tilman After He Had Taken Orders

Friedman, Herbert

To the astronomer of today, probing ever deeper with mind and telescope, the universe is more than beautiful: it is amazing, violent, and endlessly mysterious. The revelations of recent research have been so dramatic and so extreme as to leave both scientists and laymen bewildered. Modern astronomy deals with the birth and death of stars; with exotic matter and fantastic energies; with near-infinities of space and time; with creation, evolution, and the ultimate destiny of the universe. As the sum of knowledge grows, the astronomer continues to seek answers to man's most profound questions: what is the grand design of the universe? How was it created? How did we get here? What are we? Are we alone?

The Amazing Universe
Chapter 1 (p. 10)

Gibran, Kahlil
The astronomer may speak to you of his understanding of space, but he cannot give you his understanding.

The Prophet
On Teaching (p. 56)

Grondal, Florence Armstrong
How thrilling to read of great hunts for treasure! Yet the pirates who dug their spades into the earthy loam never cached such jewels as are hidden along the dark slopes of the sky. Armed with a chart of the heavens, the fledgling astronomer prods about in the depths of the gloom, shovels the dark with the aid of his telescope, and discovers, even more surely than the pirate his chest, some wonderful treasure. Sometimes the find is a star-like diamond, a twinkling emerald, a fire-filled ruby or a cluster star gems of colorful hues, but it may be, too, a profusion of riches, heaped in a magnificence that leaves one breathless.

The Music of the Spheres
Chapter I (p. 3)

Halley, Edmond
We therefore recommend again and again, to the curious investigators of the stars to whom, when our lives are over, these observations are entrusted, that they, mindful of our advice, apply themselves to the understanding of these observations vigorously. And for them we desire and pray for all good luck, especially that they not be deprived of this coveted spectacle by the unfortunate obscuration of cloudy heavens, and that the immensities of the celestial spheres, compelled to more precise boundaries, may at last yield to their glory and eternal fame.

Philosophical Transactions of the Royal Society
A Unique Method by Which the Parallax of the Sun,
or its Distance from the Earth, may be Securely Determined
by Means of Observing Venus Against the Sun (p. 460)
Number 348, April, May, June, 1716

Herbert, George
The fleet Astronomer can bore,
And thred the spheres with his quick-piercing mind:
He views their stations, walks from dore to dore,
Surveys, as if he had design'd

To make a purchase there: hee sees their dances,
And knoweth long before,
Both their full ey'd aspects, & secret glances.

The Temple
The Church
Vanity, L. 1–7 (p. 126)

Hoyle, Fred

The astronomer seems at first sight to be the most helpless of all scientists. He cannot experiment with the Universe. It is a significant matter of nomenclature that whereas we speak of experimental work in other sciences we speak of observational work in astronomy.

Frontiers of Astronomy
Prologue (p. xv)

Jeans, Sir James

The task of the observational astronomer is to survey and explore the universe, and to describe and classify the various types of objects which it is constituted, discovering what law and order he may in their observed arrangement and behavior. But only the dullest of human minds can rest content with a mere catalogue of observed facts; an alert mind asks always for the why and the wherefore.

Astronomy and Cosmogony
Chapter I (p. 1)

Jeffers, Robinson

The learned astronomer
Analyzing the light of most remote star-swirls
Has found them—or a trick of distance deludes his prism—
All at incredible speeds fleeing outward from ours.

The Selected Poetry of Robinson Jeffers
Margrave

Jones, Sir Harold Spencer

The task of the astronomer is to learn what he can about the universe as he finds it. To endeavor to understand the purpose behind it and to explain why the universe is built as it is, rather than on some different pattern which might have accorded better with our expectations, is a more difficult task; for this the astronomer is no better qualified than anybody else.

Life On Other Worlds
Chapter X (p. 253)

Keats, John

Then I felt like some watcher of the skies
When a new planet swims into his keen.

The Complete Poetical Works of Keats
On First Looking Into Chapman's Homer

Kühnert, Franz

Probably another reason why many Europeans consider the Chinese such barbarians is on account of the support they give to their Astronomers—people regarded by our cultivated Western mortals as completely useless. Yet there they rank with Heads of Departments and Secretaries of State. What frightful barbarism!

In Joseph Needham
Science and Civilisation in China
Volume 3
Science and Civilisation in China (p. iii)

Mackay, Charles

Upon thy lofty tower,
O lonely Sage,
Reading at midnight hour
Heaven's awful page!

The Collected Songs of Charles Mackay
The Astronomer

Milton, John

Hereafter, when they come to model Heav'n
And calculate the Starrs, how they will wield
The mightie frame, how build, unbuild, contrive,
To save appearances, how grid the Sphere
With Centric and Eccentric scribl'd o'er,
Cycle and Epicycle, Orb in Orb.

Paradise Lost
Book VIII, L. 78

Mitchell, Maria

I cannot expect to make astronomers, but I do expect that you will invigorate your minds by the effort at healthy modes of thinking.

In Phebe Mitchell Kendall
Maria Mitchell: Life, Letters, and Journals
Chapter VII (p. 138)

The Astronomer breaks up the starlight just as the geologist breaks up the rock with his hammer, and with similar results, he finds copper, sodium and other elements in sun and stars... If you look at the beautiful ribbon of colors which a ray of sunlight gives when passed through a prism, you see that it is crossed by dark bands, sometimes single, sometimes crowded close together—each of these is a black-lettered message from the sun.

In Helen Wright
Sweeper in the Sky
Chapter 10 (pp. 188–9)

Osiander, Andrew

...it is the job of the astronomer to use painstaking and skilled observation in gathering together the history of the celestial movements, and then—since he cannot by any line of reasoning reach the true causes of these movements—to think up or construct whatever causes or hypotheses he pleases such that, by the assumptions of these causes, those same movements can be calculated from the principles of geometry for the past and for the future too...

In Nicholas Copernicus
On the Revolutions of the Heavenly Spheres
Introduction
To the Reader Concerning the Hypothesis of this Work (p. 505)

Rees, Martin

Everything astronomers observe turns out to be a small and atypical fraction of what exists.

Before the Beginning
Chapter 6 (p. 103)

Sayers, Dorothy L.
Eustace, R.

The biologist can push it back to the original protist, and the chemist can push it back to the crystal, but none of them touch the real question of why or how the thing began at all. The astronomer goes back untold million of years and ends in gas and emptiness, and then the mathematician sweeps the whole cosmos into unreality and leaves one with mind as the only thing of which we have any immediate apprehension. *Cogito ergo sum, ergo omnia esse videntur.* All this bother, and we are no further than Descartes. Have you noticed that the astronomers and mathematicians are much the most cheerful people of the lot? I suppose that perpetually contemplating things on so vast a scale makes them feel either that it doesn't matter a hoot anyway, or that anything so large and elaborate must have some sense in it somewhere.

The Documents in the Case
Letter 22, John Munting to Elizabeth Drake (p. 70)

Shakespeare, William

These earthly godfathers of heaven's lights,
That give a name to every fixed star
Have no more profit of their shining nights
Than those that walk, and wot not what they are.

Love's Labor's Lost
Act I, scene i, L. 86–9

Stoll, Clifford

The astronomer's rule of thumb: if you don't write it down, it didn't happen.

The Cuckoo's Egg
Chapter 5 (p. 28)

Thompson, Francis

Starry amorist, starward gone,
Thou art—what thou didst gaze upon!
Passed through thy golden garden's bars,
Thou seest the Gardner of the Stars.

Complete Poetical Works of Francis Thompson
A Dead Astronomer
Stanza 1

Twain, Mark

I do not see how astronomers can help feeling exquisitely insignificant, for every new page of the Book of the Heavens they open reveals to them more & more that the world we are so proud of is to the universe of careering globes as is one mosquito to the winged & hoofed flocks & herds that darken the air & populate the plains & forests of all the earth. If you killed the mosquito, would it be missed? Verily, What is Man, that he should be considered of God?

The Mark Twain Papers
Mark Twain's Letters
Volume 4
1870–1871
Letter to Olivia L. Langdon (p. 12)
8 January 1870

For three hundred years now, the Christian astronomer has known that his Deity didn't make the stars in those tremendous six days; but the Christian astronomer doesn't enlarge upon that detail. Neither does the priest.

Letters from the Earth
Letter III (p. 16)

Unknown

Astronomers seem to be able to predict everything more and more precisely—except the end of the century.

Source unknown

Walcott, Derek

I try to forget what happiness was,
and when that didn't work, I study the stars.

The Star-Apple Kingdom
The Schooner Flight
Section 11

ASTRONOMICAL

Boethius

Think then thus upon it, and see that it is but a slight thing of no weight. As you have learnt from astronomers' shewing, the whole circumference of the earth is but as a point compared with the size of the heavens. That is, if you compare the earth with the circle of the universe, it must be reckoned as of no size at all.

On the Consolation of Philosophy
Book II, Prose VII (pp. 48–9)

Cook, Joseph

Bye Baby Bunting,
Father's gone star-hunting;
Mother's at the telescope
Casting baby's horoscope.
Bye Baby Buntoid,
Father's found an asteroid;
Mother takes by calculation
The angle of its inclination.

In Sara and John E. Brewton and John Brewton Blackburn
Of Quarks, Quasars, and Other Quirks
Boston Nursery Rhymes
Rhyme for Astronomical Baby (p. 40)

Hoyle, Fred

No literary genius could have harvested a story one-hundredth part as fantastic as the sober facts that have been unearthed by astronomical science.

The Nature of the Universe
Chapter 7 (p. 133)

Paracelsus

All this you should know exists in man and realize that the firmament is within man, the firmament with its great movements of bodily planets and stars which result in exaltations, conjunctions, oppositions and the like, as you call these phenomena as you understand them. Everything which astronomical theory has searched deeply and gravely by aspects, astronomical tables and so forth,—this self-same knowledge should be a lesson and teaching to you concerning the bodily firmament. For, none among you who is devoid of astronomical knowledge may be filled with medical knowledge.

In Allen G. Debus
The French Paracelsians
Chapter 1 (p. 9)

ASTRONOMICAL SONGS

Krisciunas, Kevin
Give me a supernova,
A bright one in the northern sky.
Something like minus seventh magnitude.
I beg you don't be shy.
Zap me with your neutrinos.
Produce a little lead and gold.
Give me a bright new star at night
Before I'm wrinkled, grey, and old.

Give Me a Supernova
Sung to "Give My Regards to Broadway"

Jedicke, Peter
Bigellow, David
The stars go nova one by one. KA-BOOM, KA-BOOM!
Nucleosynthesis is done, KA-BOOM, KA-BOOM!
The supernovas dissipate
What fusion energy helped create
And the stars go nova in the galaxy.

The heavy elements are born. KA-BOOM, KA-BOOM!
And from the stellar cores are torn. KA-BOOM, KA-BOOM!
Shells of gas are strewn through space
Distributing matter all over the place
And the spiral arms are littered with debris.

Do galaxies in space collide? KA-BOOM, KA-BOOM!
It seems they might, 'cause they're so wide. KA-BOOM, KA-BOOM!
Computer simulations show,
Five hundred million years or so,
Is what it takes for galaxies to merge.

As the years go by the remnants spread. KA-BOOM, KA-BOOM!

But the universe is far from dead. KA-BOOM, KA-BOOM!
To eliminate the tedium
Of the interstellar medium
Come the molecules that make up you and me.

The Stars Go Nova
Sung to "When Johnny Comes Marching Home"

Unknown
Whoe'er would search the starry sky,
Its secrets to divine, sir,
Should take his glass—I mean, should try
A glass or two of wine, sir!
True virtue lies in golden mean,
And man must wet his clay, sir;
Join these two maxims, and 'tis seen
He should drink his bottle a day, sir!

Old Archimedes, reverend sage!
By trump of fame renowned, sir,
Deep problems solved in every page,
And the sphere's curved surface found, sir:
Himself he would have far outshone,
And borne a wider sway, sir,
Had he our modern secret known,
And drank a bottle a day, sir!

When Ptolemy, now long ago,
Believed the Earth stood still, sir,
He never would have blundered so,
Had he but drunk his fill, sir:
He'd then have felt it circulate,
And would have learnt to say, sir,
The true way to investigate
Is to drink your bottle a day, sir!

Copernicus, that learned weight,
The glory of his nation,
With draughts of wine refreshed his sight,
And saw the Earth's rotation
Each planet then its orb described,
The Moon got under way, sir;
These truths from nature he imbibed
For he drank his bottle a day, sir!

The noble Tycho placed the stars,
Each in its due location;

He lost his nose by spite of Mars,
But that was no privation:
Had he but lost his mouth, I grant
He would have felt dismay, sir,
Bless you! he knew what he should want
To drink his bottle a day, sir!

Cold water makes no lucky hits;
On mysteries the head runs:
Small drink let Kepler time his wits
On the regular polyhedrons:
He took to wine, and it changed the chime,
His genius swept away, sir,
Through area varying as the time
At the rate of a bottle a day, sir!

Poor Galileo, forced to rat
Before the Inquisition,
E pur si muove was the pat
He gave them in addition:
He meant, whate'er you think you prove,
The Earth must go its way, sirs;
Spite of your teeth I'll make it move,
For I'll drink my bottle a day, sirs!

Great Newton, who was never beat
Whatever fools may think, sir;
Though sometimes he forgot to eat,
He never forgot to drink, sir:
Descartes took nought but lemonade,
To conquer him was play, sir;
The first advance that Newton made
Was to drink his bottle a day, sir!

D'Alembert, Euler, and Clairaut,
Though they increased our store, sir,
Much further had been seen to go
Had they tippled a little more, sir!
Lagrange gets mellow with Laplace,
And both are wont to say, sir,
The philosopher who's not an ass
Will drink his bottle a day, sir!

In Augustus De Morgan
A Budget of Paradoxes
Volume I
The Astronomer's Drinking Song (p. 382)

ASTRONOMY

Bennett, Arnold

He knew not how to look at a landscape nor at a sky. Of plants and trees he was as exquisitely ignorant as of astronomy. It had not occurred to him to wonder why the days are longer in summer, and he vaguely supposed that the cold of winter was due to an increased distance of the earth from the sun. Still, he had learnt that Saturn had a ring and sometimes he unconsciously looked for it in the firmament, as for a tea-tray.

Clayhanger
Book I
Chapter II, Section III (p. 14)

Bichat, Xavier

To say that physiology is made up of the physics of animals, is to give a very inaccurate idea of it; as well might we say that astronomy is the physiology of the stars.

Physiological Researches on Life and Death
Chapter VII
Section I (p. 81)

Bronowski, Jacob

Astronomy is not the apex of science or of invention. But it is a test of the cast of temperament and mind that underlies a culture.

The Ascent of Man
Chapter 6 (p. 190)

Burnham, Robert Jr

No one can date that remote epoch when astronomy "began"—we can say only that the fascination of the heaven is as old as man's ability to think; as ancient as his capacity to wonder and to dream. And in company with most of the special enchantments of human life, the unique appeal of astronomy is incommunicable; easily understood through direct experience, but not to be precisely defined or explained. Nor should any explanation be thought necessary. The area of astronomy is both intellectual and aesthetic; it combines the thrill of exploration and discovery, the fun of sight-seeing, and the sheer pleasure of firsthand acquaintance with incredibly wonderful and beautiful things.

Burnham's Celestial Handbook
Chapter 1 (p. 5)

Chargaff, Erwin
Thus, astronomy was probably the first exact science, practiced long before the concept of science as such had been formulated. (Mathematics may have been earlier, but I do not consider it a natural science: the mother of many kings is not necessarily a queen.)

Serious Questions
Nature (p. 153)

Clerke, Agnes M.
[Astronomy] is a science of hairbreadths and fractions of a second. It exists only by the rigid enforcement of arduous accuracy and unwearying diligence. Whatever secrets the universe still has in store for man will only be communicated on these terms.

A Popular History of Astronomy during the Nineteenth Century
Part I
Chapter VI (p. 123)

Conrad, Joseph
The demonstration must be against learning-science. But not every science will do. The attack must have all the shocking senselessness of gratuitous blasphemy. Since bombs are your means of expression, it would be really telling if one could throw a bomb into pure mathematics. But that is impossible... What do you think of having a go at Astronomy?

The Secret Agent: A Simple Tale
Chapter II (p. 38)

Emerson, Ralph Waldo
It is noticed, that the consideration of the great periods and spaces of astronomy induces a dignity of mind, and an indifference to death.

Essays and Lectures
The Conduct of Life
Culture (p. 1030)

Emerson, William
Astronomy is the science which treats of the motions of the heavenly bodies, and all the phenomena arising therefrom.

A System of Astronomy
Astronomy (p. 1)

Huxley, Julian
I turn the handle and the story starts:
Reel after reel is all astronomy,
Till life, enkindled in a niche of sky,
Leaps on the stage to play a million parts.

The Captive Shrew
Evolution: At the Mind's Cinema

Keill, John
Among all the mathematical sciences which have been continually improved, and are daily improving in the world, the first place has, as it were, by general consent, been always given to Astronomy.

An Introduction to the True Astronomy
To His Grace Jones Duke of Chandos

Laplace, Pierre Simon
Astronomy, from the dignity of the subject, and the perfection of its theories, is the most beautiful monument of the human mind—the noblest record of its intelligence. Seduced by the illusions of the senses, and of self-love, man considered himself, for a long time, as the centre about which the celestial bodies revolved, and his pride was justly punished by the vain terrors they inspired. The labour of many ages has at length withdrawn the veil which covered the system. And man now appears, upon a small planet, almost imperceptible in the vast extent of the solar system, itself only an insensible point in the immensity of space. The sublime results to which this discovery has led, may console him for the limited place assigned to the Earth, by showing him his proper magnitude, in the extreme smallness of the base which he made use of to measure the heavens.

The System of the World
Volume II
Book V, Chapter VI (p. 342)

Long, Roger
Astronomy is a science which, in all ages and countries flourishing in arts and politeness, has engaged the attention of the curious: it has not only employed the pens of the most eloquent orators and embellished the writings of poets of the most elevated genius; but has also been cultivated by the greatest princes, the ablest statesmen, and the wisest philosophers...

Astronomy, in Five Books
Volume I
Preface (p. iii)

Mitchell, Maria

But star-gazing is not science. The entrance to astronomy is through mathematics.

In Phebe Mitchell Kendall
Maria Mitchell: Life, Letters, and Journals
Chapter IX (pp. 184–5)

Murdin, Paul

The aims of astronomy are nothing less than to search for the origins of the Universe and of its constituent stars and galaxies.

In Derek McNally
The Vanishing Universe
The Aims of Astronomy in Science and the Humanities:
Why Astronomy Must Be Protected (p. 16)

Newcomb, Simon

[astronomy] seems to have the strongest hold on minds which are not intimately acquainted with its work. The view taken by such minds is not distracted by the technical details which trouble the investigator, and its great outlines are seen through an atmosphere of sentiment, which softens out the algebraic formulae with which the astronomer is concerned into those magnificent conceptions of creation which are the delight of all minds, trained or untrained.

Harper's Magazine
February 1885

Penrose, Roger

Yet nature does not always prefer conventional explanations, least of all in astronomy.

Scientific American
Black Holes (p. 46)
Volume 226, Number 5, May 1972

Penzias, Arno

Astronomy leads us to a unique event, a universe which was created out of nothing, one with the very delicate balance needed to provide exactly the conditions required to permit life, and one which has an underlying (one might say 'supernatural') plan.

In H. Margenau and R.A. Varghese
Cosmos, Bios, and Theos
Chapter 16 (p. 83)

Plato

...astronomy tells us about the motions of the stars and sun and moon, and their relative swiftness.

Gorgias
Section 451

...in astronomy, as in geometry, we should employ problems, and let the heavens alone if we would approach the subject in the right way and so make the natural gift of reason to be of any real use.

The Republic
Book VII, 530c

Sagan, Carl

It has been said that astronomy is a humbling and character building experience. There is perhaps no better demonstration of the folly of human conceits than this distant image of our tiny world. To me, it underscores our responsibility to deal more kindly with one another, and to preserve and cherish the pale blue dot, the only home we've ever known.

Pale Blue Dot
Chapter 1 (p. 9)

Shapley, Harlow

...the most interesting feature of this science astronomy (and of all science) is our eager ignorance.

Scientific American
Astronomy (pp. 25–6)
Volume 183, Number 3, September 1950

Sherrod, P. Clay

Above us, the sparkling stars of the night skies stretch out like thousands of diamonds suspended on the curtain of space. Unfolding through the beauty and the mysteries of this seemingly endless expanse are patterns and answers familiar to those willing to study them...There is an affinity for the eternity of space experienced by all mankind, a kind of motherhood in the stars to those who study space.

A Complete Manual of Amateur Astronomy
Introduction (p. 1)

Sillman, Benjamin

Astronomy is not without reason. Regarded, by mankind, as the sublimest of the natural sciences. Its objects, so frequently visible, and therefore familiar, being always remote and inaccessible, do not lose their dignity.

Elements of Chemistry
Volume I
Introduction (p. 11)

Struve, Otto

Astronomy has had three great revolutions in the past four hundred years: The first was the Copernican revolution that removed the earth from the center of the solar system and placed it 150 million kilometers away from it; the second occurred between 1920 and 1930 when, as a result of the work of H. Shapley and R.J. Trumpler, we realized that the solar system is not at the center of the Milky Way but about 30,000 light years away from it, in a relatively dim spiral arm; the third is occurring now, and, whether we want it or not, we must take part in it. This is the revolution embodied in the question: Are we alone in the universe?

The Universe
Chapter VI (p. 157)

Twain, Mark

I love to revel in philosophical matters—especially astronomy. I study astronomy more than any other foolishness there is. I am a perfect slave to it. I am at it all the time. I have got more smoked glass than clothes. I am as familiar with the stars as the comets are. I know all the facts and figures and I have all the knowledge there is concerning them. I yelp astronomy like a sun-dog, and paw the constellations like Ursa Major.

Letter from Mark Twain
San Francisco Alta California
August 1, 1869

Unknown

It may indeed appear extraordinary that no mention should yet have been made of the great desiderata of astronomy,—those questions which have exercised the curiosity and employed the time and attention of astronomers ever since the science has assumed its present character—such as the parallax of the fixed stars, their proper motion, the motion or rest of our own system, and its connection with the rest of the universe. But these and many other points are too obviously suggested by their importance to need any distinction which this society can bestow:

the applause of the human race attends his labours; and no additional stimulus can be offered to those by which he is impelled.

Memoirs of the Astronomical Society of London
Report to the First Annual General Meeting
9 February 1821 (pp. 24–5)
Volume I, 1822–25

Virgil
Give me the ways of wandering stars to know,
The depths of heaven above, and earth below;
Teach me the various labours of the moon,
And whence proceed the eclipses of the sun.

Georgics
Book 2

Whitehead, Hal
Studying the behavior of large whales has been likened to astronomy. The observer glimpses his subjects, often at long range; he cannot do experiments, and he must continually try to infer from data that are usually inadequate.

Scientific American
Why Whales Leap (p. 86)
Volume 252, Number 3, March 1985

ASTRONOMY MNEMONICS

The first letter of each word gives you the first letter of the planets, in order: Mercury, Venus, Earth, Mars, Jupiter, Saturn, Uranus, Neptune, Pluto.

Matilda Visits Every Monday, Just Stays Until Noon, Period.

My Very Easy Method—Just Set Up Nine Planets.

My Very Educated Mother Just Served Us Nine Pizzas.

Mary's Violet Eyes Make John Stay Up Nights Praying.

My Very Energetic Mother Just Sat Upon (the) North Pole.

My Very Educated Mother Just Sent Us Nine Pickles.

My Very Educated Mother Just Served Us Nine Peas.

My Very Excellent Memory Just Served Up Nine Planets.

Source unknown

Galilean Satellites of Jupiter (Io, Europa, Ganymede, Callisto):

I Eat Good Cake.

I Expect God Cries.

Source unknown

Saturnian Satellites (Mimas, Enceladus, Tethys, Dione, Rhea, Titan, Hyperion, Iapetus, Phoebe):

MET DR THIP.

Source unknown

Uranian Satellites (Miranda, Ariel, Umbriel, Titania, Oberon):

MAUTO.

Source unknown

Harvard Spectral Classification Scheme

The Traditional

Oh Be A Fine [Guy/Gal/Girl] Kiss Me (Right Now [Smack/Sweetheart]).

Oh Begone, A Friend's Gonna Kiss Me (Right Now Smack).

Only Boys Accepting Feminism Get Kissed Meaningfully.

Political

Official Bureaucrats At Federal Government Kill Many Researchers' National Support

Only Big Astronomy Federal Grants Keep Money. Research Needs Support!

The Joys of College

Oh Boy! Another Failing Grade Keeps Me Reconsidering Night School.

Oh Bother, Astronomers Frequently Give Killer Midterms.

Oh Bother, Another F's Gonna Kill Me.

Old Boring Astronomers Find Great Kicks Mustily Regaling Napping Students.

Obese Balding Astronomer Found Guilty Killing Many Reluctant Nonscience Students.

Appeals to Physics and Astronomy

Observationalists Basically Are Fine Generous Kind Men (Really Not Sexist)

Oh Backward Astronomer, Forget Geocentricity;
Kepler's Motions Reveal Nature's Simplicity.

Organs Blaring and Fugues Galore,
Kepler's Music Reads Nature's Score.

Out Beyond Andromeda, Fiery Gases Kindle Many Radiant New Stars.

Orbs, Bright And Fair, Generate Kinder Memories: Revolving Night-time Skies.

Only Bright Astral Fires Going Kaput Make Real Neutron Stars.

For the old film buffs

Oven-Baked Ants, when Fried Gently, and Kept Moist, Retain Natural Succulence.

Ecology
When Obstreperous Beasts Approach, Fragrant Geraniums Knowingly May Receive Night's Stigmata.

Old Bottles And Filthy Garbage Kill Many Rare Natural Species.

Oregon Beavers Attack Famous Gardens, Killing Many, Rangers Now Shooting.

One Bug Ate Five Green Killer Moths.

OBAFGKMRNS
Star classes on a stellar evolution diagram ranging from hot, early type stars (O) to cool, late type stars (M) class. (Three additional classes (RNS) are no longer used.)

Oh, Be A Fine Girl, Kiss Me Right Now Sweetheart.

Oh Be A Fine Girl Kiss Me Right Now Smack (the classic!).

Oh Big And Ferocious Gorilla, Kill My Roommate Next Saturday.

Oh Boy A F Grade Kills Me.

Only Bored Astronomers Find Gratification Knowing Mnemonics.

Oh Bloody Astronomy! F Grades Kill Me.

Odd Ball Astronomers Find Generally Kooky Mnemonics Really Nifty Stuff.

Source unknown

The order of the phases of the moon (and thus, whether the current appearance of the moon indicates waxing or waning) can be remembered with the phrase "What's up, DOC?": D indicates a (waxing) half-moon with the curve on the right, O indicates a full moon, and C indicates a (waning) crescent moon with the curve on the left, heading toward a new moon.

Source unknown

To remember the constellations of the zodiac (Aries, Taurus, Gemini, Cancer, Leo, Virgo, Libra, Scorpio, Sagittarius, Capricorn, Aquarius, Pisces):

A Tense Gray Cat Lay Very Low, Sneaking Slowly, Contemplating A Pounce.

Source unknown

ASTROPHYSICS

Douglas, Vibert

On the uplifting wings of imagination the astrophysicist roams the universe from atom to atom, from star to star, from star to atom, from atom to star.

Atlantic Monthly
From Atoms to Stars (p. 165)
Volume 144, Number 2, August 1929

Greenstein, J.L.

Theory may often delay understanding of new phenomena observed with new technology unless theorists are quite open-minded as to what types of physical laws may need to be applied: conservatism is unsafe... In astrophysics, historically, theories have only seldom had predictive usefulness as guides to experimenters.

In W.T. Sullivan III
The Early Years of Radio Astronomy
Optical and Radio Astronomers in the Early Years (p. 77)

Luminet, Jean-Pierre

Astrophysicists have the formidable privilege of having the largest view of the Universe; particle detectors and large telescopes are today used to study distant stars, and throughout space and time, from the infinitely large to the infinitely small, the Universe never ceases to surprise us by revealing its structures little by little.

Black Holes
Forward to the French Edition (p. xv)

Spenser, Edmund

For who so list into the heavens looke,
And search the courses of the rowling spheares,
Shall find that from the point, where first they tooke
Their setting forth, in these few thousand yeares

They all are wandred much; that plaine appears.

The Complete Works in Verse and Prose of Edmund Spenser
Volume 8
The Faerie Queene
The Fifth Book
Introduction

ATOM

Davies, Paul
Measure for measure, we are to an atom, what a star is to us.

Superforce
Chapter 1 (p. 18)

Eddington, Sir Arthur
The atom is as porous as the solar system.

The Nature of the Physical World
Chapter I (p. 1)

The physical atom is, like everything else in physics, a schedule of pointer readings.

The Nature of the Physical World
Chapter XII (p. 259)

Esar, Evan
[Atom] The smallest thing in the world which, when split or fused, becomes the biggest thing.

Esar's Comic Dictionary
Atom

Feynman, Richard P.
I, a universe of atoms, an atom of the universe.

What Do You Care What Other People Think?
The Value of Science (p. 243)

...the atoms that are in the brain are being replaced: the ones that were there before have gone away.

So what is this mind of ours: what are these atoms with consciousness? Last week's potatoes! They now can remember what was going on in my mind a year ago—a mind which has long ago been replaced.

To note that the thing I call my individuality is only a pattern or dance, that is what it means when one discovers how long it takes for the atoms of the brain to be replaced by other atoms. The atoms come into my brain, dance a dance, and then go out—there are always new atoms, but always doing the same dance, remembering what the dance was yesterday.

What Do You Care What Other People Think?
The Value of Science (p. 244)

Mach, Ernst
The atomic theory plays a part in physics similar to that of certain auxiliary concepts in mathematics: it is a mathematical *model* for facilitating the mental reproduction of facts. Although we represent vibrations by the harmonic formula, the phenomena of cooling by exponentials, falls by squares of time, etc, no one would fancy that vibrations in *themselves* have anything to do with circular functions, or the motion of falling bodies with squares.

The Science of Mechanics
Chapter IV
Section IV (p. 590)

Nabokov, Vladimir
But the individual atom is free: it pulsates as it wants, in high or low gear; it decides itself when to absorb and when to radiate energy.

Bend Sinister (p. 159)

Rowland, Henry
The round hard atom of Newton which God alone could break into pieces has become a molecule composed of many atoms and each of these smaller atoms has become so elastic that after vibrating 100,000 times its amplitude of vibration is scarcely diminished. It has become so complicated that it can vibrate with as many thousand notes. We cover the atom with patches of electricity here and there and make of it a system compared with which the planetary system, nay the universe itself, is simplicity.

The Physical Papers of Henry Augustus Rowland
The Highest Aime of the Physicist (p. 671)

Rukeyser, Muriel
The universe is made of stories,
not of atoms.

The Speed of Darkness
The Speed of Darkness
Stanza IX

Unknown
I remember, I remember,

When an atom was so small
It really hardly paid you
To think of one at all.
It was so small that anywhere
An atom safe could be
And pass his time in molecules
In elemental glee.

Industrial and Engineering Chemistry: News Edition
Past and Present (p. 161)
Volume 12, Number 3, April 20, 1934

ATOMIC POWER

Church, Peggy Pond

We had thought the magicians were all dead, but this was the blackest of magic.

There was even the accompaniment of fire and brimstone,
The shape of evil, towering leagues high into the heaven
In terrible, malevolent beauty, and, beneath, the bare trees
Made utterly leafless in one instant, and the streets where no one
Moved, and some wall still standing
Eyeless, and as silent as before Time.

Ultimatum for Man
The Nuclear Physicist

Laurence, William L.

And just at that instant there rose as if from the bowels of the earth a light not of this world, the light of many suns in one.

It was a sunrise such as the world had never seen, a great green super-sun climbing in a fraction of a second to a height of more than 8,000 feet, rising ever higher until it touched the clouds, lighting up earth and sky all around with dazzling luminosity.

Up it went, a great wall of fire about a mile in diameter, changing colors as it kept shooting upward, from deep purple to orange, expanding, growing bigger, rising as it was expanding, an elemental force freed from its bonds after being chained for billions of years.

The New York Times
Drama of the Atomic Bomb Found Climax in New Mexico Test
A16, column 5
September 26, 1945

ATOMISM

Democritus
By convention are sweet and bitter, hot and cold, by convention is color; in truth are atoms and the void... In reality we apprehend nothing exactly, but only as it changes according to the condition of our body and the things that impinge on or offer resistance to it.

In G.S Kirk and J.E. Raven
The Presocratic Philosophers
Fragment 589 (p. 422)

Lucretius
Again after the revolution of many of the sun's years a ring on the finger is thinned on the under side by wearing, the dripping from the eaves hollows a stone, the bent ploughshare of iron imperceptibly decreases in the fields, and we behold the stone-paved streets worn down by the feet of the multitude... These things then we see are lessened, after they are thus worn down; but what bodies depart at any given time nature has jealously shut out the means of seeing.

On The Nature of Things
Book 1, 265 (p. 5)

AURORA BOREALIS

Aytoun, William
All night long the northern streamers
Shot across the trembling sky:
Fearful lights, that never beckon
Save when kings or heroes die.

Harper's New Monthly Magazine
Edinburgh after Flodden (p. 337)
Volume 28, Number 165, February 1864

Burns, Robert
The cauld blae North was streaming forth
Her lights, wi' hissing eerie din.

The Poems and Songs of Robert Burns
A Vision

Haliburton, T.C.
The sun has scarcely set behind the dark wavy outline of the western hills, ere the aurora borealis mimics its setting beams, and revels with wild delight in the heavens, which it claims as its own, now ascending with meteor speed to the zenith, then dissolving into a thousand rays of variegated light, that vie with each other which shall first reach the horizon; now flashing bright, brilliant and glowing, as emanations of the sun, then slowly retreating from view pale and silvery white like wandering moonbeams.

The Old Judge
The Seasons (p. 210)

Kingsley, Charles
Night's son was driving
His golden-haired horses up;
Over the eastern firths

High flashed their manes.

Poems
The Longbeards' Saga

Scott, Robert F.
The eastern sky was massed with swaying auroral light, the most vivid and beautiful display that I have ever seen—fold on fold the arches and curtains of vibrating luminosity rose and spread across the sky, to slowly fade and yet again spring to glowing life...

It is impossible to witness such a beautiful phenomenon without a sense of awe, and yet this sentiment is not inspired by its brilliancy but rather by its delicacy of light and colour, its transparency, and above all by its tremulous evanescence of form. There is no glittering splendour to dazzle the eye...rather the appeal is to the imagination by the suggestion of something wholly spiritual, something instinct with a fluttering ethereal life, serenely confident yet restlessly mobile...To the little silent group which stood at gaze before such enchantment it seemed profane to return to the mental and physical atmosphere of our house.

Scott's Last Expedition
Chapter XI (p. 227)

Scott, Sir Walter
He knew, by streamers that shot so bright,
That spirits were riding the northern light.

The Complete Poetical Works of Sir Walter Scott
The Lay of the Last Minstrel
Canto Second
VIII, L. 91–2

Service, R.W.
And the Northern Lights in the crystal nights came forth with a mystic gleam.
They danced and they danced the devil-dance over the naked snow;
And soft they rolled like a tide upshoaled with a ceaseless ebb and flow.
They rippled green with a wondrous sheen, they fluttered out like a fan;
They spread with a blaze of rose-pink rays never yet seen of man.

Collected Poems of Robert Service
The Ballad of the Northern Lights

Some say that the Northern Lights are the glare of the Arctic ice and snow;
And some that it's electricity, and nobody seems to know.

Collected Poems of Robert Service
The Ballad of the Northern Lights

Taylor, Bayard
The amber midnight smiles in dreams of dawn.

The Poetical Works of Bayard Taylor
From the North

Wilde, Oscar
"What are fireworks like?" she asked...
"They are like the Aurora Borealis," said the King,... "only much more natural. I prefer them to stars myself, as you always know when they are going to appear..."

Oscar Wilde Selected Writing
The Remarkable Rocket (p. 196)

AXIAL TILT

Milton, John
Some say he bid his angels turn askance
The poles of Earth twice ten degrees and more
From the sun's axle; they with labor pushed
Oblique the centric globe: some say the sun
Was bid turn reins from the equinoctial road
Like distant breadth to Taurus with the seven
Atlantic Sisters, and the Spartan Twins,
Up to the Tropic Crab; thence down again
By Leo and the Virgin and the Scales,
As deep as Capricorn, to bring in change
Of seasons to each clime; else had the spring
Perpetual smiled on Earth with vernant flowers,
Equal in days and nights,...
The sun, as from Thyestean banquet, turned
His course intended; else how had the World
Inhabited, through sinless, more than now
Avoided pitching cold and scorching heat?

Paradise Lost
Book X, L. 679–95

AXIOM

Doyle, Sir Arthur Conan
It has long been an axiom of mine that the little things are infinitely the most important.

The Adventures of Sherlock Holmes
A Case of Identity (p. 61)

Frayn, Michael
For hundreds of pages the closely-reasoned arguments unroll, axioms and theorems interlock. And what remains with us in the end? A general sense that the world can be expressed in closely-reasoned arguments, in interlocking axioms and theorems.

Constructions
Number 277

Planck, Max
Axioms are instruments which are used in every department of science, and in every department there are purists who are inclined to oppose with all their might any expansion of the accepted axioms beyond the boundary of their logical application.

Where is Science Going?
Chapter VI (p. 179)

BEAUTY

Cayley, Arthur
It is difficult to give an idea of the vast extent of modern mathematics. The word 'extent' is not the right one: I mean extent crowded with beautiful detail—not an extent of mere uniformity such as an objectless plain, but of a tract of beautiful country seen at first in the distance, but which will bear to be rambled through and studied in every detail of hillside and valley, stream, rock, wood, and flower. But, as for every thing else, so for mathematical theory—beauty can be perceived but not explained.

The Collected Mathematical Papers of Arthur Cayley
Volume XI
Presidential Address to the British Association
September 1883 (p. 449)

Chandrasekhar, Subrahmanyan
This "shuddering before the beautiful", this incredible fact that a discovery motivated by a search after the beautiful in mathematics should find its exact replica in Nature, persuades me to say that beauty is that to which the human mind responds at its deepest and most profound.

Truth and Beauty: Aesthetics and Motivations in Science
Shakespeare, Newton, and Beethoven, or Patterns of Creativity (p. 54)

Collins, Wilkie
Admiration of those beauties of the inanimate world, which modern poetry so largely and so eloquently describes, is not, even in the best of us, one of the original instincts of our nature. As children, we none of us possess it. No uninstructed man or woman possesses it. Those whose lives are exclusively passed amidst the ever changing wonders of sea and land are also those who are most universally insensible to every aspect of Nature not directly associated with the human interest of their calling. Our capacity of appreciating the beauties of the earth we live on is, in truth, one of the civilized accomplishments which we all learn, as an art;

and, more, that very capacity is rarely practised by any of us except when our minds are most indolent and most unoccupied.

The Woman in White
The First Epoch
The Story Begun by Walter Hartright
Chapter VIII (pp. 41–2)

Copernicus, Nicolaus

Among the many and varied literary and artistic studies upon which the natural talents of man are nourished, I think that those above all should be embraced and pursued with the most loving care which have to do with things that are very beautiful and very worthy of knowledge.

On the Revolutions of the Heavenly Spheres
Book One (p. 510)

Dirac, P.A.M.

It is quite clear that beauty does depend on one's culture and upbringing for certain kinds of beauty, pictures, literature, poetry and so on... But mathematical beauty is of a rather different kind. I should say perhaps it is of a completely different kind and transcends these personal factors. It is the same in all countries and at all periods of time.

In Helge Kragh
Dirac: A Scientific Biography
Chapter 14 (p. 288)

We may try to make progress by following in Hamilton's footsteps, taking mathematical beauty as our guiding beacon, and setting up theories which are of interest, in the first place, only because of the beauty of their mathematics. We may then hope that such equations will ultimately prove their value in physics, basing this hope on the belief that Nature demands mathematical beauty in her laws.

Proceedings of the Royal Irish Academy
Hamiltonian Methods and Quantum Mechanics (p. 59)
Volume 63, Section A, Number 3, January 1964

Duhem, Pierre

It is impossible to follow the march of one of the greatest theories of physics, to see it unroll majestically its regular deductions starting from initial hypotheses, to see its consequences represent a multitude of experimental laws down to the smallest detail, without being charmed by the beauty of such a construction, without feeling keenly that such a creation of the human mind is truly a work of art.

The Aim and Structure of Physical Theory
Chapter II (p. 24)

Emerson, Ralph Waldo

I do not wonder at a snow-flake, a shell, a summer landscape, or the glory of the stars; but at the necessity of beauty under which the universe lies.

Essays and Lectures
The Conduct of Life
Fate (p. 967)

For the world is not painted, or adorned, but is from the beginning beautiful; and God has not made some beautiful things, but Beauty is the creator of the Universe.

Essays and Lectures
Essays
Second Series
The Poet (p. 449)

Hilbert, David

Our Science, which we loved above everything, had brought us together. It appeared to us as a flowering garden. In this garden there were well-worn paths where one might look around at leisure and enjoy oneself without effort, especially at the side of a congenial companion. But we also liked to seek out hidden trails and discovered many an unexpected view which was pleasing to our eyes; and when the one pointed it out to the other, and we admired it together, our joy was complete.

In S. Chandrasekhar
Truth and Beauty: Aesthetics and Motivation in Science
Chapter 3, Section VI (p. 52)

Leibniz, Gottfried Wilhelm

The beauty of nature is so great and its contemplation so sweet... whoever tastes it, can't help but view all other amusements as inferior.

In Ernst Peter Fischer
Beauty and the Beast
Chapter 2 (p. 47)

Misner, Charles W.
Thorne, Kip S.
Wheeler, John

Some day a door will surely open and expose the glittering central mechanism of the world in all its beauty and simplicity.

Gravitation
Chapter 44 (p. 1197)

Poincaré, Henri

...what are the mathematical entities to which we attribute this character of beauty and elegance, and which are capable of developing in us a kind of aesthetic emotion? They are those whose elements are harmoniously disposed so that the mind without effort can embrace their totality while realizing the details. This harmony is at once a satisfaction of our aesthetic needs and an aid to the mind, sustaining and guiding.

The Foundations of Science
Science and Method
Chapter III (p. 391)

Steensen, Niels

Beautiful are the things we see
More beautiful those we understand
Much the most beautiful those we do not comprehend.

Introductory Lecture
Copenhagen Anatomical Theater 1673
Source unknown

BIG BANG

Guth, Alan

...the big bang theory is not really a theory of a bang at all. It is only a theory of the *aftermath* of a bang...the standard big bang theory says nothing about what banged, why it banged, or what happened before it banged.

The Inflationary Universe
Preface (p. xiii)

Hoyle, Fred

Big-bang cosmology is a form of religious fundamentalism, as is the furore over black holes, and this is why these peculiar states of mind have flourished so strongly over the past quarter century. It is in the nature of fundamentalism that it should contain a powerful streak of irrationality and that it should not relate, in a verifiable, practical way, to the everyday world. It is also necessary for a fundamentalist belief that it should permit the emergence of gurus, whose pronouncements can be widely reported and pondered on endlessly—endlessly for the reason that they contain nothing of substance, so that it would take an eternity of time to distill even one drop of sense from them. Big-bang cosmology refers to an epoch that cannot be reached by any form of astronomy, and, in more than three decades, it has not produced a single successful prediction.

Home is Where the Wind Blows
Part III
Chapter 28 (pp. 413–14)

Maddox, John

The microwave background radiation, which fills even the corners of the universe, would psychologically have been more compelling evidence for the Big Bang if it had been predicted before its discovery in 1965. That it was not is something of a surprise, which is nevertheless now irrelevant.

Nature
The Best Cosmology There Is (p. 15)
Volume 372, Number 6501, 3 November 1994

Poe, Edgar Alan

I am fully warranted in announcing that *the Law which we have been in the habit of calling Gravity exists on account of Matter's having being irradiated, at its origin, atomically, into a limited sphere of Space, from one, individual, unconditional, irrelative, and absolute Particle Proper, by the sole process in which it is possible to satisfy, at the same time, the two conditions, irradiation and generally-equable distribution throughout the sphere, that is to say, by a force varying in direct proportion with the squares of the distances between the irradiated atoms, respectively, and the Particular centre of Irradiation.*

Eureka (p. 67)

BLACK HOLE

Asimov, Isaac
Since 1960 the universe has taken on a wholly new face. It has become more exciting, more mysterious, more violent, and more extreme as our knowledge concerning it has suddenly expanded. And the most exciting, most mysterious, most violent, most extreme phenomena of all has the simplest, plainest, calmest, and mildest name—nothing more than a "black hole."

The Collapsing Universe
Chapter 1 (p. 1)

Gardner, Martin
The healthy side of the black-hole craze is that it reminds us of how little science knows, and how vast is the realm about which science knows nothing.

Science: Good, Bad, and Bogus
Seven Books on Black Holes (p. 343)

Our entire universe may slowly stop expanding, go into a contracting phase, and finally disappear into a black hole, like an acrobatic elephant jumping into its anus.

Science: Good, Bad, and Bogus
Seven Books on Black Holes (p. 336)

Koestler, Arthur
Young Archie, the intrepid mole,
Went down to explore a Black Hole.
A stark singularity,
Devoid of all charity,
Devoured the mole as a whole.

In Bernard Dixon (ed.)
From Creation to Chaos
Cosmic Limerick (p. 108)

Lasota, Jean-Pierre
Black holes may still be black, but they can no longer hide in disguise. We are learning how to unmask them.

Scientific American
Unmasking Black Holes (p. 47)
Volume 280, Number 5, May 1999

Milne, A.A.
A huge great big thing, like—like nothing. A huge big—well, like a—I don't know—like an enormous big nothing.

Winnie-the-Pooh
Piglet Meets a Heffalump

Thorne, Kip
Of all the conceptions of the human mind, from unicorns to gargoyles to the hydrogen bomb, the most fantastic, perhaps, is the black hole.

Black Holes and Time Warps
Prologue (p. 23)

BOOK

Bernstein, Jeremy

...a physics book, unlike a novel, not only has no happy ending, but has no real ending at all.

Elementary Particles and Their Currents
Chapter 15 (p. 318)

Eco, Umberto

Books are not made to be believed, but to be subjected to inquiry. When we consider a book, we must not ask ourselves what it says but what it means...

The Name of the Rose
Fourth Day
After Compline (p. 316)

Mitchell, Maria

A book is a very good institution! To read a book, to think it over, and to write out notes is a useful exercise; a book which will not repay some hard thought is not worth publishing.

In Phebe Mitchell Kendall
Maria Mitchell: Life, Letters, and Journals
Chapter IX (p. 180)

Newton rolled up the cover of a book; he put a small glass at one end, and a large brain at the other—it was enough.

In Phebe Mitchell Kendall
Maria Mitchell: Life, Letters, and Journals
Chapter IX (p. 180)

CELESTIAL MOTION

Milton, John

That day, as other solemn dayes, they spent
In song and dance about the sacred Hill,
Mystical dance, which yonder starrie Sphere
Of Planets and of fixt in all her Wheeles
Resembles nearest, mazes intricate,
Eccentric, intervovl'd, yet regular
Then most, when most irregular they seem:
And in their motions harmonic Divine
So smooths her charming tones, that Gods own ear
Listens delighted.

Paradise Lost
Book V, L. 618–27

CHANCE

Eddington, Sir Arthur Stanley
We have swept away the anti-chance from the field of our current physical problems, but we have not got rid of it. When some of us are so misguided as to try to get back millards of years into the past we find the sweepings piled up high like a high wall, forming a boundary—a beginning of time—which we cannot climb over.

New Pathways in Science
Chapter III (p. 60)

Jevons, W. Stanley
Happily the Universe in which we dwell is not the result of chance, and where chance seems to work it is our own deficient faculties which prevent us from recognising the operating of Law and of Design.

The Principles of Science
Book I
Chapter I (p. 2)

John of Salisbury
Chance blows together the atoms into an immense heap so that this very globe of the world comes into being, and so that the elements are fixed in their places under an eternal law...

In John van Laarhoven (ed.)
Entheticus Maior and Minor
Volume I
Part II
Section I, Notes from Epicuris, L. 567–9

CHANGE

Burns, Robert
Look abroad through Nature's range
Nature's mighty law is change.

The Poems and Songs of Robert Burns
Inconstancy in Love

Emerson, Ralph Waldo
There are no fixtures in nature. The universe is fluid and volatile. Permanence is but a word of degrees. Our globe seen by God is a transparent law, not a mass of facts. The law dissolves the fact and holds it fluid.

Essays and Lectures
Essays
First Series
Circles (p. 403)

Mitchell, Maria
We may turn our gaze [to other stars] as we turn a kaleidoscope, and the changes are infinitely more startling, the combinations infinitely more beautiful; no flower garden presents such a variety and such delicacy of shades.

In Phebe Mitchell Kendall
Maria Mitchell: Life, Letters, and Journals
Chapter XI (p. 235)

Ovid
All things are fluent; every image forms,
Wandering through change. Time is itself a river
In constant movement, and the hours flow by
Like water, wave on wave, pursued, pursuing,
Forever fugitive, forever new.

Metamorphosis
Book 15, L. 178–82

CHAOS

Bradley, John Hodgdon Jr
From the smallest satellite slave of the smallest star to the largest super-galaxy of worlds in space, everything bows to the first law of nature. Chaos and caprice do not exist.

Parade of the Living
Chapter I (p. 3)

Miller, Henry
...chaos is the score upon which reality is written.

Tropic of Cancer (p. 2)

COMET

Babylonian Inscription

A comet arose whose body was bright like the day, while from its luminous body a tail extended, like the sting of a scorpion.

In John Brandt and Robert Chapman
Introduction to Comets
Chapter 1 (p. 4)

Byron, George Gordon

The angels all were singing out of tune,
And hoarse with having little else to do,
Excepting to wind up the sun and moon,
Or curb a runaway young star or two,
Or wild colt of a comet, which too soon
Broke out of bounds o'er the ethereal blue,
Splitting some planet with its beautiful tail,
As boats are sometimes by a wanton whale.

The Complete Poetical Works of Byron
The Vision of Judgment

de Fontenelle, Bernard Le Bovier

These foreign planets, with their tails and their beards, have a terrible menacing countenance, it may be they are sent to affront us...

Conversations on the Plurality of Worlds
The Fifth Evening (p. 173)

We think ourselves unhappy when a comet appears; but it is the comet itself which is unfortunate.

Conversations on the Plurality of Worlds
The Fifth Evening (p. 175)

Dick, Thomas

Whatever opinions we may adopt as to the physical constitution of comets, we must admit that they serve some grand and important purpose in the economy of the universe; for we cannot suppose that the Almighty has created such an immense number of bodies, and set them in rapid motion according to established laws, without an end worthy of his perfection, and, on the whole, beneficial to the inhabitants of the system through which they move.

The Sidereal Heavens and Other Subjects Connected with Astronomy, As Illustrative of the Character of the Deity and of an Infinity of Worlds
Chapter XX (p. 342)

...what I conceive to be one of the main designs of the Creator in the formation of such a vast number of splendid bodies is, that they may serve as habitations for myriads of intelligent beings...If this position be admitted, then we ought to contemplate the approach of a comet, not as an object of terror or harbinger of evil, but as a splendid world, of a different construction from ours, conveying millions of happy beings to survey a new region of the Divine empire...

The Sidereal Heavens and Other Subjects Connected with Astronomy, As Illustrative of the Character of the Deity and of an Infinity of Worlds
Chapter XX (p. 345)

Donne, John

Who vagrant transitory comets sees,
Wonders because they're rare; but a new star
Whose motion with the firmament agrees,
Is miracle; for there no new things are.

In Charles M. Coffin (ed.)
The Complete Poetry and Selected Prose of John Donne
To the Countess of Huntingdon (p. 169)

Halley, Edmond

Aristotle's opinion...that comets were nothing else than sublunary vapors or airy meteors...prevailed so far amongst the Greeks, that this sublimest part of astronomy lay altogether neglected; since none could think it worthwhile to observe, and to give an account of the wandering and uncertain paths of vapours floating in the Aether.

Transactions of the Royal Society of London
Astronomiae Cometicae Synopsis (p. 1882)
Volume 24, Number 297, March 1705

Holmes, Oliver Wendell

The Comet! He is on his way,
And singing as he flies;
The whizzing planets shrinks before

The spectre of the skies...

The Complete Poetical Works of Oliver Wendell Holmes
The Comet
Stanza 1

Lee, Oliver Justin

Dynamically it is quite possible that great numbers of comets were once well-behaved members of the solar system and that they have been bullied and kicked around gravitationally by the great planets and by possible dark bodies in space that they have become the pariahs they are.

Measuring Our Universe: From the Inner Atom to Outer Space
Chapter 7 (p. 93)

Levy, David H.

Comets are like cats: they have tails, and they do precisely what they want.

Comets
Preface (p. 13)

Maunder, E. Walter

Comets cannot be homes of life; they are not sufficiently condensed; indeed, they are probably but loose congeries of small stones. But even if comets were of planetary size it is clear that life could not be supported on them; water could not remain in the liquid state on a world that rushed from one such extreme of temperature to another.

Are the Planets Inhabited?
Chapter IX (pp. 119–20)

Peltier, Leslie C.

I had watched a dozen comets, hitherto unknown, slowly creep across the sky as each one signed its sweeping flourish in the guest book of the Sun.

Starlight Nights
Chapter 6 (p. 43)

Time has not lessened the age-old allure of the comets. In some ways their mystery has only deepened with the years. At each return a comet brings with it the questions which were asked when it was here before, and as it rounds the sun and backs away toward the long, slow night of its aphelion, it leaves behind with us those questions still unanswered.

Starlight Nights
Chapter 27 (p. 231)

Seneca

If one of these fires of unusual shape have made its appearance, everybody is eager to know what it is. Blind to all other celestial bodies, each asks about the newcomer; one is not quite sure whether to admire or to fear it. Persons there are who seek to inspire terror by forecasting its grave import.

Physical Science in the Time of Nero
Book VII
Chapter I (p. 272)

...how many other bodies besides [these comets] move in secret, never dawning upon human eyes? Nor is it for man that God has made all things.

Physical Science in the Time of Nero
Book VII
Chapter XXX (p. 305)

Shakespeare, William
Hung by the heavens with black, yield day to night!
Comets, importing change of time and states,
Brandish your crystal tresses in the sky...

The First Part of King Henry the Sixth
Act I, scene i, L. 1–3

Thomson, James
Lo! from the dread immensity of space,
Returning, with accelerated course,
The rushing comet to the sun descends;
And, as he sinks below the shading earth,
With awful train projected o'er the heavens,
The guilty nations tremble.

Seasons
Summer, L. 1705-10

Tolstoy, Leo
...at this bright comet which, after traveling in its orbit with inconceivable velocity through infinite space, seemed suddenly—like an arrow piercing the earth—to remain fixed in a chosen spot, vigorously holding its tail, shining and displaying its white light amid countless other scintillating stars.

War and Peace
Book VIII, Chapter 22 (p. 341)

von Humboldt, Alexander
Since scientific knowledge, although frequently blended with vague and superficial views, has been more extensively diffused through wider circles of social life, apprehensions of the possible evils threatened by

comets have acquired more weight, as their direction has become more definite.

Cosmos
Volume I
Chapter I (p. 96)

COMMUNICATION

Archytas

That tho' a Man, were admitted into Heaven to view the wonderful Fabrick of the World, and the Beauty of the Stars, yet what would otherwise be Rapture and Extasie, would be but a melancholy Amazement if he had not a Friend to communicate it to.

Attributed
In Christianus Huygens
The Celestial Worlds Discover'd
Book the First (p. 4)

Mach, Ernst

Science is communicated by instruction, in order that one man may profit by the experience of another and be spared the trouble of accumulating it for himself; and thus, to spare posterity, the experiences of whole generations are stored up in libraries.

The Science of Mechanics
Chapter IV
Section IV (p. 578)

Neal, Patricia

Gort, Klaatu berada nikto!

The Day the Earth Stood Still
20th Century Fox, 1951

COMPULSORY

Bilaniuk, O.
Sudarshan, E.C.
There is an unwritten precept in modern physics, often facetiously referred to as Gell-Mann's totalitarian principle, which states that in physics "anything which is not prohibited is compulsory." Guided by this sort of argument we have made a number of remarkable discoveries, from neutrinos to radio galaxies.

Physics Today
Particles Beyond the Light Barrier (p. 44)
Volume 22, Number 5, May 1969

CONCEPT

Einstein, Albert
Infeld, Leopold
Physical concepts are free creations of the human mind, and are not, however it may seem, uniquely determined by the external world.

The Evolution of Physics
Chapter I (p. 31)

Weaver, Warren
With the extremely small or the extremely large, with inconceivably brief or extended phenomena, science has a difficult time. It is by no means

clear that our present concepts or even our existing language is suitable for these ranges.

Science and Imagination
Chapter 2 (p. 50)

CONSTELLATION

Aratus

Some man of ages past
Observed their goings; and devised their titles,
Forming the constellations. For the name
Of each star singly none could tell or learn;—
So numerous are they everywhere, and many
Of the same size and color, as they roll.
Thus he bethought him to combine them so,
That, ranged in neighborhood, they might present
Images,—each taking his proper name,
And henceforth none rising to doubt or guess at.

In N.L. Frothingham
Metrical Pieces
The Appearances of the Stars (pp. 39–40)

Burns, Robert

. . . O, had I power like inclination,
I'd heeze thee up a constellation!
To canter with the Sagitare,
Or loup the Ecliptic like a bar,
Or turn the Pole like any arrow;. . .

The Poems and Songs of Robert Burns
Epistle to Hugh Parker

de Cervantes, Miguel

". . . Even if your worship will not altogether give up this exploit, put it off at least till morning. For by the science I learned when I was a shepherd, it can't be more than three hours till dawn, since the muzzle of the Bear is at the top of his head, and at midnight it is in line with the left paw."

"How can you see, Sancho, where the line is, or the muzzle, or the top of the head you speak of? The night is so dark that there is not a star to be seen in the whole sky."

"That's true," said Sancho, "but fear has many eyes, and can see things underground. So it'll easily see things up above in the sky. Besides, it's reasonable to suppose that it won't be long till dawn."

Don Quixote
Part 1, Chapter 20

Donne, John

And in these Constellations then arise
New Starres, and old doe vanish from our eyes.

In Charles M. Coffin (ed.)
The Complete Poetry and Selected Prose of John Donne
First Anniversary

Frost, Robert

You'll wait a long, long time for anything much
To happen in heaven...

Complete Poems of Robert Frost
On Looking Up By Chance at the Constellations

Homer

He wroght the earth, the heavens, and the sea; the moon also at her full and the untiring sun, with all the signs that glorify the face of heaven—the Pleiads, the Hyads, huge Orion, and the Bear, which men also call the Wain and which turns around ever in one place, facing Orion, and alone never dips into the stream of Oceanus.

Illiad
Book 18, L. 483–9

Noyes, Alfred

Night after night, among the gabled roofs,
Climbing and creeping through a world unknown
Save to the roosting stork, he learned to find
The constellations, Cassiopeia's throne,
The Plough still pointing to the Pole star,
The Sword-belt of Orion. There he watched
The movement of the planets, hours and hours,
And wondered at the mystery of it all.

The Torch-Bearers
Volume I
Watchers of the Sky
Tycho Brahe (p. 40)

Sagan, Carl

In the night sky, when the air is clear, there is a cosmic Rorschach test awaiting us. Thousands of stars, bright and faint, near and far, in a glittering variety of colors, are peppered across the canopy of night. The eye, irritated by randomness, seeking order, tends to organize into patterns these separate and distinct points of light.

The Cosmic Connection
Chapter 2 (p. 9)

You'll wait a long, long time for anything much
To happen in heaven...
Robert Frost – (See p. 68)

Twain, Mark
Constellations have always been troublesome things to name. If you give one of them a fanciful name, it will always persist in not resembling the thing it has been named for.

Following the Equator
Volume I
Chapter 5 (p. 52)

Whitman, Walt

The earth, that is sufficient,
I do not want the constellations any nearer,
I know they are very well where they are,
I know they suffice for those who belong to them.

In James E. Miller, Jr (ed.)
Complete Poetry and Selected Prose
Song of the Open Road

ANDROMEDA

Keats, John

Andromeda! Sweet woman! Why delaying
So timidly among the stars: come hither!
Join this bright throng, and nimbly follow whither
They all are going.

The Complete Poetical Works of Keats
Endymion

Kingsley, Charles

High for a star in the heavens, a sign and a hope for the seamen,
Spreading thy long white arms all night in the heights of the aether,
Hard by thy sire and the hero, thy spouse, while near thee thy mother
Sits in her ivory chair, as she plaits ambrosial tresses;
All night long thou wilt shine.

Poems
Andromeda

ARCTURUS

Dickinson, Emily

Arcturus is his other name,—
I'd rather call him star.
It's so unkind of Science
To go and interfere!

The Complete Poems of Emily Dickinson
Poem 70
Old-Fashioned

Teasdale, Sara

When, in the gold October dusk, I saw you near to setting,
Arcturus, bringer of spring,
Lord of the summer nights, leaving us now in autumn,

Having no pity on our withering;...

The Collected Poems of Sara Teasdale
Arcturus in Autumn

ARIES

Longfellow, Henry Wadsworth
Now the zephyrs diminish the cold, and the year being ended,
Winter Maeotian seems longer than ever before;
And the Ram that bore unsafely the burden of Helle,
Now makes the hours of the day equal with those of the night.

The Poetical Works of Henry Wadsworth Longfellow
Ovid in Exile
Tristia
Book III, Elegy XII

CANIS MAJOR

Frost, Robert
The Great Overdog,
That heavenly beast
With a star in one eye,
Gives a leap in the East.

Complete Poems of Robert Frost
Canis Major

CAPRICORNUS

Aratus
Then blow the fearful south-winds, when the Goat
With the sun rises; and then Jove's sharp cold,
Still worse, besets the stiffening mariner.

In N.L. Frothingham
Metrical Pieces
The Appearances of the Stars (p. 33)

DRACO

Darwin, Erasmus
With vast convolutions Draco holds
Th' ecliptic axis in his scaly folds.
O'er half the skies his neck enormous rears,

And with immense meanders parts the Bears.

The Botanic Garden
Part I, Canto I, XI, L. 517

LIBRA

Longfellow, Henry Wadsworth
I hear the Scales, where hang in equipoise
The night and day...

The Poetical Works of Henry Wadsworth Longfellow
The Poet's Calendar
September

LOST PLEIAD

Hemans, Felicia
And is there glory from the heavens departed?
Oh! Void unmarked!—Thy sisters of the sky
Still hold their place on high,
Though from its rank thine orb so long hath started,
Thou, that no more art seen of mortal eye.

The Complete Works of Mrs Hemans
Volume I
The Lost Pleiad

ORION

Teasdale, Sara
But when I lifted up my head
From the shadows shaken on the snow,
I saw Orion in the east
Burn steadily as long ago.

The Collected Poems of Sara Teasdale
Winter Stars

Tennyson, Alfred
...those three stars of the airy Giants' zone
That glitter burnished by the frosty dark.

The Complete Poetical Works of Tennyson
Princess

PLEIADES

Tabb, John Banister
"Who are ye with clustered light,
Little Sisters seven?"
"Crickets, chirping all the night
On the hearth of heaven."

In Francis A. Litz (ed.)
The Poetry of Father Tabb
Humorous Verse
The Pleiads

Tennyson, Alfred
Many a night I saw the Pleiads, rising thro' the mellow shade,
Glitter like a swarm of fireflies in a silver braid.

The Complete Poetical Works of Tennyson
Locksley Hall

SAGITTARIUS

Longfellow, Henry Wadsworth
The Centaur, Sagittarius, am I,
Born of Ixion and the cloud's embrace:
With sounding hoofs across the earth I fly,
A steed Thessalian with a human face.

The Poetical Works of Henry Wadsworth Longfellow
Poet's Calendar
November

SCORPIO

Longfellow, Henry Wadsworth
Though on the frigid scorpion I ride,
The dreamy air is full, and overflows
With tender memories of the summer-tide
And mingled voices of the doves and crows.

The Poetical Works of Henry Wadsworth Longfellow
Poet's Calendar
October

SOUTHERN CROSS

Meredith, Owen
Then did I feel as one who, much perplext,
Led by strange legends and the light of stars
Over long regions of the midnight sand
Beyond the red tract of the Pyramids,
Is suddenly drawn to look upon the sky,
From sense of unfamiliar light, and sees,
Reveal'd against the constellated cope,
The great cross of the South.

The Poetical Works of Owen Meredith
Queen Guenevere

VIRGO

Longfellow, Henry Wadsworth
I am the Virgin, and my vestal flame
Burns less intensely than the Lion's rage;
Sheaves are my only garments, and I claim
A golden harvest as my heritage.

The Poetical Works of Henry Wadsworth Longfellow
Poet's Calendar
August

COSMOCHEMISTRY

Frost, Robert
Say something to us we can learn
By heart and when alone repeat.
Say something! And it says, "I burn."
But say with what degree of heat.
Talk Fahrenheit, Talk Centigrade.
Use language we can comprehend.
Tell us what elements you blend.

Complete Poems of Robert Frost
Take Something Like a Star

Fuller, Buckminster
Universe has no pollution.
All the chemistries of the Universe are essential
To its comprehensive self regulation.

In John S. Lewis
Physics and Chemistry of the Solar System
Chapter II (p. 43)

Marcet, Jane
Nature also has her laboratory, which is the universe, and there she is incessantly employed in chemical operations.

Conversations on Chemistry
Conversation I (p. 2)

COSMOGONY

Bridgman, Percy

...the most striking thing about cosmogony is the perfectly hair raising extrapolations which it is necessary to make. We have to extend the times of the order of 10^{13} years and distances of the order of 10^9 light years laws which have been checked in a range of not more than 3×10^2 years, and certainly in distances not greater than the distance which the solar system has traveled in that time, or about 4×10^{-2} light years. It seems to me that one cannot take such extrapolations seriously unless one subscribes to a metaphysics that claims that laws of the necessary mathematical precision really control the actual physical universe.

The Nature of Physical Theory
Chapter VIII (p. 109)

COSMOLOGY

Chaisson, Eric

Exploring the whole universe requires large thoughts. There are no larger thoughts than cosmological ones.

Cosmic Dawn
Prologue (p. 3)

Görtniz, Thomas

Modern cosmology is myth which does not know itself to be myth.

International Journal of Theoretical Physics
Connections between Abstract Quantum Theory and Space–Time Structure II,
A Model of Cosmological Evolution (p. 659)
Volume 27, No 6, June 1988

Hawking, Stephen
Penrose, Roger

Cosmology used to be considered a pseudoscience and the preserve of physicists who might have done some useful work in their earlier years but who had gone mystic in their dotage.

The Nature of Space and Time
Chapter Five (p. 75)

Hogan, John

...cosmologists—and the rest of us—may have to forego attempts at understanding the universe and simply marvel at its infinite complexity and strangeness.

Scientific American
Universal Truths (p. 117)
Volume 263, Number 4, October 1990

Popper, Karl R.

I, however, believe that there is at least one philosophical problem in which all thinking men are interested. It is the problem of cosmology: *the problem of understanding the world—including ourselves and our knowledge, as part of the world.* All science is cosmology, I believe, and for me the interest of philosophy lies solely in the contributions which it has made to it.

The Logic of Scientific Discovery
Preface to English edition (p. 15)

Tolman, R.C.
It is appropriate to approach the problems of cosmology with feelings of respect for their importance, of awe for their vastness, and of exultation for the temerity of the human mind in attempting to solve them. They must be treated, however, by the detailed, critical, and dispassionate methods of the scientist.

Relativity, Thermodynamics and Cosmology
Part IV, section 187 (p. 488)

Turok, Neil G.
...Maybe the problems cosmology has set for itself will turn out to be just too difficult to solve scientifically. After all, we've got a lot of gall to suppose that the universe can be described by some simple theory.

Scientific American
In John Hogan
Universal Truths (p. 117)
Volume 263, Number 4, October 1990

COSMOS

Ferris, Timothy

The history of the cosmos is arrayed in the sky for those who care to read it.

Galaxies
Chapter VI (p. 161)

Lawrence, D.H.

We and the cosmos are one. The cosmos is a vast body, of which we are still parts. The sun is a great heart whose tremors run through our smallest veins. The moon is a great gleaming nerve-centre from which we quiver forever. Who knows the power that Saturn has over us or Venus? But it is a vital power, rippling exquisitely through us *all the time*... Now all this is *literally* true, as men knew in the great past and as they will know again.

Apocalypse
Five (p. 45)

Mencken, H.L.

1. The cosmos is a gigantic fly-wheel making 10,000 revolutions a minute.
2. Man is a sick fly taking a dizzy ride on it.
3. Religion is the theory that the wheel was designed and set spinning to give him the ride.

Prejudices: Third Series
Chapter V, Section 5 (p. 132)

Plato

...this universe is called Cosmos, or order, not disorder or misrule.

Gorgias
Section 508

Reeves, Hubert

Knowledge of the cosmos is much more than a luxury for cultivated souls. It is the foundation of a cosmic consciousness. It casts light on the heavy responsibilities that have fallen upon us.

Atoms of Silence
Chapter 13 (p. 146)

Santayana, George

... the cosmos has its own way of doing things, not wholly rational nor ideally best, but patient, fatal, and fruitful. Great is this organism of mud and fire, terrible this vast, painful glorious experiment.

In Logan Pearsall Smith
Little Essays
Piety (p. 86)

Shapley, Harlow

Cosmography is to the Cosmos what geography is to the earth.

Of Stars and Men
Introduction (fn, p. 4)

Tomlinson, C.

He sees in Nature's laws a code divine,
A living Presence he must first adore,
Ere he the sacred mysteries explore,
Where Cosmos is his temple, Earth his shrine.

The Graphic
Michael Faraday (p. 183)
Volume XX, Number 508, 23 August 1879

CREATION

Bush, Vannevar
With all our wide vision we may be looking at only a small part of a grand creation. Our universe with its billions of galaxies may be only one among many.

Science is Not Enough
Chapter IX (p. 168)

Epictetus
No great thing is created suddenly, any more than a bunch of grapes or a fig. If you tell me that you desire a fig, I answer you that there must be time. Let it first blossom, then bear fruit, then ripen.

The Discourses and Enchiridions
Discourses
Book I
Chapter 15 (p. 43)

Fitzgerald, Edward
There was a Door to which I found no Key:
There was a Veil past which I could not see...

The Rubaiyat of Omar Khayyam
XXXII

Hoyle, Fred
Without continuous creation, the Universe must evolve toward a dead state in which all the matter is condensed into a vast number of dead stars.

The Nature of the Universe

CURIOSITY

Amaldi, Ginestra Giovene

Man in his universe is like a baby in a strange room. Just as a baby reaches out to finger or taste all the mysterious objects in the room, so man's curiosity is excited by the wonderful sights, sounds, and smells that greet him whichever way he turns.

Our World and the Universe Around Us
Volume I
Introduction (p. ix)

Haber, Heinz
The curiosity of man must forever find its greatest challenge in the magnificent riddle of the universe.

Stars, Men and Atoms
Chapter 11 (p. 188)

Pittendreigh, W. Maynard Jr
I burn with curiosity about what lies beyond the sky.

Sky and Telescope
Pittendreigh's Law of Planetary Motion (p. 6)
Volume 87, Number 2, February 1994

Weisskopf, Victor
Curiosity without compassion is inhuman; compassion without curiosity is ineffectual.

Physics in the Twentieth Century: Selected Essays
Marie Curie and Modern Science (p. 325)

Wright, Helen
The curiosity of Alice to see what lives behind the looking glass may be likened to the desire of the astronomer to see beyond the range of his vision. By each addition to the light-gathering power of his instrument he soon yearns for a glimpse of things farther away and plans for a larger telescope.

Palomar: The World's Largest Telescope
A 200 Inch Mirror (p. 91)

DARK MATTER

Browning, Robert
. . . greet the unseen with a cheer!

The Complete Poetical Works of Browning
Asolando: Fancies and Facts
Epilogue

De Saint-Exupéry, Antoine
. . . what is essential is invisible to the eye.

The Little Prince
Part XXI (p. 70)

Reeve, F.D.
Because all things balance—as on a wheel—and we cannot see nine-tenths of what is real, our claims of self-reliance are pieced together by unpanned gold. The whole system is a game: the planets are the shells; our earth, the pea. *May there be no moaning of the bar.* Like ships at sunset in a reverie, we are shadows of what we are.

The American Poetry Review
Coasting (p. 38)
Volume 24, Number 4, July–August 1995

DATA

Greenstein, George
Data in isolation are meaningless, a collection of numbers. Only in context of a theory do they assume significance...

Frozen Star
Chapter 1 (pp. 3-4)

Hoyle, Fred
To the optical astronomer, radio data serves like a good dog on a hunt.

Galaxies, Nuclei and Quasars (p. 43)

Lowell, Percival
All deduction rests ultimately upon the data derived from experience. This is the tortoise that supports our conception of the cosmos.

Mars
Chapter I (p. 6)

Russell, Bertrand
...there is more difficulty in stating our principle so as to be applicable when our data are confined to a finite part of the universe. Things from outside may always crash in and have unexpected effects.

Religion and Science
Determinism (p. 149)

DEPLETION

Jenkins, Edward B.

Figuratively, when we study depletions, it is as if we were looking at the crumbs left on the plate after the grains have eaten their dinner.

In L.J. Allamandola and A.G.G.M. Tielens (eds)
Interstellar Dust
Proceedings of the 135th Symposium of the International Astronomical Union
Insights on Dust Grain Formation and Destruction
Provided by Gas-Phase Element Abundances
Section 2 (p. 24)

DESIGN

Davies, Paul

I belong to the group of scientists who do not subscribe to a conventional religion but nevertheless deny that the universe is a purposeless accident... [T]he physical universe is put together with an ingenuity so astonishing that I cannot accept it merely as a brute fact. There must... be a deeper level of explanation.

The Mind of God
Preface (p. 16)

Russell, Bertrand

You all know the argument from design: everything in the world is made just so that we can manage to live in the world, and if the world was ever so little different, we could not manage to live in it.

Why I Am Not a Christian
Why I Am Not A Christian (p. 9)

DETERMINISM

Einstein, Albert

Everything is determined, the beginning as well as the end, by forces over which we have no control. It is determined for the insect as well as the star. Human beings, vegetables, or cosmic dust, we all dance to a mysterious tune, intoned in the distance by an invisible piper.

The Saturday Evening Post
26 October 1929 (p. 17)

Popper, K.

The intuitive idea of determinism may be summed up by saying that the world is like a motion-picture film: the picture or still which is just being projected is *the present*. Those parts of the film which have already been shown constitute *the past*. And those which have not yet been shown constitute *the future*.

In the film, the future co-exists with the past; and the future is fixed, in exactly the same sense as the past. Though the spectator may not know the future, every future event, without exception, might in principle be known with certainty, exactly like the past, since it exists in the same sense in which the past exists. In fact, the future will be known to the producer of the film—to the Creator of the world.

The Open Universe
Chapter I (p. 5)

DIFFERENTIAL EQUATIONS

Eddington, Sir Arthur Stanley

We should suspect an intention to reduce God to a system of differential equations...

The Nature of the Physical World
Chapter XIII (p. 282)

Haldane, J.B.S.

Men have fallen in love with statues and pictures. I find it easier to imagine a man falling in love with a differential equation, and I am inclined to think that some mathematicians have done so. Even in a non-mathematician like myself, some differential equations evoke fairly violent physical sensations similar to those described by Sappho and Catullus when viewing their mistresses. Personally, however, I obtain an even greater kick from finite difference equations, which are perhaps more like those which an up-to-date materialist would use to describe human behavior.

The Inequality of Man and Other Essays
Scientific Calvinism (p. 39)

Lanczos, Cornelius

Our symbolic mechanism is eminently useful and powerful, but the danger is ever-present that we become drowned in a language which has its well-defined grammatical rules but eventually loses all content and becomes a nebulous sham.

Linear Differential Operators
Preface (p. vii)

Whitehead, Alfred North

Matter-of-fact is an abstraction, arrived at by confining thought to purely formal relations which then masquerade as the final reality. This is why science, in its perfection, relapses into the study of differential equations. The concrete world has slipped through the meshes of the scientific net.

Modes of Thought
Chapter I
Lecture 1 (p. 25)

DIMENSION

Einstein, Albert

Imagine a bedbug completely flattened out, living on the surface of a globe. This bedbug may be gifted with analysis, he may study physics, he may even write a book. His universe will be two-dimensional. He may even intellectually or mathematically conceive of a third dimension, but he cannot visualize it. Man is in the same position as the unfortunate bedbug, except that he is three-dimensional. Man can imagine a fourth dimension mathematically, but he cannot see it, he cannot visualize it, he cannot represent it physically. It exists only mathematically for him. The mind cannot grasp it.

Cosmic Religion
On Science (pp. 102–3)

DISCOVERY

Bolyai, John

Mathematical discoveries, like springtime violets in the woods, have their season which no human can hasten or retard.

Mathematics Teacher
In Israel Kleiner
Thinking the Unthinkable: The Story of Complex Numbers (with a Moral) (p. 590)
Volume 81, Number 7, October 1988

Bruner, Jerome

First, I should be clear about what the act of discovery entails. It is rarely on the frontier of knowledge or elsewhere, that new facts are "discovered" in the sense of being encountered, as Newton suggests, in the form of islands of truth in an uncharted sea of ignorance. Or if they appear to be discovered in this way, it is almost always thanks to some happy hypothesis about where to navigate. Discovery, just like surprise favors the well-prepared mind.

On Knowing—Essays for the Left Hand
The Act Of Discovery (p. 82)

Clerke, Agnes M.

It is impossible to follow with intelligent interest the course of astronomical discovery without feeling some curiosity as to the means by which such surpassing results have been secured. Indeed, the bare acquaintance with *what* has been achieved, without any corresponding knowledge of *how* it has been achieved, supplies food for barren wonder rather than for fruitful and profitable thought.

A Popular History of Astronomy during the Nineteenth Century
Part I
Chapter VI (p. 108)

Curie, Marie
A great discovery does not leap completely achieved from the brain of the scientist, as Minerva sprang, all panoplied, from the head of Jupiter; it is the fruit of accumulated preliminary work.

Pierre Curie
Chapter VII (p. 144)

Eddington, Sir Arthur Stanley
...we have not to discover the properties of a thing which we have recognized in nature, but to discover how to recognize in nature a thing whose properties we have assigned.

The Mathematical Theory of Relativity
Introduction (p. 6)

Glass, Bentley
We are like the explorers of a great continent who have penetrated to its margins in most points of the compass and have mapped the major mountain chains and rivers. There are still innumerable details to fill in, but the endless horizons no longer exist.

Science
Science: Endless Horizons or Golden Age? (p. 24)
Volume 171, Number 3966, January 8, 1971

Holton, G.
...it is precisely because the drive toward discovery is in a sense irrational that it is so powerful.

Thematic Origins of Scientific Thought: Kepler to Einstein
Chapter 11 (p. 390)

Kepler, Johannes
Here it is a question not only of leading the reader to an understanding of the subject matter in the easiest way, but also, chiefly, of the arguments, meanderings, or even chance occurrences by which I the author first came upon that understanding. Thus, in telling of Christopher Columbus, Magellan, and of the Portuguese, we do not simply ignore the errors by which the first opened up America, the second, the China Sea, and the last, the coast of Africa; rather, we would not wish them omitted, which would indeed be to deprive ourselves of an enormous pleasure in reading.

New Astronomy
Summaries of the individual chapters (p. 78)

Körner, T.W.

It is sometimes said that the great discovery of the nineteenth century was that the equations of nature were linear, and the great discovery of the twentieth century is that they are not.

Fourier Analysis
Chapter 24 (p. 99)

Lowell, Percival

The road to discovery is not an easy one to travel...There is to add to its forbiddingness no warm compensating reception at its end, except in one's own glow of attainment. For progress is first obstructed by the reticence of nature and then opposed by the denunciation of man. Nature does not help and humanity hinders. If nature abhors a vacuum, mankind abhors filling it. A really new idea is a foundling without friends. Indeed a doorstep acquisition is welcome compared with the gift of a brand new upsetting thought. The undesired outsider is ignored, pooh-poohed, denounced, or all three according to circumstances. A generation or more is needed to secure it a hearing and more time still before its worth is recognized.

In William Graves Hoyt
Lowell and Mars
Chapter 15 (p. 299)

Milne, A.A.

"Oh!" said Pooh again. "What is the North Pole?" he asked.
"It's just a thing you discover," said Christopher Robin carelessly, not being quite sure himself.

Winnie the Pooh
Christopher Robin Leads an Expotition to the North Pole

Thomson, J.J.

...in the distance tower still higher peaks, which will yield to those who ascend them, still wider prospects.

Nature
Inaugural Address (p. 257)
Volume 81, Number 2078, 26 August, 1909

Thoreau, Henry

Do not engage to find things as you think they are.

The Writings of Henry David Thoreau
Volume 6
Letter, August 9, 1850 to Harrison Blake (p. 186)

Unknown

Pioneers occupy new land. Only later, one comes to understand that the cabins they built were really cathedrals.

In Jeremy Bernstein
Experiencing Science
Chapter 1(p. 3)

von Lenard, Philipp E.A.

. . . I have by no means always been numbered among those who pluck the fruit; I have been repeatedly only one of those who planted or cared for the trees. . .

In Nobel Foundation
Nobel Lecture
Physics 1901–1921
Nobel Lecture of Philipp E.A. von Lenard
May 28, 1906 (p. 105)

DISTANCE

Coblentz, Stanton

Our race has want of sages such as these,
Whose measuring-rods are light-years, and who say
That points a million trillion leagues away
Are only as our next-door galaxies.

Sky and Telescope
Astronomers (p. 12)
Volume III, Number 10, August 1940

Heidmann, Jean

The distances we are going to embrace are so enormous that galaxies will appear as tiny toys, infinitesimal as the dust-specks dancing in the sunbeam in the crack of the curtain.

Extragalactic Adventure
Chapter 2 (p. 20)

DOGMA

Abbey, Edward

Let our practice form our doctrine, thus assuring precise theoretical coherence.

The Monkey Wrench Gang
Chapter 5 (p. 68)

Huxley, Julian

'The undevout astronomer is mad.'
Thus Young declared a century ago.
To-day we'd like to know what right he had
To dogmatize and lay the law down so.

The Captive Shrew
Undevout Astronomers

DUST

Greenberg, J. Mayo

Astronomers no longer consider interstellar dust a nuisance. Rather it is a major source of information about the birth of stars, planets and comets, and it may even hold clues to the origin of life itself.

Scientific American
The Secrets of Stardust (p. 75)
Volume 283, Number 6, December 2000

Kapteyn, A.J.

Undoubtedly one of the greatest difficulties, if not the greatest of all, in the way of obtaining an understanding of the real distribution of the stars in space, lies in our uncertainty about the amount of loss suffered by the light on its way to the observer.

The Astrophysical Journal
On the Absorption of Light in Space (p. 46)
Volume 29, Number 1, January 1909

ECLIPSE

Amos 8:9
On that day, says the Lord God, I will make the sun go down at noon and darken the earth in broad daylight.

The Bible

Archilochus
Nothing can be surprising any more or impossible or miraculous, now that Zeus, father of the Olympians has made night out of noonday, hiding the bright sunlight, and... fear has come upon mankind. After this, men can believe anything, expect anything. Don't any of you be surprised in future if land beasts change places with dolphins and go to live in their salty pastures, and get to like the sounding waves of the sea more than the land, while the dolphins prefer the mountains.

In F. Richard Stephenson
Historical Eclipses and Earth's Rotation
Chapter 10 (p. 338)

Caithness, James Balharrie
I watched the shadow of our globe
Pass sheer across the moon,
She sadly donned the somber robe,
But glad emerging, soon
Cast all its dismal folds aside,
Bright through the heavens again to ride.

Pastime Poems
An Eclipse of the Moon (Second Version)

Flammarion, Camille
Eclipses, like comets, have always been interpreted as the indication of inevitable calamities. Human vanity sees the finger of God making signs

to us on the least pretext, as if we were the end and aim of universal creation.

Popular Astronomy
Book II, Chapter IX (p. 180)

Hardy, Thomas
Thy shadow, Earth, from Pole to Central Sea,
Now steals along upon the Moon's meek shine
In even monochrome and curving line
Of imperturbable serenity.

How shall I like such sun-cast symmetry
With the torn troubled form I know as thine,
That profile, placid as a brow divine,
With continents of moil and misery?

And can immense mortality but throw
So small a shade, and Heaven's high human scheme
Be hemmed within the coasts yon arc implies?

Is such a stellar gauge of earthly show,
Nation at war with nation, brains that teem,
Heroes, and women fairer than the skies?

Collected Poems of Thomas Hardy
At a Lunar Eclipse

Joel 2:30, 31
I will show portents in the sky and on earth, blood and fire and columns of smoke; the sun shall be turned into darkness and the moon into blood before the great and terrible day of the Lord comes.

The Bible

Plato
...people may injure their bodily eye by observing and gazing on the sun during an eclipse, unless they take the precaution of only looking at the image reflected in the water, or in some similar medium.

Phaedo
Section 99

...people may injure their bodily eye by observing and gazing on the sun during an eclipse, unless they take the precaution of only looking at the image reflected in the water, or in some similar medium.
Plato – (See p. 101)

Poincaré, Henri
Why do the rains, the tempests themselves seem to us to come by chance, so that many persons find it quite natural to pray for rain or shine, when they would think it ridiculous to pray for an eclipse?

The Foundations of Science
Science and Method
Book I
Chapter IV, section II (p. 398)

Shakespeare, William
And now the house of York, thrust from the crown
By shameful murder of a guiltless king
And lofty proud encroaching tyranny,
Burns with revenging fire; whose hopeful colours
Advance our half-faced sun, striving to shine,
Under the which is writ "Invitis nubibus."

The commons here in Kent are up in arms:
And, to conclude, reproach and beggary
Is crept into the palace of our king.
And all by thee. Away! convey him hence.

The Second Part of King Henry the Sixth
Act IV, scene I, L. 94–103

Unknown
The sky is an immense place where everything moves in its own way. Some go faster than others, some turn around themselves while they are going around the sun and each other. Sometimes the fast ones catch up with the slow ones and that is an eclipse. Also, when the moon gets in the way we can't see the sun, and that is an eclipse too and everybody talks about it.

That is an important eclipse.

The Sky
Reported by little girl who visited the Hayden Planetarium (p. 25)
December 1939

Wordsworth, William
High on her speculative tower
Stood Science waiting for the hour
When Sol was destined to endure
That darkening of his radiant face
Which Superstition strove to chase,
Erewhile, with rites impure.

The Complete Poetical Works of Wordsworth
The Eclipse of the Sun

ELECTRON

Bragg, Sir William
. . . an electron springs into existence.

Scientific Monthly
Electrons and Ether Waves (p. 156)
Volume XIV, February 1922, Number 8

Frankel, Felice
Whitesides, George M.
Electrons know two verbs: *seek* and *avoid*. They seek the positive charges of atomic nuclei; they avoid the negative charges of other electrons. That is almost all they know.

On the Surface of Things
Microelectrodes (p. 99)

Lederman, Leon
The "naked" electron is an imaginary object cut off from the influences of the field, whereas a "dressed" electron carries the imprint of the universe.

The God Particle

Sullivan, J.W.N.
The electron is not. . . an enduring something that can be tracked through time. Its mathematical description does not involve that degree of definiteness. Any picture we form of the atom errs, as it were, by excess of solidity. The mathematical symbols refer to entities more indefinite than our pictorial imagination, limited as it is by experience of "gross matter," can construct.

The Bases of Modern Science
Chapter XI (pp. 252–3)

ELLIPSE

Hardy, Thomas
His world was an ellipse, with a dual centrality, of which his own was not the major.

Collected Poems of Thomas Hardy
For Conscience' Sake
Chapter III

ENERGY

Feynman, Richard
Leighton, R.B.
Sands, M.

There is a fact, or if you wish, a law governing all natural phenomena that are known to date. There is no known exception to this law—it is exact as far as we know. The law is called the conservation of energy. It states that there is a certain quantity, which we call energy, that does not change in the manifold changes which nature undergoes. That is a most abstract idea, because it is a mathematical principle; it says that there is a numerical quantity which does not change when something happens.

The Feynman Lectures on Physics
Volume I
Chapter 4 (p. 4-1)

It is important to realize that in physics today, we have no knowledge of what energy *is*.

The Feynman Lectures on Physics
Volume I
Chapter 4 (p. 4-2)

Huxley, Julian

I am Energy. Sublime and meaningless Energy.
I stream in floods across the empty ocean
Of Space, where island-universes float,
Each like a little lonely boat.

The Captive Shrew
Matter, Energy, Time and Space

Meyerson, Emile

Energy is really only an integral; now, what we want to have is a *substantial* definition, like that of Leibniz, and this demand is justifiable to a certain degree, since our very conviction of the conservation of energy rests in great part on this foundation... And so the manuals of physics contain really two discordant definitions of energy, the first which is verbal, intelligible, capable of establishing our conviction, and false; and the second which is mathematical, exact, but lacking verbal expression.

Identity & Reality
Chapter VIII (p. 280)

It is important to realize that in physics today, we have no knowledge of what energy *is*.
Richard Feynman, R.B. Leighton and M. Sands – (See p. 106)

Moulton, Forest Ray
The innumerable members of our galaxy assure us, however, that though the stars may be evaporating, so to speak, like the dew, their energies will in some way be integrated again.

Astronomy
Chapter XV (p. 471)

Soddy, Frederick
Energy, someone may say, is a mere abstraction, a mere term, not a real thing. As you will. In this, as in many another respects, it is like an abstraction no one would deny reality to, and that abstraction is wealth. Wealth is the power of purchasing, as energy is the power of working. I cannot show you energy, only its effects... Abstraction or not, energy is as

real as wealth—I am not sure that they are not two aspects of the same thing.

Science and Life
Physical Force—Man's Servant or His Master (p. 27)

Unknown
Don't bother me, I'm busy conserving energy, momentum, and angular momentum.

Source unknown

EQUATION

Hawking, Stephen
What is it that breathes fire into the equations and makes a universe for them to describe? The usual approach of science of constructing a mathematical model cannot answer the questions of why there should be a universe for the model to describe. Why does the universe go to all the bother of existing?

A Brief History of Time
Chapter 11 (p. 174)

Holton, Gerald
Roller, Duane H.D.
Without the clear understanding that equations in physical science always have hidden limitations, we cannot expect to interpret or apply them successfully. For instance, we would continually be tempted to make unwarranted extrapolations and interpolations. We would be in the catastrophic position of a navigator who has to negotiate a rocky channel without having any idea of the length, width, and draft of his ship.

Foundations of Modern Physical Science
Chapter I (pp. 4–5)

London, Jack
The difference between the sun's position and the position where the sun ought to be if it were a decent, self-respecting sun is called the Equation of Time.

The Cruise of the Snark
Chapter 14 (p. 244)

Saaty, Thomas L.
Equations are the lifeblood of applied mathematics and science.

Modern Nonlinear Equations
Preface (p. vii)

ERROR

van de Kamp, Peter

...should we not come to the rescue of a cosmic phenomenon trying to reveal itself in a sea of errors?

Vistas in Astronomy
The Planetary System of Barnard's Star (p. 157)
Volume 26, 1982

Whitehead, Alfred North

The results of science are never quite true. By a healthy independence of thought perhaps we sometimes avoid adding other people's errors to our own.

The Aims of Education
Chapter X (p. 233)

Wright, Wilbur

If a man is in too big a hurry to give up an error he is liable to give up some truth with it, and in accepting the arguments of the other man he is sure to get some error with it.

In Fred C. Kelly (ed.)
Miracle at Kitty Hawk
Chapter III
Letter from Wilbur Wright to George A. Spratt
April 27, 1903 (p. 89)

ETERNITY

Harrison, Edward
If eternity is silliness, then infinity of space is sheer madness.

Masks of the Universe
Chapter 12 (p. 202)

Paine, Thomas
It is difficult beyond description to conceive that space can have no end; but it is more difficult to conceive an end. It is difficult beyond the power of man to conceive an eternal duration of what we call time; but it is more impossible to conceive a time when there shall be no time.

The Age of Reason
Part First
Chapter X (p. 25)

Vaughan, Henry
I saw eternity the other night,
Like a great ring of pure and endless light,
All calm, as it was bright:—
And round beneath it, Time, in hours, days, years,
Driven by the spheres,
Like a vast shadow moved; in which the World
And all her train were hurl'd.

Poetry and Selected Prose
A Vision

Young, Edward
Eternity is written in the skies.

Night Thoughts
9

Eternity is written in the skies.
Edward Young – (See p. 111)

EVENT

Ferguson, Kitty

Events in the heavens happen in their own good time and not before, and they are often not repeatable. Astronomers have learned to take what's on offer and make the best of it.

Measuring the Universe
Prologue (p. 4)

Milne, Edward

Not only the laws of nature, but also the events occurring in nature, the world itself, must appear the same to all observers, wherever they may be.

Zeitschrift für Betriebswirtschaft
Volume 6, 1933 (p. 1)

EXPERIMENT

Ehrlich, Paul

Much testing; accuracy and precision in experiment; no guesswork or self-deception.

In Martha Marquardt
Paul Ehrlich
Chapter XIII (p. 134)

Gore, G.

In scientific study also, as in other abstruse meditations, the mind soon becomes exhausted by intense thinking, but is usually relieved by preparing and making experiments.

The Art of Scientific Discovery
Chapter XXXIII (p. 313)

Rutherford, E.

Experiment without imagination or imagination without recourse to experiment, can accomplish little, but, for effective progress, a happy blend of these two powers is necessary.

Science
The Electrical Structure of Matter (p. 221)
Volume 58, Number 1499, September 21, 1923

Sagan, Carl

When theory is not adequate in science, the only realistic approach is experimental. Experiment is the touchstone of science on which the theories are framed. It is the court of last resort.

The Cosmic Connection
Chapter 5 (p. 37)

EXTRATERRESTRIAL LIFE

Butler, Samuel

Quoth he—Th' Inhabitants o' the *Moon*,
Who when the Sun shines hot at Noon,
Do live in Cellars underground
Of eight Miles deep and eighty round
(In which at once they fortify
Against the Sun and th' Enemy)
Which they count towns and Cities there,
Because their People's civiler
Than those rude Peasants, that are found
To live upon the upper Ground,
Call'd Privolvans, with whom they are
Perpetually at open War.

In René Lamar (ed.)
Satires and Miscellaneous Prose
The Elephant in the Moon

de Saint-Exupéry, Antoine

So you, too, come from the sky! Which is your planet?

The Little Prince
Part III (p. 14)

Diamond, Jared

Think again of those astronomers who beamed radio signals into Space from Arecibo, describing Earth's location and its inhabitants. In its suicidal folly that act rivaled the folly of the last Inca emperor, Atahualpa, who described to his gold-crazy Spanish captors the wealth of his capital and provided them with guides for the journey. If there really are any radio civilizations within listening distance of us, then for heaven's sake let's turn off our own transmitters and try to escape detection, or we are doomed.

Fortunately for us, the silence from Outer Space is deafening... What woodpeckers teach us about flying saucers is that we're unlikely to ever see one.

The Third Chimpanzee
Part 3
Chapter 12 (pp. 214–15)

So you, too, come from the sky! Which is your planet?
Antoine de Saint-Exupéry – (See p. 115)

Dick, Steven

As we stand on the threshold of the new millennium, we may conjecture that 1,000 years from now we will have had our answer to this age-old question. Humanity 3,000 will know whether or not it is alone in the universe, at least within our galaxy.

Extraterrestrial Life and Our World at the Turn of the Millennium (p. 44)

Dickinson, Terence

... the lure of backyard exploration of the universe: the chilling realization that Earth is but a mote of dust adrift in the ocean of space. The fact that Earth harbours creatures who are able to contemplate their place in the cosmic scheme must make our dust speck a little special. But wondering who else is out there only deepens the almost mystical enchantment of those remote celestial orbs.

Nightwatch

Dietrich, Marlene
Until they come to see us from their planet, I wait patiently. I hear them saying: Don't call us, we'll call you.

Marlene Dietrich's ABC
Venus

Eddington, Sir Arthur Stanley
I do not think that the whole purpose of the Creation has been staked on the one planet where we live; and in the long run we cannot deem ourselves the only race that has been or will be gifted with the mystery of consciousness. But I feel inclined to claim that *at the present time* our race is supreme; and not one of the profusion of stars in their myriad clusters looks down on scenes comparable to those which are passing beneath the rays of the sun.

The Nature of the Physical World
Chapter VIII (p. 178)

Eiseley, Loren
So deep is the conviction that there must be life out there beyond the dark, one thinks that if they are more advanced than ourselves they may come across space at any moment, perhaps in our generation. Later, contemplating the infinity of time, one wonders if perchance their messages came long ago, hurtling into the swamp muck of the steaming coal forests, the bright projectile clambered over by hissing reptiles, and the delicate instruments running mindlessly down with no report.

The Immense Journey
Little Men and Flying Saucers (p. 144)

In a universe whose size is beyond human imagining, where our world floats like a dust mote in the void of night, men have grown inconceivably lonely. We scan the time scale and the mechanisms of life itself for portents and signs of the invisible. As the only thinking mammals on the planet—perhaps the only thinking animals in the entire sidereal universe—the burden of consciousness has grown heavy upon us. We watch the stars, but the signs are uncertain. We uncover the bones of the past and seek for our origins. There is a path there, but it appears to wander. The vagaries of the road may have a meaning, however; it is thus we torture ourselves.

The Immense Journey
Little Men and Flying Saucers (pp. 161-2)

. . . nowhere in all space or on a thousand worlds will there be men to share our loneliness. There may be wisdom; there may be power; somewhere across space great instruments, handled by strange, manipulative organs, may stare vainly at our floating cloud wrack, their owners yearning as we yearn. Nevertheless, in the nature of life and in the principles of evolution we have had our answer. Of men, elsewhere, and beyond, there will be none forever.

The Immense Journey
Little Men and Flying Saucers (p. 162)

Fuller, R. Buckminster
Sometimes I think we're alone. Sometimes I think we're not. In either case, the thought is quite staggering.

In James A. Haught (ed.)
2000 Years of Disbelief
Chapter 71 (p. 290)

Giraudoux, Jean
COUNTESS:... are you so stupid as to think that just because we're alone here, there's nobody else in the room? Do you consider us so boring or so repulsive that of all the millions of beings, imaginary or otherwise, who are prowling around in space looking for a little company, there is not one who might possibly enjoy spending a moment with us? On the contrary, my dear—my house is full of guests...

The Madwoman of Chaillot
Act II (p. 94)

Huygens, Christianus
A Man that is of *Copernicus*'s Opinion, that this Earth of ours is a planet, carry'd round and enlighten'd by the Sun, like the rest of them, cannot but sometimes have a fancy, that it's not improbable that the rest of the Planets have their Dress and Furniture, nay and their Inhabitants too as well as this Earth of ours.

The Celestial Worlds Discover'd
Book the First (pp. 1–2)

Jones, Sir Harold Spencer
We see the Earth as a small planet, one member of a family of planets revolving round the Sun; the Sun, in turn, is an average star situated somewhat far out from the centre of a vast system, in which the stars are numbered by many thousands of millions; there are many millions of such systems, more or less similar to each other, peopling space to the farthest limits to which modern exploration has reached.

Can it be that throughout the vast deeps of space nowhere but on our own little Earth is life to be found?

Life on Other Worlds
Chapter I (p. 19)

Koch, Howard

Good heavens, something's wriggling out of the shadow like a grey snake. Now it's another one, and another. They look like tentacles to me. There, I can see the thing's body. It's large as a bear and glistens like wet leather. But that face. It—it's indescribable. I can hardly force myself to keep looking at it. The eyes are black and gleam like a serpent's. The mouth is V-shaped with saliva dripping from its rimless lips that seem to quiver and pulsate.

In Isabel S. Gordon and Sophie Sorkin (eds)
The Armchair Science Reader
Part I
Man Among the Stars
Invasion from Mars (p. 9)

Metrodorus of Chios

...it would be strange if a single ear of corn grew in a large plain or there were only one world in the infinite.

In F.M. Cornford
The Classical Quarterly
Innumerable Worlds in Presocratic Philosophy (p. 13)
January 1934

Milton, John

Dream not of other Worlds; what Creatures there
Live, in what state, condition or degree...

Paradise Lost
Book VIII, L. 175–6

Oparin, A.I.

...there is every reason now to see in the origin of life not a "happy accident" but a completely regular phenomenon, an inherent component of the total evolutionary development of our planet. The search for life beyond Earth is thus only a part of the more general question which confronts science, of the origin of life in the universe.

In M. Calvin and O.G. Gazenko (eds)
Foundations of Space Biology and Medicine
Volume I
Theoretical and Experimental Prerequisites of Exobiology
Chapter 7 (p. 321)

Pallister, William

No one can yet show proof that there exists
A single planet save the solar ones.
But space is wide and high, and time is long,

And there are millions more of other suns.

So men imagine why they do not know
And they assume that surely there must be
Some other planets, peopled like our own;
Some other worlds with creatures such as we.

Poems of Science
Other Worlds and Ours
Life on Other Planets (p. 210)

Pope, Alexander
He, who through vast immensity can pierce,
See worlds on worlds compose one universe,
Observe how system into system runs,
What other planets circle other suns,
What varied Being peoples every star,
May tell why Heaven has made us as we are...

The Complete Poetical Works of Pope
Essay on Man
Epistle I

Sagan, Carl
...there are a million other civilizations, all fabulously ugly, and all a lot smarter than us. Knowing this seems to me to be a useful and character-building experience for mankind.

In Richard Berendzen (ed.)
Life Beyond Earth & the Mind of Man
Sagan (p. 64)

After centuries of muddy surmise, unfettered speculation, stodgy conservatism, and unimaginative disinterest, the subject of extraterrestrial life has finally come of age.

Cosmic Connections
Preface (p. viii)

Occasionally, I get a letter from someone who is in "contact" with extraterrestrials. I am invited to "ask them anything." And over the year's I've prepared a little list of questions. The extraterrestrials are very advanced, remember. So I ask things like, "Please provide a short proof of Fermat's Last Theorem."...I write out the simple equation with the exponents...It's a stimulating exercise to think of questions to which no human today knows the answers, but where a correct answer would immediately be recognized as such. It's even more challenging to formulate such questions in fields other than mathematics. Perhaps we should hold a contest and collect the best responses in "Ten Questions to Ask an Alien."

The Demon-Haunted World
Chapter 6 (p. 100, fn)

We are like the inhabitants of an isolated valley in New Guinea who communicate with societies in neighboring valleys (quite different societies, I might add) by runner and by drum. When asked how a very advanced society will communicate, they might guess by an extremely rapid runner or by an improbably large drum. They might not guess a technology beyond their ken. And yet, all the while, a vast international cable and radio traffic passes over them, around them, and through them...

We will listen for the interstellar drums, but we will miss the interstellar cables. We are likely to receive our first messages from the drummers of the neighboring galactic valleys—from civilizations only somewhat in our future. The civilizations vastly more advanced than we, will be, for a long time, remote both in distance and in accessibility. At a future time of vigorous interstellar radio traffic, the very advanced civilizations may be, for us, still insubstantial legends.

The Cosmic Connection
Chapter 31 (pp. 224, 224–5)

Sagan, Carl
Newman, William I.
We think it possible that the Milky Way Galaxy is teeming with civilizations as far beyond our level of advance as we are beyond the ants, and paying us about as much attention as we pay to the ants.

Quarterly Journal of the Royal Astronomical Society
The Solipsist Approach to Extraterrestrial Intelligence (p. 120)
Volume 24, Number 3, June 1983

Sakharov, Andrei
In infinite space many civilizations are bound to exist, among them societies that may be wiser and more "successful" than ours. I support the cosmological hypothesis which states that the development of the universe is repeated in its basic characteristics an infinite number of times... Yet this should not minimize our sacred endeavors in this world of ours, where, like faint glimmers in the dark, we have emerged for a moment from the nothingness of dark unconscious into material existence.

Nobel Peace Prize Lecture
December 11, 1975

Shakespeare, William

Glendower: I can call spirits from the vasty deep.
Hotspur: Why, so can I, or so can any man;
But will they come when you do call them?

The First Part of King Henry the Fourth
Act III, scene i, L. 53–5

Horatio: O day and night, but this is wondrous strange!
Hamlet: And therefore as a stranger give it welcome.
There are more things in heaven and earth, Horatio,
Than are dreamt of in your philosophy.

Hamlet, Prince of Denmark
Act I, scene v, L. 164–7

Shaw, George Bernard

Drier: Mr. Shaw, do you believe in life on other planets?
Shaw: Indeed I do.
Drier: But, Mr. Shaw, what proof do you have?
Shaw: The proof is that they're using us for an insane asylum.

Chemistry
Volume 42, Number 4, April 1969 (p. 2)

Tsiolkovsky, Konstantin

Is it probable for Europe to be inhabited and not the other parts of the world? Can one island have inhabitants and numerous other islands have none? Is it conceivable for one apple-tree in the infinite orchard of the Universe to bear fruit, while innumerable other trees have nothing but foliage?

In Adam Starchild (ed.)
The Science Fiction of Konstantin Tsiolkovsky
Dreams of the Earth and Sky (p. 154)

von Braun, Wernher

Our sun is one of 100 billion stars in our galaxy. Our galaxy is one of billions of galaxies populating the universe. It would be the height of presumption to think that we are the only living things in that enormous immensity.

The New York Times
Text of the Address by von Braun Before the Publishers' Group Meeting Here
29 April 1960
L. 20, column 2

Wells, H.G.

No one would have believed in the last years of the nineteenth century that this world was being watched keenly and closely by intelligences greater than man's and yet as mortal as his own...

Seven Famous Novels by H.G. Wells
The War of the Worlds
Book I
Chapter 1 (p. 265)

Those who have never seen a living Martian can scarcely imagine the strange horror of its appearance.

Seven Famous Novels by H.G. Wells
The War of the Worlds
Book I
Chapter 4 (p. 276)

FACT

Bridgman, P.W.

...the *fact* has always been for the physicist the one ultimate thing from which there is no appeal, and in the face of which the only possible attitude is a humility almost religious.

The Logic of Modern Physics
Chapter I (pp. 2–3)

Chesterton, G.K.

Facts as facts do not always create a spirit of reality, because reality is a spirit.

Come to Think of It
On the Classics (p. 49)

Collins, Wilkie

"Facts?" he repeated. "Take a drop more grog, Mr. Franklin, and you'll get over the weakness of believing in facts! Foul play, Sir!"

The Moonstone
Chapter IV (p. 275)

Faraday, Michael

...it is always safe and philosophic to distinguish, as much as is in our power, fact from theory; the experience of past ages is sufficient to show us the wisdom of such a course; and considering the constant tendency of the mind to rest on an assumption, and, when it answers every present purpose, to forget that it is an assumption, we ought to remember that it, in such cases, becomes a prejudice, and inevitably interferes, more or less, with a clear-sighted judgment. I cannot doubt but that he who, as a wise philosopher, has most power of penetrating the secrets of nature, and guessing by hypothesis at her mode of working, will also be most careful, for his own safe progress and that of others, to distinguish that knowledge which consists of assumption, by which I mean theory and hypothesis, from that which is the knowledge of facts and laws; never raising the former to the dignity or authority of the latter, nor confusing the latter more than is inevitable with the former.

Philosophical Magazine
A Speculation Touching Electric Conduction and the Nature of Matter (p. 136)
Volume XXIV, January–June, 1844

Huxley, Thomas
Men of science do not pledge themselves to creeds; they are bound by articles of no sort; there is not a single belief that it is not a bounden duty with them to hold with a light hand and to part with cheerfully, the moment it is really proved to be contrary to any fact, great or small.

Collected Essays
Volume II
Darwiniana
On Our Knowledge of the Causes of the Phenomena of Organic Nature
Lecture VI (pp. 468–9)

Jacks, L.P.
Facts are popularly regarded as antidotes to mysteries. And yet, in sober earnest, there is nothing so mysterious as a fact.

The Atlantic Monthly
Is There a Foolproof Science? (p. 229)
Volume 133, Number 2, February 1924

Krough, A.
Facts are necessary, of course, but unless fertilized by ideas, correlated with other facts, illuminated by thought, I consider them as material only for science.

Science
The Progress of Physiology (p. 203)
Volume 70, 1929

Mayer, J.R.
If a fact is known on all its sides, it is, by that knowledge, explained, and the problem of science is ended.

In Ernst Mach
History and Root of the Principle of the Conservation of Energy
Chapter III (p. 58)

Michelson, A.A.
The more important fundamental laws and facts of physical science have all been discovered, and these are now so firmly established that the possibility of their ever being supplanted in consequence of

new discoveries is exceedingly remote... Our future discoveries must be looked for in the sixth place of decimals.

Light Waves and Their Uses
Lecture II (pp. 23–4, 24)

Poincaré, Henri
The historian, the physicist, even, must make a choice among facts; the head of the scientist, which is only a corner of the universe, could never contain the universe entire; so that among the innumerable facts nature offers, some will be passed by, others retained.

The Foundations of Science
Science and Method
Book I
Chapter II (p. 369)

Shaw, George Bernard
But an Englishman was not daunted by facts. To explain why all the lines in his rectilinear universe were bent, he invented a force called gravitation and thus erected a complete British universe and established it as a religion which was devoutly believed in for 300 years. The book of this Newtonian religion was not that oriental magic thing, the Bible. It was that British and matter-of-fact thing, a Bradshaw[a British railway timetable]. It gives the stations of all the heavenly bodies, their distances, the rates at which they are traveling, and the hour at which they reach eclipsing points or crash into the earth like Sirius. Every time is precise, ascertained, absolute and English.

In B. Patch
Thirty Years with G.B.S.
Chapter 12 (p. 235)

Snow, C.P.
A fact is a fact is a fact.

The Two Cultures: And a Second Look
Chapter 4 (p. 45)

Tyndall, John
It is as fatal as it is cowardly to blink facts because they are not to our taste.

Fragments of Science
Volume II
Science and Man

Whewell, William
When we inquire what Facts are to be made the materials of Science, perhaps the answer which we should most commonly receive would be,

that they must be *True Facts*, as distinguished from any mere inferences or opinions of our own.

Novum Organon Renovatum
Chapter III (pp. 50–1)

FORCE

Moleschott, Jakob

Force is not an impelling God, not an essence separate from the material substratum of things. A force not united to matter, but floating freely above it, is an idle conception. Nitrogen, carbon, hydrogen, oxygen, sulphur, and phosphorus, possess their inherent qualities from eternity.

In Ludwig Büchner
Force and Matter
Chapter I (p. 1)

Weil, Simone

Two forces rule the universe: light and gravity.

Gravity and Grace
Gravity and Grace (p. 45)

Whitman, Walt

You unseen force, centripetal, centrifugal, through space's spread,
Rapport of sun, moon, earth, and all the constellations,
What are the messages by you from distant stars to us?

In James E. Miller, Jr (ed.)
Complete Poetry and Selected Prose
Fancies at Navesink

FORMULA

Emerson, Ralph Waldo

The formulas of science are like the papers in your pocketbook, of no value to any but their owner.

Essays and Lectures
The Conduct of Life
Beauty (p. 1100)

Mitchell, Maria

Every formula which expresses a law of nature is a hymn to praise of God.

In Phebe Mitchell Kendall
Maria Mitchell: Life, Letters, and Journals
Chapter IX (p. 185)

Planck, Max

Ever since the observation of nature has existed, it has held a vague notion of its ultimate goal as the composition of the colorful multiplicity of phenomena in a uniform system, where possible, in a single formula.

In Ernest Peter Fischer
Beauty and the Beast
Chapter 2 (p. 47)

Stoppard, Tom

THOMASINA: If you could stop every atom in its position and direction, and if your mind could comprehend all the actions thus suspended, then if you were really, *really*, good at algebra you could write the formula for all the future; and although nobody can be so clever to do it, the formula must exist just as if one could.

Arcadia
Act I, Scene I (p. 5)

FUSION

Pauli, Wolfgang
Let no man join together what God hath put asunder.

The Atlantic Monthly
In Robert P. Crease and Charles C. Mann
How the Universe Works (p. 68)
August 1984

FUTURE

Albran, Kehlog
I have seen the future and it is very much like the present, only longer.

The Profit (p. 89)

Einstein, Albert
The scientist is possessed by the sense of universal causation. The future, to him, is every whit as necessary and determined as the past.

Ideas and Opinions
The Religious Spirit of Science (p. 40)

Hilbert, David
Who of us would not be glad to lift the veil behind which the future lies hidden; to cast a glance at the next advances of our science and at the secrets of its development during future centuries?

Bulletin of the American Mathematical Society
Mathematical Problems (p. 437)
Volume 8, July 1902

Poincaré, Lucien
It would doubtless be exceedingly rash, and certainly very presumptuous, to seek to predict the future which may be reserved for physics. The role of prophet is not a scientific one, and the most firmly established precisions of to-day may be overthrown by the reality of to-morrow.

The New Physics and its Evolution
Chapter XI (p. 322)

Tennyson, Alfred
When I dipt into the Future, far as human eye could see;
Saw the vision of the world, and all the wonder that would be.

The Complete Poetical Works of Tennyson
Locksley Hall

GALAXY

Eddington, Arthur

Let us first understand what a galaxy is. The following is a recipe for making galaxies: Take about ten thousand million stars. Spread them so that on the average light takes three or four years to pass from one to the next. Add about the same amount of matter in the form of diffuse gas between the stars. Roll it all out flat. Set it spinning in its won plane. Then you will obtain an object which, viewed from a sufficient distance, will probably look more or less like a spiral nebula.

New Pathways in Science
Chapter X (p. 206)

Hoyle, Fred

Think of the stars as ordinary household specks of dust. Then we must think of a galaxy as a collection of specks a few miles apart from each other, the whole distribution filling a volume about equal to the Earth. Evidently one such collection of specks could pass almost freely through another.

Frontiers of Astronomy
Chapter 16 (p. 278)

Jeffers, Robinson

Galaxy on galaxy, innumerable swirls of innumerable
stars, endured as it were forever and humanity
Came into being, its two or three million years are a
moment, in a moment it will certainly cease out from being
And galaxy on galaxy endure after that as it were forever...

The Selected Poetry of Robinson Jeffers
Margrave

Wolf, Fred Alan
Stars, like little lost children seeking shelter on a cold night, tend to cluster, via gravitationally induced starlight, into galaxies.

Parallel Universes
Chapter 6 (p. 71)

GEOMETRY

Chern, Shiing Shen

Physics and geometry are one family.
Together and holding hands they roam to the limits of outer space...
Surprisingly, Math. has earned its rightful place for man and in the sky;
Fondling flowers with a smile—just wish nothing is said!

Notices of the American Mathematical Society
Interview with Shiing Shen Chern (p. 865)
Volume 45, Number 7, August 1998

de Fontenelle, Bernard Le Bovier

A work of morality, politics, criticism will be more elegant, other things being equal, if it is shaped by the hand of geometry.

L'utilité des Mathématiques et de la Physique
Preface

Manning, Henry Parker

...the greatest advantage to be derived from the study of geometry of more than three dimensions is a real understanding of the great science of geometry. Our plane and solid geometries are but the beginning of this science. The four-dimensional geometry is far more extensive than the three-dimensional, and all the higher geometries are more extensive than the lower.

Geometry of Four Dimensions
Introduction (p. 13)

Suspended o'er geometry,
I am a fish-worm dangling—
A creature too obtuse to see
What is acute in angling.

In Francis A. Litz (ed.)
The Poetry of Father Tabb
Humorous Verse
A Problem in Mathematics

Morgan, Frank

We are just beginning to understand how geometry rules the universe.

American Mathematical Monthly
Review: The Parsimonious Universe (p. 376)
Volume 104, Number 4, April, 1997

Russell, Bertrand

It was formerly supposed that Geometry was the study of the nature of the space in which we live, and accordingly it was urged, by those who held that what exists can only be known empirically, that Geometry should really be regarded as belonging to applied mathematics. But it has gradually appeared, by the increase of non-Euclidean systems, that Geometry throws no more light upon the nature of space than Arithmetic throws upon the population of the United States.

Mysticism and Logic and Other Essays
Chapter V (p. 92)

GOD

Card, Orson Scott
All the universe is just a dream in God's mind, and as long as he's asleep, he believes in it, and things stay real.

Seventh Son
Chapter 10 (p. 126)

Dawkins, Richard
If God is a synonym for the deepest principles of physics, what word is left for a hypothetical being who answers prayers, intervenes to save cancer patients or helps evolution over difficult jumps, forgives sins or dies for them?

Forbes ASAP
Snake Oil and Holy Water (p. 236)
October 4, 1999

Feynman, Richard
God was invented to explain mystery. God is always invented to explain those things that you do not understand. Now when you finally discover how something works, you get some laws which you're taking away from God; you don't need him anymore. But you need him for the other mysteries. So therefore you leave him to create the universe because we haven't figured that out yet; you need him for understanding those things which you don't believe the laws will explain, such as consciousness, or why you only live to a certain length of time—life and death—stuff like that. God is always associated with those things that you do not understand. Therefore I don't think that the laws can be considered to be like God because they have been figured out.

In P.C.W. Davies and J. Brown (eds)
Superstrings: A Theory of Everything
Chapter 9 (pp. 208–9)

Greenstein, George

As we survey all the evidence, the thought insistently arises that some supernatural agency—or, rather, Agency—must be involved. Is it possible that suddenly, without intending to, we have stumbled upon scientific proof of the existence of a Supreme Being? Was it God who stepped in and so providentially crafted the cosmos for our benefit?

The Symbiotic Universe
Prologue (p. 27)

Hawking, Stephen

The idea that space and time may form a closed surface without boundary also has profound implications for the role of God in the affairs of the universe. With the success of scientific theories in describing events, most people have come to believe that God allows the universe to evolve according to a set of laws and does not intervene in the universe to break these laws. However, the laws do not tell us what the universe should have looked like when it started—it would still be up to God to wind up the clockwork and choose how to start it off. So long as the universe had a beginning, we could suppose it had a creator. But if the universe is really completely self-contained, having no boundary or edge, it would have neither beginning nor end: it would simply be. What place, then, for a creator?

A Brief History of Time
Chapter 8 (pp. 140–1)

Herrick, Robert

Science in God, is known to be
A Substance, not a Qualitie.

In J. Max Patrick (ed.)
The Complete Poetry of Robert Herrick
Science in God

Jastrow, Robert

When an astronomer writes about God, his colleagues assume he is either over the hill or going bonkers.

God and the Astronomers
Chapter 1 (p. 11)

Keillor, Garrison

We wondered if there is a God or is the universe only one seed in one apple on a tree in another world where a million years of ours is only one of their moments and what we imagine as our civilization is only a tiny charge of static electricity and the great truth that our science is slowly grasping is the fact the apple in which we are part of one seed is falling, has been falling for a million years and in one one-millionth of a second it will hit hard-frozen ground in that other world and split open and lie on the ground and a bear will come along and gobble it up, everything, the Judeo-Christian heritage, science, democracy, the Renaissance, art, music, sex, sweet corn—all disappear into that black hole of a bear.

The Atlantic Monthly
Leaving Home (p. 48)
Volume 260, Number 3 September 1987

Kepler, Johannes

...if there are globes in the heaven similar to our Earth, do we vie with them over who occupies the better portion of the universe? For if their globes are nobler, we are not the noblest of rational creatures. Then how can all things be for man's sake? How can we be the master of God's handiwork?

Conversations with Galileo's Sidereal Messenger
Section VIII (p. 43)

Lambert, Johann Heinrich

If we admit the existence of a Supreme Disposer, who brought order out of Chaos, and gave form to the universe, it will follow that the universe is a perfect work, the impression, the character, the reflected image of the perfections of its author.

The System of the World
Part I
Chapter II (p. 9)

Orgel, Irene

"But before Man," asked Jonah, shocked out of his wits, "do you mean you understood nothing at all? Didn't you exist?"
"Certainly," said God patiently. "I have told you how I exploded in the stars. Then I drifted for aeons in clouds of inchoate gas. As matter stabilized, I acquired the knowledge of valency. When matter cooled, I lay sleeping in the insentient rocks."

The Odd Tales of Irene Orgel
Jonah (pp. 17–18)

Polyakov, Alexander

We know that nature is described by the best of all possible mathematics because God created it.

In S. Gannes
Fortune
Alexander Polyakov; 40: Probing the Forces of the Universe (p. 57)
Volume 114, Number 8, October 13, 1986

Reade, Winwood

When we have ascertained, by means of Science, the methods of nature's operations, we shall be able to take her place to perform them for ourselves...men will master the forces of nature; they will become themselves architects of systems, manufacturers of worlds. Man will then be perfect; he will be a creator; he will therefore be what the vulgar worship as God.

Martyrdom of Man
Chapter IV (pp. 513, 515)

Thomson, J. Arthur

The heavens are telling the glory of God.

Concerning Evolution
Chapter I, section 6 (p. 12)

Twain, Mark

If I were going to construct a God I would furnish Him with some ways and qualities and characteristics which the Present (Bible) One lacks...He would spend some of His eternities in trying to forgive Himself for making man unhappy when He could have made him happy with the same effort and He would spend the rest of them in studying astronomy.

In Albert Bigelow Paine (ed.)
Mark Twain's Notebook
Chapter XXVI (pp. 301, 302)

von Braun, Wernher

The more we learn about God's creation, the more I am impressed with the orderliness and unerring perfection of the natural laws that govern it.

In Erik Bergaust
Wernher von Braun
The Starry Sky Above Me (p. 113)

Ziman, John

As has been said of some experiments in high-energy physics: the process to be observed has never occurred before in the history of the Universe; God himself is waiting to see what will happen!

Reliable Knowledge
Chapter 3 (fn 11, p. 62)

GRAIN

Heiles, Carl

...needle-like grains tend to spin end-over-end, like a well-kicked American football.

In D.J. Hollenbach and H.A. Thronson (eds)
Interstellar Processes
Section III (p. 171)

Seab, C.G.

Once the newly formed grains are injected into the interstellar medium, they are subject to a variety of indignities...

In M.E. Bailey and D.A. Williams (eds)
Dust in the Universe
Chapter 32, section 32.1 (p. 304)

GRAVITATIONAL LENS

Drake, Frank
Sobel, Dava
"I know perfectly well that at this moment the whole universe is listening to us," Jean Giraudoux wrote in *The Madwoman of Chaillot*, "and that every word we say echoes to the remotest star." That poetic paranoia is a perfect description of what the Sun, as a gravitational lens, could do for the Search for Extraterrestrial Intelligence.

Is Anyone Out There?
Chapter 10 (p. 232)

GRAVITY

Arnott, Neil

Attraction, as gravitation, is the muscle and tendon of the universe, by which its mass is held together and its huge limbs are wielded. As cohesion and adhesion, it determines the multitude of physical features of its different parts. As chemical or interatomic action, it is the final source to which we trace all material changes.

In J. Dorman Steele
Popular Physics
Chapter III (p. 41)

Bierce, Ambrose

Gravitation, n. The tendency of all bodies to approach one another with a strength proportional to the quantity of matter they contain—the quantity of matter they contain being ascertained by the strength of their tendency to approach one another. This is a lovely and edifying illustration of how science, having made A the proof of B, makes B the proof of A.

The Devil's Dictionary

Blake, William

God keep me... from supposing Up and Down to be the same thing as all experimentalists must suppose.

The Complete Prose and Poetry of William Blake
Letter to George Cumberland
12 April 1827

Feynman, Richard P.

But I would like *not* to underestimate the value of the world view which is the result of scientific effort. We have been led to imagine all sorts of things infinitely more marvelous than the imaginings of poets and dreamers of the past. It shows that the imagination of nature is far, far greater than the imagination of man. For instance, how much more remarkable it is for us all to be stuck—half of us upside down—by a mysterious attraction to a spinning ball that has been swinging in space for billions of years than to be carried on the back of an elephant supported on a tortoise swimming in a bottomless sea.

What Do You Care What Other People Think
The Value of Science (p. 242)

Einstein, Albert
Falling in love is not at all the most stupid thing that people do—but gravitation cannot be held responsible for it.

Quoted in Helen Dukas and Banesh Hoffman
Albert Einstein: The Human Side (p. 56)

Lockyer, Joseph Norman
The force of gravity on their surfaces must be very small. A man placed on one of them would spring with ease 60 feet high, and sustain no greater shock in his descent than he does on the Earth from leaping a yard. On such planets giants may exist; and those enormous animals which here require the buoyant power of water to counteract their weight, may there inhabit the land.

Elements of Astronomy
Chapter IX (p. 153)

Longfellow, Henry Wadsworth
Every arrow that flies feels the attraction of the earth.

The Poetical Works of Henry Wadsworth Longfellow
Hiawatha

Newton, Sir Isaac
...what hinders the fixed stars from falling upon one another?

Optics
Book III, Part I
Query 28 (p. 529)

HEAVENS

Addison, Joseph
The ways of Heaven are dark and intricate,
Puzzled in Mazes and perplex'd with errors:
Our understanding traces 'em in vain,
Lost and bewilder'd in the fruitless search;
Nor sees with how much art the winding run,
Nor where the regular confusion ends.

Cato
Act I, scene I

Alighieri, Dante
...Heaven calls you, and revolves around you, displaying to you its eternal beauties...

The Divine Comedy of Dante Alighieri
Purgatory
Canto XIV, L. 147

Browning, Robert
Ah, but a man's reach should exceed his grasp,
Or what's a heaven for?

The Complete Poetical Works of Browning
Andrea del Sarto

Burnham, Robert Jr
Here, in the dark unknown immensity of the heavens, we shall meet the glories beyond description and witness scenes of inexpressible splendor. In the great black gulfs of space and in the realm of the innumerable stars, we shall find mysteries and wonders undreamed of.

Burnham's Celestial Handbook
Chapter 2 (p. 13)

Chesterton, G.K.

Oh with what prayers and fasting
Shall mortal man deserve
To see that glimpse of Heaven…

The Coloured Lands
The Joys of Science (p. 209)

Cicero

If I had ascended the very heaven, and beheld completely the nature of the universe, and the beauty of the stars, the wonder of it would give me no pleasure, if I did not have you as a friendly, attentive, and eager reader to whom to tell it.

De amicita
Laelius
23

Dickinson, Emily

What once was heaven, is zenith now.
Where I proposed to go
When time's brief masquerade was done,
Is mapped, and charted too!

Collected Poems of Emily Dickinson
XLVIII
Old-Fashioned

Donne, John

And then that heaven, which spreads so farre, as that subtill men have, with some appearance of probabilitie, imagined, that in that heaven, in those manifold Sphere of the Planets and the Starres, there are many earths, many worlds, as big as this which we inhabit…

Donne's Sermons (p. 352)
The Heavens and Earth
Sermon 98 (p. 161)

Man has weav'd out a net, and this net throwne
Upon the Heavens, and now they are his owne…

In Charles M. Coffin (ed.)
The Complete Poetry and Selected Prose of John Donne
An Anatomy of the World
The First Anniversary

Grondal, Florence Armstrong

To the true lover of the stars, one universe or a million makes not a whit of difference. The silent song of the heavens is as sweet today, its mystery as alluring, its delights more marvelous, than in the days of yore when planets rolled out heavenly notes and stars shone through the seven spheres of pure, translucent crystal.

The Music of the Spheres
Chapter IX (p. 200)

Herschel, William

The Heavens...are now seen to resemble a luxuriant garden, which contains the greatest variety of productions, in different flourishing beds; and one advantage we may at least reap from it is that we can, as it were, extend the range of our experience to an immense duration. For, to continue the simile I have borrowed from the vegetable kingdom, is it not almost the same thing, whether we live successively to witness the germination, blooming, foliage, fecundity, fading, withering, and conception of a plant, or whether a vast number of specimens, selected from every change through which the plant passes in the course of its existence be brought at once to our view?

In Timothy Ferris
Galaxies
Introduction (p. 1)

Seneca

No man is so utterly dull and obtuse, with head so bent on Earth, as never to lift himself up and rise with all his soul to the contemplation of the starry heavens, especially when some fresh wonder shows a beacon-light in the sky. As long as the ordinary course of heaven runs on, custom robs it of its real size. Such is our constitution that objects of daily occurrence pass us unnoticed even when most worthy of our admiration. On the other hand, the sight even of trifling things is attractive if their appearance is unusual. So this concourse of stars, which paints with beauty the spacious firmament on high, gathers no concourse of the nation. But when there is any change in the wonted order, than all eyes are turned to the sky...So natural is it to admire what is strange rather than what is great.

Physical Science in the Time of Nero
Book VII
Chapter I (pp. 271, 272)

Shelley, Percy Bysshe

Heaven's ebon vault
Studded with stars unutterably bright,
Through which the moon's unclouded grandeur rolls,
Seems like a canopy which love has spread
To curtain her sleeping world.

The Complete Poetical Works of Shelley
Queen Mab, IV

Simes, James

The prose of the heavens surpasses the brightest poetry of earth.

William Herschel and His Work
Chapter V
Ocean of Ether: Star-Dust (p. 153)

HYPOTHESIS

Huxley, Thomas

Every hypothesis is bound to explain, or, at any rate, not be inconsistent with, the whole of the facts which it professes to account for; and if there is a single one of these facts which can be shown to be inconsistent with (I do not merely mean inexplicable by, but contrary to) the hypothesis, the hypothesis falls to the ground—it is worth nothing.

Collected Essays
Volume II
Darwiniana
On Our Knowledge of the Causes of the Phenomena of Organic Nature
Lecture VI (p. 463)

Osiander, Andrew

There is no need for these hypotheses to be true or even to be at all like the truth: rather one thing is sufficient for them—that they yield calculations which agree with the observations.

In Nicholas Copernicus
On the Revolutions of the Heavenly Spheres
Introduction
To the Reader Concerning the Hypothesis of this Work

Steinbeck, John

When a hypothesis is deeply accepted it becomes a growth which only a kind of surgery can amputate.

The Log from the Sea of Cortez
Chapter 17 (p. 183)

Sterne, Laurence

It is in the nature of a hypothesis when once a man has conceived it, that it assimilates everything to itself, as proper nourishment, and from the first moment of your begetting it, it generally grows stronger by everything you see, hear or understand.

The Life & Opinions of Tristram Shandy
Book II, Chapter XVIII (p. 100)

Whewell, William
The hypotheses which we accept ought to explain phenomena which we have observed. But they ought to do more than this: our hypotheses ought to *foretell* phenomena which have not yet been observed.

The Philosophy of the Inductive Sciences
Volume II
Part II
Book XI, Chapter V, Section III, article 10 (p. 62)

IDEA

Sagan, Carl

Someone has to propose ideas at the boundaries of the plausible, in order to so annoy the experimentalists or observationalists that they'll be motivated to disprove the idea.

The Washington Post
In J. Achenbach
The Final Frontier?, C1–C2
May 30, 1996

Toulmin, Stephen
Goodfield, June

New ideas are the tools of science, not its end-product. They do not *guarantee* deeper understanding, yet our grasp of Nature will be extended only if we are prepared to welcome them and give them a hearing. If at the outset exaggerated claims are made on their behalf, this need not matter. Enthusiasm and deep conviction are necessary if men are to explore all possibilities of any new idea, and later experience can be relied on either to confirm or to moderate the initial claims—for science flourishes on a double programme of speculative liberty and unsparing criticism.

The Architecture of Matter
Chapter 2 (p. 41)

Tucker, Abraham

...an idea, on being displaced by another, does not wholly vanish, but leaves a spice and tincture of itself behind, by which it operates with a kind of attraction upon the subsequent ideas, determining which of their associates they shall introduce, namely such as carry some conformity with itself...This regular succession of ideas, all bearing a reference to some one purpose retained in view, is what we call a train; and daily experience testifies how readily they follow one another in this manner of themselves, without any pains or endeavor of ours to introduce them.

The Light of Nature Pursued
Volume I
Chapter X
Trains
Section 2 (pp. 147–8)

Whewell, William

Facts are the materials of science, but all Facts involve Ideas. Since, in observing Facts, we cannot exclude Ideas, we must, for the purposes of science, take care that the Ideas are clear and rigorously applied.

The Philosophy of the Inductive Sciences
Volume II
Aphorisms
Aphorisms Concerning Science, IV (p. 467)

Whitehead, Alfred North

In the study of ideas, it is necessary to remember that insistence on hard-headed clarity issues from sentimental feeling, as it were a mist, cloaking the perplexities of fact. Insistence on clarity at all costs is based on sheer superstition as to the mode in which human intelligence functions. Our reasonings grasp at straws for premises and float on gossamers for deductions.

Adventures of Ideas
Part I
Chapter V, Section II (p. 79)

IGNORANCE

Hellman, C. Doris
There are many things whose existence we allow, but whose character we are still in ignorance of... Why should we be surprised, then, that comets, so rare a sight in the universe, are not embraced under definite laws, or that their return is at long intervals?... The day will yet come when the progress of research through long ages will reveal to sight the mysteries of nature that are now concealed... The day will yet come when posterity will be amazed that we remained ignorant of things that will to them seem so plain.

The Comet of 1577: Its Place in the History of Astronomy
Chapter I (p. 33)

Hilbert, David
... in mathematics there is no *ignorabimus*.

Bulletin of the American Mathematical Society
Mathematical Problems (p. 445)
Volume 8, July 1902

Pratchett, Terry
Because the universe was full of ignorance all around and the scientist panned through it like a prospector crouched over a mountain stream, looking for the gold of knowledge among the gravel of unreason, the sand of uncertainty and the little whiskery eight-legged swimming things of superstition.

Witches Abroad (p. 7)

IMAGINATION

Douglas, Vibert

To every investigator there come moments when his thought is baffled, when the limits of experimental possibility seem to have been reached and he faces a barrier which defies his curiosity. Then it is that imagination, like a glorious greyhound, comes bounding along, leaps the barrier, and a vision is flashed before the mind—a vision no doubt that is partly false, but a vision that may be partly true.

Atlantic Monthly
From Atoms to Stars (p. 158)
Volume 144, Number 2, August 1929

Einstein, Albert

Imagination is more important than knowledge. Knowledge is limited. Imagination encircles the world.

The Saturday Evening Post
What Life Means to Einstein: An Interview by George Sylvester Viereck
October 26, 1929

Herschel, William

If we indulge a fanciful imagination and build worlds of our own, we must not wonder at our going wide from the path of truth and nature; but these will vanish like the Cartesian vortices, that soon give way when better theories were offered.

Philosophical Transactions
The Construction of the Heavens (p. 213)
Volume LXXV, February 3, 1785

Jeans, Sir James

...are there any limits at all to the extent of space?

Even a generation ago I think most scientists would have answered this question in the negative. They would have argued that space could be limited only by the presence of something which is not space. We, or rather our imaginations, could only be prevented from journeying for ever through space by running against a wall of something different from space. And, hard though it may be to imagine space extending for ever, it is far harder to imagine a barrier of something different from space which could prevent our imaginations from passing into a further space beyond.

The Universe Around Us
Chapter I (p. 70)

Lowell, Percival

Imagination is the single source of the new... reason, like a balance wheel, only keeping the action regular. For reason... compares what we imagine with what we know, and gives us the answer in terms of the here and now, which we call the actual. But the actual... does not mark the limit of the possible.

In William Graves Hoyt
Lowell and Mars
Chapter 2 (p. 20)

A good education is indispensable, one as broad as it is long; without it he runs the risk of becoming a crank. Then enters the important quality of imagination. This word to the routine rabble of science is a red rag to a bull; partly because it is beyond their conception, partly because they do not comprehend how it is used. To their thinking to call a man imaginative is to damn him; when, did they but know it, it is admitting the very genius they would fain deny. For all great work imagination is vital; just as necessary in science and business as it is in novels and art... The difference between the everyday and scientific use of it is in that in science every imagining must be tested to see whether it explains the facts. Imagination harnessed to reason is the force that pulls an idea through. Reason, too, of the most complete, uncompromising kind. Imagination supplies the motive power, reason the guiding rein.

In William Graves Hoyt
Lowell and Mars
Chapter 2 (p. 21)

Mitchell, Maria

We especially need imagination in science. It is not all mathematics, nor all logic, but it is somewhat beauty and poetry.

In Phebe Mitchell Kendall
Maria Mitchell: Life, Letters, and Journals
Chapter IX (p. 187)

Thoreau, Henry
...the imagination, give it the least license, dives deeper and soars higher than Nature goes.

Walden
Chapter XVI (p. 286)

Tombaugh, Clyde
You have to have the imagination to recognize a discovery when you make one. When they examined Voyager images and saw for the first time the volcanic eruptions on Io, that called for some intuitive imagination. I would suggest that above everything else, in observing you have to be very alert to everything. You have to be able to recognize a discovery as such. There are so many people who don't seem to have that talent. A research astronomer cannot afford to be in such a rut. I might say that different types of personalities in astronomy make certain types of discoveries that are in line with their personalities.

In David H. Levy
Clyde Tombaugh: Discoverer of Planet Pluto
Chapter 5 (p. 61)

Velikovsky, Immanuel
Imagination coupled with skepticism and an ability to wonder—if you possess these, bountiful nature will hand you some of the secrets out of her inexhaustible store. The pleasure you will experience in discovering truth will repay you for your work; don't expect other compensation, because it may not come. Yet dare.

Earth In Upheaval
Supplement
Worlds in Collision in the Light of Recent Finds in
Archaeology, Geology, and Astronomy
Address
Princeton University
October 14, 1953 (p. 279)

Wheeler, John Archibald
The vision of the Universe that is so vivid in our minds is framed by a few iron posts of true observation—themselves resting on theory for their meaning—but most of all the walls and towers in the vision are of paper-maché, plastered in between those posts by an immense labor of imagination and theory.

In John Archibald Wheeler and Wojciech Hubert Zurek (eds)
Quantum Theory and Measurement
Law Without Law (p. 203)

Whitrow, G.J.

Our idea of the universe as a whole remains a product of the imagination.

The Structure and Evolution of the Universe
Chapter 8 (p. 197)

IMPOSSIBLE

von Braun, Wernher

...the past few decades should have taught us to use the word 'impossible' with utmost caution.

In Erik Bergaust
Wernher von Braun
Reaching for the Straws (p. 2)

INFINITE

Aristotle

...it is impossible that the infinite should move at all. If it did...there is another place, infinite like itself, to which it will move. But that is impossible.

On the Heavens
Book 1, Chapter 7, 274^b [30]

Blake, William

The nature of infinity is this: That every thing has its
Own Vortex, and when once a traveller thro' Eternity
Has pass'd that Vortex, he percieves [sic] it roll backward behind
His path, into a globe itself unfolding like a sun;
Or like a moon, or like a universe of starry majesty,
While he keeps onwards in his wondrous journey on the earth,
Or like a human form, a friend (with) whom he lived benevolent.

The Complete Prose and Poetry of William Blake
Milton
L. 21–7

Descartes, Rene

We call infinite that thing whose limits we have not perceived, and so by that word we do not signify what we understand about a thing, but rather what we do not understand.

Isis
In P. Mancosu and E. Vailati
Torricelli's Infinitely Long Solid and Its Philosophical Reception in the Seventeenth Century (p. 62)
Volume 82, Number 311, 1991

Emerson, Ralph Waldo
For it is only the finite that has wrought and suffered; the infinite lies stretched in smiling repose.

Essays and Lectures
Essays
First Series
Spiritual Laws (p. 305)

Greene, Brian
Like a sharp rap on the wrist from an old-time schoolteacher, an infinite answer is nature's way of telling us that we are doing something that is quite wrong.

The Elegant Universe

Harrison, Edward
Only a cosmic jester could perpetrate eternity and infinity...

Masks of the Universe
Chapter 12 (p. 201)

Kasner, Edward
Newman, James
The infinite in mathematics is always unruly unless it is properly treated.

Mathematics and the Imagination
Paradox Lost and Paradox Regained (p. 210)

Mitchell, Maria
Do not forget the infinite in the infinitesimal.

In Helen Wright
Sweeper in the Sky
Chapter 9 (p. 164)

von Haller, Albrecht
Infinity! What measures thee?
Before the worlds as days, and men as moments flee!

In W. Hastie
Kant's Cosmogony
Seventh Chapter (p. 146)

Whitehead, Alfred North
From what science has discovered about the infinitely small and the infinitely vast, the size of our bodies is almost totally irrelevant. In this little mahogany stand...may be civilizations as complex and diversified in scale as our own; and up there, the heavens, with all their vastness,

may be only a minute strand of tissue in the body of a being in the scale of which all our universes are as a trifle.

As recorded by Lucien Price
Dialogues of Alfred North Whitehead as Recorded by Lucien Price
Dialogue XLIII (pp. 367–8)

Wolf, Fred Alan
... infinity is just another name for mother nature.

Parallel Universes
Chapter 6 (p. 70)

Zebrowski, George
Science, when it runs up against infinities, seeks to eliminate them, because a proliferation of entities is the enemy of explanation.

OMNI
Time Is Nothing But A Clock (p. 144)
Volume 17, Number 1, October 1994

INSTRUMENT

Bacon, Francis
The unassisted hand and the understanding left to itself possess but little power. Effects are produced by the means of instruments and helps, which the understanding requires no less than the hand; and as instruments either promote or regulate the motion of the hand, so those that are applied to the mind prompt or protect the understanding.

Novum Organum
Aphorism II

Bridgman, P.W.
Not only do we use instruments to give us fineness of detail inaccessible to direct sense perception, but we also use them to extend qualitatively the range of our senses into regions where our senses no longer operate...

The Way Things Are
Chapter V (p. 149)

Egler, Frank E.
Dazed with this brightness of our technology, I wonder if some are not inclined to forget that the most important instrument in science must always be the mind of man.

The Way of Science
Methodology and Instrumentation (p. 59)

Eisenhart, Churchill
Of wonders of science and feats of design
Has many a scribe writ the praise;
And if I now mention the subject again
It's distinctly a relative phase.
For while science and gadgets are fine in their ways
One worries at times 'bout their clutch,
Especially when science, design, and math'matics

Combine to get us in Dutch.

Science
Operational Aspects of Instrument Design (p. 343)
Volume 110, October 7, 1949

Kuhn, Thomas

...scientists see new and different things when looking with familiar instruments in places they have looked before...It is as elementary prototypes for these transformations of the scientist's world that the familiar demonstrations of a switch in visual gestalt prove so suggestive. What were ducks in the scientist's world before the revolution are rabbits afterwards.

The Structure of Scientific Revolutions
Chapter X (p. 111)

Lavoisier, Antoine

As the usefulness and accuracy of chemistry depend entirely upon the determination of the weights of the ingredients and products, too much precision cannot be employed in this part of the subject; and for this purpose, we must be provided with good instruments.

Elements of Chemistry in a New Systematic Order
Part III, Chapter I (p. 88)

In the present advanced state of chemistry, very expensive and complicated instruments are become indispensably necessary for ascertaining the analysis and synthesis of bodies with the requisite precision as to quantity and proportion; it is certainly proper to endeavor to simplify these, and to render them less costly; but this ought by no means to be attempted at the expense of their convenience of application, and much less of their accuracy.

Elements of Chemistry in a New Systematic Order
Part III, Chapter II, Section II (p. 94)

von Goethe, Johann Wolfgang

Ye instruments, ye surely jeer at me,
With handle, wheel and cogs and cylinder.
I stood beside the gate, ye were to be the key.
True, intricate your ward, but no bolts do ye stir.
Inscrutable upon a sunlit day,
Her veil will Nature never let you steal,
And what she will not to your mind reveal,
You will not wrest from her with levers and with screws.

Faust
The First Part
Night, L. 668–75

Whitehead, Alfred North

The reason why we are on a higher imaginative level is not because we have finer imagination, but because we have better instruments.

Science and the Modern World
Chapter VII (p. 166)

INTERACTION

Hugo, Victor

Nothing is small, in fact; any one who is subject to the profound and penetrating influence of nature knows this. Although no absolute satisfaction is given to philosophy, either to circumscribe the cause or to limit the effect, the contemplator falls into those unfathomable ecstasies caused by these decompositions of force terminating in unity. Everything toils at everything.

Algebra is applied to the clouds; the radiation of the star profits the rose; no thinker would venture to affirm that the perfume of the hawthorn is useless to the constellations. Who then, can calculate the course of a molecule? How do we know that the creation of worlds is not determined by the fall of grains of sand? Who knows the reciprocal ebb and flow of the infinitely great and the infinitely little, the reverberations of causes in the precipices of being, and the avalanches of creation? The tiniest worm is of importance; the greatest is little, the little is great; everything is balanced in necessity; alarming vision for the mind.

Les Miserables
St Denis
The House in the Rue Plumet
Foliis ac frondibus (pp. 61–2)

KNOWLEDGE

Addison, Joseph

The utmost extent of man's knowledge, is to know that he knows nothing.

Interesting Anecdotes, Memoirs, Allegories, Essays, and Poetical Fragments
Volume 3 & 4
Essay on Pride (p. 230)

Alighieri, Dante

... to have heard without retaining does not make knowledge.

The Divine Comedy of Dante Alighieri
Paradise
Canto V, L. 41–2

Clerke, Agnes M.

... our knowledge will, we are easily persuaded, appear in turn the merest ignorance to those who come after us. Yet it is not to be despised, since by it we reach up groping fingers to touch the hem of the garment of the Most High.

A Popular History of Astronomy during the Nineteenth Century
Part II
Chapter VIII (p. 442)

Confucius

When you know a thing, to hold that you know it; and when you do not know a thing, to allow that you do not know it;—this is knowledge.

In James Legge
The Chinese Classics
Volume I
The Confucian Analects
Book 2:17

Gauss, Carl Friedrich

The knowledge whose content makes up astronomy is the gain from more than 2,000 years' work on one of the most abundant objects of human knowledge, in which the foremost minds of all times have summoned up all the resources of genius and diligence.

In G. Waldo Dunnington (ed.)
Inaugural Lecture on Astronomy and Papers on the Foundations of Mathematics
Inaugural Lecture on Astronomy (p. 49)

Gore, George
New knowledge is not like a cistern, soon emptied, but is a fountain of almost unlimited power and duration.

The Art of Scientific Discovery
Chapter III (p. 27)

Hinshelwood, C.N.
To some men knowledge of the universe has been an end possessing in itself a value that is absolute: to others it has seemed a means of useful application.

The Structure of Physical Chemistry (p. 2)

James, William
...our science is a drop, our ignorance a sea. Whatever else be certain, this at least is certain—that the world of our present natural knowledge *is* enveloped in a larger world of *some* sort of whose residual properties we at present can frame no positive idea.

The Will to Believe and other Essays in Popular Philosophy
Is Life Worth Living(p. 54)

Jeans, Sir James
...our knowledge of the external world must always consist of numbers, and our picture of the universe—the synthesis of our knowledge—must necessarily be mathematical in form. All the concrete details of the picture, the apples, the pears and bananas, the ether and atoms and electrons, are mere clothing that we ourselves drape over our mathematical symbols—they do not belong to Nature, but to the parables by which we try to make Nature comprehensible. It was, I think, Kronecker who said that in arithmetic God made the integers and man made the rest; in the same spirit, we may add that in physics God made the mathematics and man made the rest.

Supplement to *Nature*
The New World—Picture of Modern Physics (p. 356)
Volume 134, Number 3384, September 1934

Latham, Peter
There is nothing so captivating as NEW knowledge.

In William B. Bean
Aphorisms from Latham (p. 38)

SAYS HERE -
"THERE IS NOTHING SO CAPTIVATING AS NEW KNOWLEDGE"
TELL ME SOMETHING I DON'T KNOW..!

There is nothing so captivating as NEW knowledge.
Peter Latham – (See p. 166)

Rabi, I.I.
We are the inheritors of a great scientific tradition and of a beautiful structure of knowledge. It is the duty of our generation to add to the perfection of this structure and to pass on to the next generation the best traditions of our science for the edification and entertainment of all mankind.

The Atlantic Monthly
The Physicist Returns from the War (p. 114)
Volume 176, Number 4, October 1945

Russell, Bertrand
With equal passion I have sought knowledge. I have wished to understand the hearts of men. I have wished to know why the stars shine. And I have tried to apprehend the Pythagorean power by which numbers holds sway above the flux. A little of this, but not much, I have achieved.

The Autobiography of Bertrand Russell
Prologue (pp. 3–4)

Smith, Theobald

It is incumbent upon us to keep training and pruning the tree of knowledge without looking to the right or the left.

Journal of Pathology and Bacteriology
Obituary Notice of Deceased Member (p. 630)
Volume 40, Number 3, May 1935

Tennyson, Alfred

And this grey spirit yearning in desire, To follow knowledge like a sinking star, beyond the utmost bound of human thought.

The Complete Poetical Works of Tennyson
Ulysses

Thoreau, Henry

Such is always the pursuit of knowledge. The celestial fruits, the golden apples of the Hesperides, are ever guarded by a hundred-headed dragon which never sleeps, so that it is an Herculean labor to pluck them.

The Writings of Henry David Thoreau
Volume 5
Wild Apples (p. 307)

Whewell, William

The *Senses* place before us the *Characters* of the Book of Nature; but these convey no knowledge to us, till we have discovered the Alphabet by which they are to be read.

The Philosophy of the Inductive Sciences
Volume II
Aphorisms
Aphorisms Concerning Ideas, II (p. 443)

LAWS

Holton, Gerald
Roller, Duane H.D.
If we liken the facts to be explained to fish in a pond, then the law or set of laws is the net with which we make the catch. It may turn out that our particular net is not fine enough to haul in *all* fish, large and small, but it may still be quite satisfactory for supplying our ordinary needs. We may go even further and maintain that to be useful at all, our conceptual schemes, like our nets, *must* contain holes; if it were otherwise (if, so to speak, we were to go fishing with large buckets instead of nets), we should not be able to distinguish between the significant and the trivial, the fish and the water.

Foundations of Modern Physical Science
Chapter 15 (p. 260)

Hoyle, Fred
It is not only the smallest features of the Universe that are controlled by the laws of physics. The behavior of matter on the very large scale that concerns us in astronomy is also determined by physics. The heavenly bodies dance like puppets on strings. If we are to understand why they dance as they do, it is necessary to find out how the strings are manipulated.

Frontiers of Astronomy
Chapter 3 (p. 40)

LaPlace, Pierre Simon
All events, even those which on account of their insignificance do not seem to follow the great laws of nature, are a result of it just as necessarily as the revolutions of the sun. In ignorance of the ties which unite such events to the entire system of the universe, they have been made to depend upon final causes or upon hazard, according as they occur and are repeated with regularity, or appear without regard to order; but these imaginary causes have gradually receded with the widening bounds of

knowledge and disappear entirely before sound philosophy, which sees in them only the expression of our ignorance of the true causes.

A Philosophical Essay on Probabilities
Chapter II (p. 3)

Mitchell, Maria

The laws of nature... are not discovered by accident; theories do not come by chance even to the greatest minds; they are not born of the hurry and worry of daily toil; they are diligently sought; they are patiently waited for, they are received with cautious reserve, they are accepted with reverence and awe. And until able women have given their lives to investigation, it is idle to discuss their capacity for original work.

In Helen Wright
Sweeper in the Sky
Chapter 10 (pp. 203–4)

The laws which regulate the influence of sun and planets are complex; the nature of the influence is not yet understood. The telescope, the spectroscope, and the camera are all at work, and although the unknown must always be infinite, Nature yields one truth after another to the earnest seeker.

In Helen Wright
Sweeper in the Sky
Chapter 11 (p. 219)

The immense spaces of creation cannot be spanned by our finite powers; these great cycles of time cannot be lived even by the life of a race. And yet, small as is our whole system compared with the infinitude of creation, brief as is our life compared with cycles of time, we are tethered to all by the beautiful dependencies of law, that not only the sparrow's fall is felt to the outermost bound, but the vibrations set in motion by the words that we utter reach through all space and the tremor is felt through all time.

In Helen Wright
Sweeper in the Sky
Chapter 11 (p. 227)

Pagels, Heinz

The fact that the universe is governed by simple natural laws is remarkable, profound and on the face of it absurd. How can the vast variety in nature, the multitude of things and processes all be subject to a few simple, universal laws.

Perfect Symmetry
Part 2
Chapter 1 (p. 160)

Rowland, Henry

He who makes two blades of grass grow where one grew before is the benefactor of mankind; but he who obscurely worked to find the laws of such growth is the intellectual superior as well as the greater benefactor of the two.

The Physical Papers of Henry Augustus Rowland
The Highest Aim of the Physicist (p. 669)

Russell, Bertrand

The discovery that all mathematics follows inevitably from a small collection of fundamental laws is one which immeasurably enhances the intellectual beauty of the whole; to those who have been oppressed by the fragmentary and incomplete nature of most existing chains of deduction this discovery comes with all the overwhelming force of a revelation; like a palace emerging from the autumn mist as the traveler ascends an Italian hill-side, the stately storeys of the mathematical edifice appear in their due order and proportion, with a new perfection in every part.

Mysticism and Logic
Chapter IV (pp. 67–8)

Schwarzschild, Martin

If simple perfect laws uniquely rule the universe, should not pure thought be capable of uncovering this perfect set of laws without having to lean on the crutches of tenuously assembled observations? True, the laws to be discovered may be perfect, but the human brain is not. Left on its own, it is prone to stray, as many past examples sadly prove. In fact, we have missed few chances to err until new data freshly gleaned from nature set us right again for the next steps. Thus pillars rather than crutches are the observations on which we base our theories; and for the theory of stellar evolution these pillars must be there before we can get far on the right track.

Structure and Evolution of the Stars
Chapter 1 (p. 1)

Unknown

The universe does not have laws, it has habits, and habits can be broken.

Source unknown

Weyl, Hermann

To gaze up from the ruins of the oppressive present toward the stars is to recognise the indestructible world of laws, to strengthen faith in reason, to realise the "harmonia mundi" that transfuses all phenomena, and never has been, nor will be, disturbed.

Space, Time, Matter
Preface to the Third Edition (p. vi)

LEARN

Dennett, Daniel
...we often learn more from bold mistakes than from cautious equivocation.

Consciousness Explained
Preface (p. xi)

Carson, Rachel
It is more important to pave the way for the child to want to know than to put him on a diet of facts he is not ready to assimilate.

The Sense of Wonder (p. 45)

LIFE

Glazkov, Yuri

The winds scatter across the planet the seeds of life to bring forth the grass and flowers and woods. The eternal winds of the universe are rushing along. What do they bring? No one knows. But I am sure that Nature has created us, endowed us with intelligence, so that we, like her servant the winds, can carry life into the vast and limitless emptiness and to its innumerable worlds. Reason should win out on Earth and then in the whole universe.

In Kevin W. Kelley
The Home Planet
With Plate 135

Jeans, Sir James

Is this, then, all that life amounts to? To stumble, almost by mistake, into a universe which was clearly not designed for life, and which, to all appearances, is either totally indifferent or definitely hostile to it, to stay clinging on to a fragment of a grain of sand until we are frozen off, to strut our tiny hour on our tiny stage with the knowledge that our aspirations are all doomed to final frustration, and that our achievement must perish with out race, leaving the universe as though we had never been?

The Mysterious Universe
Chapter I (p. 13)

Pagels, Heinz

My own view is that although we do not yet know the fundamental physical laws, when and if we find them the possibility of life in a universe governed by those laws will be written into them. The existence of life in the universe is not a selective principle acting upon the laws of nature; rather it is a consequence of them.

Perfect Symmetry
Part 4
Chapter 1 (pp. 359–60)

Wald, George

We are not alone in the universe, and do not bear alone the whole burden of life and what comes of it. life is a cosmic event—so far as we know the most complex state of organization that matter has achieved in our cosmos. It has come many times, in many places—places closed off from us by impenetrable distances, probably never to be crossed even with a signal. As men we can attempt to understand it, and even somewhat to control and guide its local manifestations. On this planet that is our home, we have every reason to wish it well. Yet should we fail, all is not lost. Our kind will try again elsewhere.

Scientific American
The Origin of Life (p. 53)
Volume 191, Number 2, August 1954

LIGHT

Feynman, Richard
Leighton, R.B.
Sands, M.

Things on a very small scale behave like nothing that you have any direct experience about. They do not behave like waves, they do not behave like particles, they do not behave like clouds, or billiard balls, or weights on springs, or anything that you have ever seen.

The Feynman Lectures in Physics
Volume III
Chapter 1 (p. 1-1)

Frankel, Felice
Whitesides, George M.

We know light best in its diluted form: a gentle rain of photons falling from the sun that illuminates and warms. More concentrated, light is a furnace and a terror.

On the Surface of Things
Silicon, Etched by Light (p. 34)

Mullaney, James

The light we see coming from celestial objects brings us into direct personal contact with remote parts of the universe as the photons end their long journey across space and time on our retinas.

Sky and Telescope
Focal Point (p. 244)
Volume 79, Number 3, 1990

Newton, Sir Isaac

Do not all fix'd Bodies, when heated beyond a certain degree, emit Light and shine; and is not this Emission perform'd by the vibrating motions of their parts?

Opticks
Query 8

Pratchett, Terry

Light thinks it travels faster than anything but it is wrong. No matter how fast light travels it finds the darkness has always got there first, and is waiting for it.

Reaper Man (p. 230)

Thomas, Dylan

Light breaks where no sun shines.

The Poems of Dylan Thomas
Light Breaks Where no Sun Shines (p. 82)

Unknown

I know the speed of light, but what's the speed of dark?

Source unknown

Light travels faster than sound, which is why some people appear bright until they speak.

Source unknown

LOGIC

Joubert, Joseph
Logic operates, metaphysics contemplates.

Pensées and Letters of Joseph Joubert
XI (p. 88)

Rexroth, Kenneth
The space of night is infinite,
The blackness and emptiness
Crossed only by thin bright fences
Of logic.

The Collected Shorter Poems
Theory of Numbers (p. 165)

Shakespeare, William
He dreweth out the thread of his verbosity finer than the staple of his argument.

Love's Labour's Lost
Act V, scene i, L. 18

Unknown
Logic is a systematic method of coming to the wrong conclusion with confidence.

Source unknown

Logic is a systematic method of coming to the wrong conclusion with confidence.
Unknown – (See p. 177)

MAGNETIC

Hale, G.E.

Thanks to Zeeman's discovery of the effect of magnetism on radiation, it appeared that the detection of such a magnetic field should offer no great difficulty, provided that it were sufficiently intense.

Astrophysical Journal
On the Possible Existence of a Magnetic Field in Sunspots (p. 315)
Volume 28, 1908

Parker, E.N.

It appears that the radical element responsible for the continuing thread of cosmic unrest is the magnetic field.

Cosmical Magnetic Fields
Chapter 1 (p. 2)

Magnetic fields (and their inevitable offspring fast particles) are found everywhere in the universe where we have the means to look for them.

Cosmical Magnetic Fields
Chapter 1 (p. 6)

Zirin, H.

If the Sun had no magnetic field, it might be a quiet, "classical" star—if such stars exist, with no corona, chromosphere, sunspots, or solar activity.

Astrophysics of the Sun
Chapter 2 (p. 39)

MAN

Bradley, John Hodgdon

...man is the only animal who can face with a thought, a dream, and a smile the mystery and the madness and the terrible beauty of the universe.

Autobiography of Earth
Chapter 12 (p. 347)

Landau, Lev

The discovery of quantum mechanics and of the principle of uncertainty has shown that man can tear himself away from deeply rooted notions, discover, and accept something that is beyond his power of visualing.

In Alexandre Dorozynski
The Man They Wouldn't Let Die
Chapter 7 (p. 95)

Lowell, Percival

If astronomy teaches anything, it teaches that man is but a detail in the evolution of the Universe, and that resemblant though diverse details are inevitably to be expected in the host of orbs around him. He learns that, though he will probably never find his double anywhere, he is destined to discover any number of cousins scattered through space.

Mars
Chapter VI (p. 212)

Miller, Perry

It is only too clear that man is not at home in this universe, and yet he is not good enough to deserve a better...

The New England Mind: The Seventeenth Century
Chapter 1 (p. 7)

Shapley, Harlow

Mankind is made of star stuff, ruled by universal laws. The thread of cosmic evolution runs through his history, as through all phases of the universe—the microcosmos of atomic structures, molecular forms, and microscopic organisms, and the macrocosmos of higher organisms, of planets, stars, and galaxies. Evolution is still proceeding in galaxies and man—to what end, we can only vaguely surmise.

The View from a Distant Star
Preface (p. 5)

Steinbeck, John

Man is related... inextricably to all reality, known and unknowable... [P]lankton, a shimmering phosphorescence on the sea and the spinning planets and an expanding universe, all bound together by the elastic string of time. It is advisable to look from the tide pool to the stars and then back to the tide pool again.

The Log from the Sea of Cortez
Chapter 21 (p. 218)

MATTER

Darling, David

You are roughly eighteen billion years old and made of matter that has been cycled through the multimillion-degree heat of innumerable giant stars. You are composed of particles that once were scattered across thousands of light-years of interstellar space, particles that were blasted out of exploding suns and that for eons drifted through the cold, starlit vacuum of the Galaxy. You are very much a child of the cosmos.

Equations of Eternity
Introduction (p. xiii)

Dyson, Freeman J.

When we examine matter in the finest detail in the experiments of particle physics, we see it behaving as an active agent rather than an inert substance. Its actions are in the strict sense unpredictable. It makes what appear to be arbitrary choices between alternative possibilities. Between matter as we observe it in the laboratory and mind as we observe it in our own consciousness, there seems to be only a difference in degree but not in kind. If God exists and is accessible to us, then his mind and ours may likewise differ from each other only in degree and not in kind. We stand, in a manner of speaking, midway between the unpredictability of matter and the unpredictability of God. Our minds may receive inputs equally from matter and from God. This view of our place in the cosmos may not be true, but it is at least logically consistent and compatible with the active nature of matter as revealed in the experiments of modern physics. Therefore, I say, speaking as a physicist, scientific materialism and religious transcendentalism are neither incompatible nor mutually exclusive. We have learned that matter is weird stuff. It is weird enough, so that it does not limit God's freedom to make it do what he pleases.

Infinite in All Directions
Part I
Chapter 1 (p. 8)

Huxley, Julian

I am Matter. I am the condensation,
The Kink in empty space that provides resistance,
Precious inertia—mine the sole foundation
On which swift Energy's flow of fluid emanation
Fraternally builds reality into existence.

The Captive Shrew
Matter, Energy, Time and Space

Reeves, Hurbert

The organization of the universe demands that matter abandon itself to the games of chance.

Atoms of Silence
Chapter 16 (p. 177)

Updike, John

There is infinitely more nothing in the universe than anything else.

The Poorhouse Fair
Chapter II (p. 90)

Weyl, Hermann

And now, in our time, there has been unloosed a cataclysm which has swept away space, time and matter, hitherto regarded as the firmest pillars of natural science, but only to make place for a view of things of wider scope, and entailing a deeper vision.

Space, Time, Matter
Introduction (p. 2)

MEASUREMENT

Bell, J.A.
The concept of 'measurement' becomes so fuzzy on reflection that it is quite surprising to have it appearing in physical theory *at the most fundamental level.*

Speakable and Unspeakable in Quantum Mechanics
Chapter 15 (p. 117)

Rankine, William John Macquorn
A party of astronomers went measuring of the earth,
And forty million meters they took to be its girth;
Five hundred million inches, though, go through from pole to pole;
So let's stick to inches, feet, and yards, and the good old three-foot rule.

Songs and Fables
The Three-Foot Rule
Stanza III

MECHANICS

Gauss, Carl Friedrich

Proper is it that in the gradual development of a science, and in the instruction of individuals, the easy should precede the difficult, the simple the complex, the special the general, yet the mind, when once it has reached a higher point of view, demands the contrary course, in which all statics shall appear simply as a special case of mechanics.

In Ernst Mach
History and Root of the Principle of the Conservation of Energy
Chapter II (p. 34)

Oliver, David

Mechanics is the wellspring from which physics flows...

The Shaggy Steed of Physics
Preface (p. ix)

Mechanics is the vehicle of all physical theory. Mechanics is the vehicle of war. The two have been inseparable.

The Shaggy Steed of Physics
Preface (p. x)

MERE

Feynman, Richard
Leighton, R.B.
Sands, M.

The stars are made of the same atoms as the earth. I usually pick one small topic like this to give a lecture on. Poets say science takes away from the beauty of the stars—mere gobs of gas atoms. Nothing is "mere." I too can see the stars on a desert night, and feel them. But do I see less or more? The vastness of the heavens stretches my imagination—stuck on this carousel my little eye can catch one-million-year-old light. A vast pattern—of which I am a part—perhaps my stuff was belched from some forgotten star, as one is belching there. Or see them with the greater eye of Palomar, rushing all apart from some common starting point when they were perhaps all together. What is the pattern, or the meaning, or the "why?" It does not do harm to the mystery to know a little about it. For far more marvelous is the truth than any artists of the past imagined! Why do the poets of the present not speak of it? What men are poets who can speak of Jupiter if he were like a man, but if he is an immense spinning sphere of methane and ammonia must be silent?

The Feynman Lectures of Physics
Volume 1
Chapter 3 (pp. 3–6)

METAPHOR

Cole, K.C.

So much of science consists of things we can never see: light "waves" and charged "particles"; magnetic "fields" and gravitational "forces"; quantum "jumps" and electron "Orbits." In fact, none of these phenomena is literally what we say it is. Light waves do not undulate through empty space in the same way that water waves ripple over a still pond; a field is only a mathematical description of the strength and direction of a force; an atom does not literally jump from one quantum state to another, and electrons do not really travel around the atomic nucleus in orbits. The words we use are merely metaphors.

Discover
On Imagining the Unseeable (p. 70)
December 1982

Moore, James R.

Clever metaphors die hard. Their tenacity of life approaches that of the hardiest micro-organisms. Living relics litter our language, their *raisons d'être* forever past, ignored if not forgotten, and their present fascination seldom impaired by the confusions they may create.

The Post-Darwinian Controversies
Chapter I (p. 19)

METEOR

Butler, Samuel
This hairy meteor did denounce
The fall of Scepters and of Crowns;...

Hudibras
First Part
Canto I, L. 245–6

Caithness, James Balharrie
Wonderful, shimmering trail of light,
Falling from whence on high!
Flooding the world in thy moment's flight
With the sense of a mystery!
Softly thy radiance works a spell,
Night is enhanced, as a note may swell
From a simple melody.

Pastime Poems
The Meteor

Darwin, Erasmus
Ethereal Powers! You chase the shooting stars,
Or yoke the vollied lightnings to your cars.

The Botanic Garden
Part I, Canto I, II, L. 115

Devaney, James
The coming of this lovely night
Lifted the world's great roof of blue
And bared the awful Infinite—
So grand an hour, so vast a view,
Abashed I stand each night anew:

When out of unimagined deeps

Spectacular you burst upon
The dark, and down the starry steeps
A trail of whitest fire you shone
One breathless moment—and were gone.

Where The Wind Goes
To A Falling Star

Frost, Robert
Did you stay up last night (the Magi did)
To see the star shower known as Leonid
That once a year by hand or apparatus
Is so mysteriously pelted at us?

Complete Poems of Robert Frost
A Loose Mountain

Hoffman, Jeffrey
Suddenly I saw a meteor go by underneath me. A moment later I found myself thinking, That can't be a meteor. Meteors burn up in the atmosphere above us; this was below us. Then, of course, the realization hit me.

In Kevin W. Kelley
The Home Planet
With Plate 10

Plum, David
Then bear us, O Earth, with our eyes upward gazing,
To the place where the Star-God his fireworks displays;
When countless as snowflakes are meteors blazing
With their red, green and orange and amber-like rays.

New York Evening Post
Meteors
November 20, 1866

Revelation 9:1–2
And the fifth angel sounded, and I saw a star fall from heaven unto the earth: and to him was given the key of the bottomless pit. And he opened the bottomless pit; and there arose a smoke out of the pit, as the smoke of a great furnace; and the sun and the air were darkened.

The Bible

Revelation 12:3–4
And behold a great red dragon, having seven heads and ten horns and seven crowns upon his heads. And his tail drew the third part of the stars of heaven and did cast them to the earth.

The Bible

Shakespeare, William

And certain stars shot madly from their spheres.

A Midsummer-Night's Dream
Act II, scene i, L. 153

Smythe, Daniel

A curve of fire traces the dark
And warns us of a visitor.
It makes an unfamiliar mark
And then is seen no more.

Nature Magazine
The Meteor (p. 493)
Volume 50, Number 9, November 1957

Tennyson, Alfred

Now slides the silent meteor on, and leaves
A shining furrow, as thy thoughts in me.

The Complete Poetical Works of Tennyson
The Princess
VII

Teasdale, Sara

I saw a star slide down the sky,
Blinding the north as it went by,
Too burning and too quick to hold,
Too lovely to be bought or sold,
Good only to make wishes on
And then forever to be gone.

The Collected Poems of Sara Teasdale
The Falling Star

Unknown

A rock from space that falls to earth is called a meteorite. However, if it lands to the left of you it's called a meteorleft.

Source unknown

Far better 'tis, to die
the death that flashes gladness,
than alone, in frigid dignity,
to live on high.
Better, in burning sacrifice,
be thrown against a world
to perish, than the sky
to circle endlessly

a barren stone.

Nature
Nature and Science in Poetry (p. 295)
Volume 132, Number 3330, August 26, 1933

A rock from space that falls to earth is called a meteorite. However, if it lands to the left of you it's called a meteorleft.
Unknown – (See p. 190)

Virgil
And oft, before tempestuous winds arise,
The seeming stars fall headlong from the skies,
And, shooting through the darkness, gild the night
With sweeping glories and long trains of light.

The Works of Virgil
Georgics
Book 1 (p. 374)

MILKY WAY

Alighieri, Dante
. . . distinct with less and greater lights, the Galaxy so whitens between the poles of the world that it makes even the wise to question. . .

The Divine Comedy of Dante Alighieri
Paradise
Canto XIV, L. 97–100

Donne, John
In that glistering circle in the firmament, which we call the *Galaxie*, the milkie way, there is not one starre of any of the six great magnitudes, which Astronomers proceed upon, belonging to that circle: it is a glorious circle, and possesseth a great part of heaven, and yet is all of so little starres, as have no name, no knowledge taken of them. . .

Donne's Sermons
Little Stars
Sermon 144 (p. 221)

Hearn, Lafcadio
In the silence of the transparent night, before the rising of the moon, the charm of the ancient tale sometimes descends upon me out of the scintillant sky, to make me forget the monstrous facts of science and the stupendous horror of space. Then I no longer behold the Milky Way, as that awful Ring of Cosmos, whose hundred million suns are powerless to lighten the abyss, but as the very Amanogwa itself—the river Celestial. I see the thrill of its shining stream, the mists that hover along the verge, and the watergrasses that bend in the winds of autumn. White Orihimé I see at her starry loom and the Ox that grazes on the farther shore—and I know that the falling dew is the spray of the Herdsman's oar.

The Writings of Lafcadio Hearn
Volume VIII
The Romance of the Milky Way (p. 257)

Herschel, William

As we are used to call the appearance of the heavens, where it is surrounded with a bright zone, the Milky Way, it may not be amiss to point out some other very remarkable Nebulae which cannot well be less, but are probably much larger than our own system; and, being also extended, the inhabitants of the planets that attend the stars which compose them must likewise perceive the same phenomena. For which reason they may also be called milky ways by way of distinction.

In Laurence A. Marschall
The Supernova Story
Chapter 2 (p. 34)

Kilmer, Joyce

God be thanked for the Milky Way that
runs across the sky.
That's the path that my feet would tread
whenever I have to die.

Some folks call it a Silver Sword, and some a
Pearly Crown.
But the only thing I think it is, is Main
Street, Heaventown.

Main Street and Other Poems
Main Street

Milton, John

...the Galaxie, that Milkie way
Which nightly as a circling Zone, thou seest
Poudered with Starrs...

Paradise Lost
Book VII, L. 579–81

Poincaré, Henri

...to the eyes of a giant for whom our suns would be as for us our atoms, the milky way would seem only a bubble of gas.

The Foundations of Science
Science and Method
Book IV
Chapter I (p. 524)

Rich, Adrienne

Driving at night I feel the Milky Way

Streaming above me like the graph of a cry.

Leaflets, Poems 1965–1968
Ghazals
7/24/68: ii

Thoreau, Henry David

This whole earth which we inhabit is but a point in space. How far apart, think you, dwell the two most distant inhabitants of yonder star, the breadth of whose disk cannot be appreciated by our instruments? Why should I feel lonely? Is not our planet in the Milky Way?

Walden
Solitude (p. 131)

MIND

Blaise, Clarke
Our minds soar with instant connection, but our feet are stuck in temporal boots.

Time Lord
Chapter 1 (p. 19)

Dyson, Freeman J.
It appears to me that the tendency of mind to infiltrate and control matter is a law of nature. The infiltration of mind into the universe will not be permanently halted by any catastrophe or by any barrier that I can imagine. If our species does not choose to lead the way, others will do so, or may have already done so. If our species is extinguished, others will be wiser or luckier. Mind is patient. Mind has waited for 3 billion years on this planet before composing its first string quartet. It may have to wait for another 3 billion years before it spreads all over the galaxy. I do not expect that it will have to wait so long. But if necessary, it will wait. The universe is like a fertile soil spread out all around us, ready for the seeds of mind to sprout and grow. Ultimately, late or soon, mind will come into its heritage.

Infinite in All Directions
Part 1
Chapter 6 (p. 118)

Einstein, Albert
The human mind is not capable of grasping the Universe. We are like a little child entering a huge library. The walls are covered to the ceiling with books in many different tongues. The child knows that something must have written these books. It does not know who or how. It does not understand the languages in which they are written. But the child notes a

definite plan in the arrangement of the books—a mysterious order which it does not comprehend, but only dimly suspects.

In M. Taube
Evolution of Matter and Energy
Chapter 1 (p. 1)

Gauss, Carl Friedrich
Astronomy and Pure Mathematics are the magnetic poles toward which the compass of my mind ever turns.

In Franz Schmidt and Paul Stäckel (eds)
Briefwechsel zwischen Carl Friedrich Gauss und Wolfgang Bolyai
Letter XXIII (p. 55)
Letter to Bolyai
June 30, 1803

Jevons, W. Stanley
Summing up, then, it would seem as if the mind of the great discoverer must combine contradictory attributes. He must be fertile in theories and hypotheses, and yet full of facts and precise results of experience. He must entertain the feeblest analogies, and the merest guesses at truth, and yet he must hold them as worthless till they are verified in experiment. When there are any grounds of probability he must hold tenaciously to an old opinion, and yet he must be prepared at any moment to relinquish it when a clearly contradictory fact is encountered.

The Principles of Science
Book IV
Chapter XXVI (p. 592)

Kepler, Johannes
A mind accustomed to mathematical deduction, when confronted with the faulty foundations [of astrology] resists a long, long time, like an obstinate mule, until compelled by beating and curses to put its foot into that dirty puddle.

In Arthur Koestler
The Sleepwalkers
Part IV
Chapter I (p. 243)

Land, Edwin
Each stage of human civilization is defined by our mental structures: the concepts we create and then project upon the universe. They not only redescribe the universe but also in so doing modify it, both for our own time and for subsequent generations. This process—the revision of old cortical structures and the formulation of new cortical structures whereby

the universe is defined—is carried on in science and art by the most creative and talented minds in each generation...

Remarks at Opening of New American Academy of Arts and Sciences
Cambridge, Massachusetts April 2, 1979
In Ansel Adams
An Autobiography
Chapter 19 (p. 306)

MODEL

Born, Max

All great discoveries in experimental physics have been due to the intuition of men who made free use of models, which were for them not products of the imagination, but representatives of real things.

Philosophical Quarterly
Physical Reality (p. 140)
Volume 3, Number 11, April 1953

Feynman, Richard

...the more you see how strangely Nature behaves, the harder it is to make a model that explains how even the simplest phenomena actually work. So theoretical physics has given up on that.

QED: The Strange Theory of Light and Matter
Chapter 3 (p. 82)

Weisskopf, Victor F.

What is a model? A model is like an Austrian timetable. Austrian trains are always late. A Prussian visitor asks the Austrian conductor why they bother to print timetables. The conductor replies: "If we did not, how would we know how late the trains are?"

In H Frauenfelder and E.M. Henley
Subatomic Physics
Part V (p. 351)

MOLECULE

Frankel, Felice
Whitesides, George M.
Molecules—like ants, lemmings, herring, people—are happiest when surrounded by their own kind.

On the Surface of Things
Introduction (p. 7)

Maxwell, James Clerk
...in the heavens we discover by their light, and by their light alone, stars so distant from each other that no material thing can ever have passed from one to another; and yet this light, which is to us the sole evidence of the existence of these distant worlds, tells us also that each of them is built up of molecules of the same kinds as those which we find on earth. A molecule of hydrogen, for example, whether in Sirius or in Arcturus, executes its vibrations in precisely the same time.

The Scientific Papers of James Clerk Maxwell
Volume II
Molecules (pp. 375–6)

...though in the course of ages catastrophes have occurred and may yet occur in the heavens, though ancient systems may be dissolved and new systems evolved out of their ruins, the molecules out of which these systems are built—the foundation-stones of the material universe—remain unbroken and unworn.

The Scientific Papers of James Clerk Maxwell
Volume II
Molecules (p. 377)

MOMENTUM

Unknown
A rolling stone gathers momentum.

Source unknown

MOON

Atwood, Margaret
I fold back the sheet, get carefully up, on silent bare feet, in my nightgown, go to the window, like a child, I want to see. The moon on the breast of the new-fallen snow. The sky is clear but hard to make out, because of the searchlight; but yes, in the obscured sky a moon does float, newly, a wishing moon, a sliver of ancient rock, a goddess, a wink. The moon is a stone and the sky is full of deadly hardware, but oh God, how beautiful anyway.

The Handmaid's Tale
Chapter 17 (p. 108)

Blake, William
The moon like a flower
In heaven's high bower,
With silent delight
Sits and smiles on the night.

The Complete Prose and Poetry of William Blake
The Moon

Brontë, Charlotte
Where, indeed, does the moon not look well? What is the scene, confined or expansive, which her orb does not fallow? Rosy or fiery, she mounted now above a not distant bank; even while we watched her flushed ascent, she cleared to gold, and in a very brief space, floated up stainless into a now calm sky.

Life and Works of The Sisters Brontë
Volume III
Villette
La Terrasse (p. 214)

Burton, Sir Richard
That gentle Moon, the lesser light, the Lover's lamp, the Swain's delight,

A ruined world, a globe burnt out, a corpse upon the road of night.

The Kasîdah (p. 10)

Carroll, Lewis

The moon was shining sulkily,
Because she thought the sun
Had got no business to be there
After the day was done—

Through the Looking-Glass
Tweedledum and Tweedledee (p. 56)

We may hope therefore to find some profit in contemplating for a few moments this land of the skies: and although we may not look for very speedy "annexation," we may possibly gather some facts and ideas which the decree of Truth will annex to the domain of Science.

American Journal of Science
On the Volcanoes of the Moon
2nd series, Volume 2, 1846 (p. 336)

Coleridge, Samuel Taylor

The moving moon went up the sky,
And no where did abide;
Softly she was going up,
And a star or two beside.

The Ancient Mariner
Part IV

Collins, Michael

It was a totally different moon than I had ever seen before. The moon that I knew from old was a yellow flat disk, and this was a huge three-dimensional sphere, almost a ghostly blue-tinged sort of pale white. It didn't seem like a very friendly place or welcoming place. It made one wonder whether we should be invading its domain or not.

In Kevin W. Kelley
The Home Planet
With Plate 39

Fry, Christopher

. . . the moon is nothing
But a circumambulating aphrodisiac
Divinely subsidized to provoke the world

Into a rising birth-rate.

The Lady's Not for Burning, A Phoenix Too Frequent
and an Essay And Experience of Critics
The Lady's Not for Burning
Act III (p. 66)

Homer

As when the moon, refulgent lamp of night,
O'er heaven's clear azure spreads her sacred light,
When not a breath disturbs the deep serene,
And not a cloud o'ercasts the solemn scene;
Around her throne the vivid planets roll,
And stars unnumbered gild the glowing pole,
O'er the dark trees a yellower verdue shed,
And tip with silver every mountain's head.

Iliad
Book VIII, L. 687

Huxley, Julian

By death the moon was gathered in
Long ago, ah long ago;
Yet still the silver corpse must spin
And with another's light must glow.
Her frozen mountains must forget
Their primal hot volcanic breath,
Doomed to revolve for ages yet,
Void amphitheatres of death.

The Captive Shrew
Cosmic Death

Jastrow, Robert
Newell, Homer E.

The moon is the Rosetta stone of the solar system, and to the student of the origin of the earth and planets, this lifeless body is even more important than Mars and Venus.

The Atlantic Monthly
Why Land on the Moon? (p. 43)
Volume 211, Number 2; August, 1963

Lear, Edward

They dined on mince, and slices of quince,
Which they ate with a runcible spoon;
And hand in hand, on the edge of the sand,
They danced by the light of the moon,

The moon, the moon,
They danced by the light of the moon.

In Tony Palazzo
Edward Lear's Nonsense Book
The Owl and the Pussycat

Lightner, Alice
Queen of Heaven, fair of face,
Undefiled by alien feet;
Where the sun's untrammeled heat
Meets the cold of outer space;
Soon no more the Queen of Night,
For your conquest is in sight.

Nature Magazine
To the Moon (p. 213)
April 1957

Longfellow, Henry Wadsworth
Saw the moon rise from the water,
Rippling, rounding from the water,
Saw the flecks and shadows on it,
Whispered, "What is that, Nokomis?"
And the good Nokomis answered,
"Once a warrior very angry,
Seized his grandmother and threw her
Up into the sky at midnight;
Right against the moon he threw her;
'Tis her body that you see there."

The Poetical Works of Henry Wadsworth Longfellow
Hiawatha

Milton, John
To behold the wandering Moon,
Riding neer her highest noon,
Like one that had bin led astray
Through the Heav'ns wide pathless way;
And oft, as if her head she bowed,
Stooping through a fleecy cloud.

Miscellaneous Poems
Il Penseroso, L. 67–72

Shelley, Percy Bysshe
Art thou pale for weariness
Of climbing heaven, and gazing on the earth,

Wandering companionless
Among the stars that have a different birth,—?

The Complete Poetical Works of Shelley
To the Moon

That orbed maiden, with white fire laden,
Whom mortals call the moon.

The Complete Poetical Works of Shelley
The Cloud
Stanza 4

Tennyson, Alfred
All night, through archways of the bridged pearl
And portals of pure silver, walks the moon.

The Complete Poetical Works of Tennyson
Sonnet

Verne, Jules
There is no one among you, my brave colleagues, who has not seen the Moon, or at least, heard speak of it.

From Earth to the Moon
Chapter II (p. 12)

MOON LANDING

Armstrong, Neil

That's one small step for a man, one giant leap for mankind.

In *The New York Times*
Men Walk on Moon
L5, column 321, July 1969

That's one small step for man, one giant leap for mankind.

Words recorded on the transmission

Crew of Apollo 11

Here Men from The Planet Earth
First Set Foot upon The Moon
July, 1969 AD
We Came in Peace for All Mankind.

Plaque left behind on the moon's surface

Hoffer, Eric

Our passionate preoccupation with the sky, the stars, and a God somewhere in outer space is a homing impulse. We are drawn back to where we came from.

The New York Times
Reactions to Man's Landing on the Moon Show Broad Variations in Opinions
A6, column 2
21 July 1969

Koestler, Arthur

Prometheus is reaching out for the stars with an empty grin on his face.

The New York Times
Reactions to Man's Landing on the Moon Show Broad Variations in Opinions
A6, column 6
21 July 1969

Here Men from The Planet Earth
First Set Foot upon The Moon
July, 1969 AD
We Came in Peace for All Mankind.
Crew of Apollo 11 – (See p. 206)

Nabokov, Vladimir

Treading the soil of the moon, palpating its pebbles, tasting the panic and splendor of the event, feeling in the pit of one's stomach the separation from terra... these form the most romantic sensation an explorer has ever known... this is the only thing I can say about the matter. The utilitarian results do not interest me.

The New York Times
Reactions to Man's Landing on the Moon Show Broad Variations in Opinions
A6, column 5
21 July 1969

MOTION

Butterfield, Herbert

Of all the intellectual hurdles which the human mind has confronted and has overcome in the last fifteen hundred years, the one which seems to me to have been the most amazing in character and the most stupendous in the scope of its consequences is the one relating to the problem of motion...

The Origins of Modern Science
Chapter 1 (p. 3)

Galilei, Galileo

...we have decided to consider the phenomena of bodies falling with an acceleration such as actually occurs in nature and to make this definition of accelerated motion exhibit the essential features of observed accelerated motions.

Concerning the Two New Sciences
Third Day
Naturally Accelerated Motion (p. 200)

Gleick, J.

The basic idea of Western science is that you don't have to take into account the falling of a leaf on some planet in another galaxy when you're trying to account for the motion of a billiard ball on a pool table on earth. Very small influences can be neglected. There's a convergence in the way things work, and arbitrarily small influences don't blow up to have arbitrarily large effects.

Chaos: Making a New Science
The Butterfly Effect (p. 15)

Meredith, George

So may we read, and little find them cold:
Not frosty lamps illuminating dead space,
Not distant aliens, not senseless Powers.

The fire is in them whereof we are born;
The music of their motion may be ours.

A Reading of Earth
Meditation under Stars

Regnault, Pére
Nothing seems more clear at first than the Idea of Motion, and yet nothing is more obscure when one comes to search thoroughly into it.

Philosophical Conversations
Volume I
Conversation VI (p. 58)

MUON

Penman, Sheldon
For the time being, however, the muon itself qualifies as a "riddle wrapped in a mystery inside an enigma."

Scientific American
The Muon (p. 55)
Volume 205, Number 1, July 1961

NATURE

Agassiz, Louis
The study of Nature is an intercourse with the highest mind. You should never trifle with Nature. At the lowest her works are the works of the highest powers, the highest something in whatever way we may look at it. A laboratory of Natural History is a sanctuary where nothing profane should be tolerated. I feel less agony at improprieties in churches than in a scientific laboratory.

In R.A. Gregory
Discovery
Chapter III (p. 43)

Bohm, David
In nature nothing remains constant. Everything is in a perpetual state of transformation, motion, and change.

Causality and Chance in Modern Physics
Chapter 1 (p. 1)

Boyle, Robert
Nature always looks out for the preservation of the universe.

A Free Enquiry into the Vulgarly Received Notions of Nature
Section IV (p. 31)

Browne, Sir Thomas
All things are artificial, for nature is the art of God.

Religio Medici
Part I, section xvi (p. 29)

Einstein, Albert
...nature is the realization of the simplest conceivable mathematical ideas. I am convinced that we can discover by means of purely mathematical constructions the concepts and the laws connecting them with each other, which furnish the key to the understanding of natural

phenomena...Experience remains, of course, the sole criterion of the physical utility of a mathematical construction. But the creative principle resides in mathematics. In a certain sense, therefore, I hold it true that pure thought can grasp reality, as the ancients dreamed.

Ideas and Opinions
On the Method of Theoretical Physics (p. 274)

Emerson, Ralph Waldo

Nature is an endless combination and repetition of a very few laws. She hums the old well-known air through innumerable variations.

Essays and Lectures
Essays
First Series
History (p. 243)

It is very odd that Nature should be so unscrupulous. She is no saint...

In Edward Waldo Emerson (ed.)
Journals of Ralph Waldo Emerson
1841–44
20 May 1843 (p. 405)

Feynman, Richard

There was a moment when I knew how nature worked. It had elegance and beauty. The goddam thing was gleaming.

In Lee Edson
The New York Times Magazine
Two Men in Search of a Quark
October 8, 1967

Ford, Kenneth W.

One of the elementary rules of nature is that, in the absence of a law prohibiting an event or phenomenon, it is bound to occur with some degree of probability. To put it simply and crudely: Anything that can happen does happen.

Scientific American
Magnetic Monopoles (p. 122)
Volume 209, Number 6, December 1963

Gould, Stephen Jay

I do not believe that nature frustrates us by design, but I rejoice in her intransigence nonetheless.

Hen's Teeth and Horses Toes
A Zebra Trilogy
What If Anything Is A Zebra? (p. 365)

Heraclitus
The real constitution of things is accustomed to hide itself.

In G.S. Kirk and J.E. Raven
The Presocratic Philosophers
Fragment 211 (p. 193)

Herschel, J.F.W.
...Nature builds up by her refined and invisible architecture, with a delicacy eluding our conception, yet with a symmetry and beauty which we are never weary of admiring.

The Cabinet of Natural Philosophy
Section 292 (p. 263)

Holton, Gerald
The study of nature is a study of the artifacts that appear during an engagement between the scientist and the world in which he finds himself.

Daedalus
The Roots of the Complementarity (p. 1019)
Number 4, Fall 1970

Hooke, Robert
...the footsteps of Nature are to be trac'd, not in her *ordinary course*, but when she seems to be put to her shifts, to make many *doublings* and *turnings*, and to use some kind of art in indeavouring to avoid our discovery.

Micrographia
Preface (Third page)

Lawrence, Louise de Kiriline
Nature is a deep reality and whether we understand it or not it is true and elemental.

The Lovely and the Wild
Chapter 3 (p. 33)

McLennan, Evan
There is a charm for man in the study of Nature. It elevates his soul to real greatness. It frees his mind from stormy life, and thrills him with the purest joy.

Cosmical Evolution
Introduction (p. 23)

Muir, John

How lavish is Nature building, pulling down, creating, destroying, chasing every material particle from form to form, ever changing, ever beautiful.

My First Summer in the Sierra
August 30 (pp. 318–19)

When we are with Nature we are awake, and we discover many interesting things and reach many a mark we are not aiming at.

In Linnie Marsh Wolfe (ed.)
John of the Mountains
Chapter VII
Section I, June, 1890 (p. 300)

Musser, George

The basic rules of nature are simple, but their consummation may never lose its ability to surprise.

The Scientific American
From the Editors (p. 6)
Volume 280, Number 1, January 1999

Petrarch

There are fools who seek to understand the secrets of nature.

In Richard Olson
Science Deified and Science Defied: The Historical Significance of Science in Western Culture
Chapter 7 (p. 210)

Sayers, Dorothy L.
Eustace, R.

Nature never worked by rule and compass.

The Documents in the Case
Letter 16, Agatha Milsom to Olive Farebrother (p. 56)

Seneca

Nature does not turn out her work according to a single pattern; she prides herself upon her power of variation...

Physical Science in the Time of Nero
Book VII
Chapter XXVII (p. 301)

Nature does not reveal all *her* secrets at once. We imagine we are initiated in her mysteries: we are, as yet, but hanging around her outer courts.

Physical Science in the Time of Nero
Book VII
Chapter XXXI (p. 306)

Tennyson, Alfred
A void was made in Nature; all her bonds
Crack'd; and I saw the flaring atom-streams
And torrents of her myriad universe
Ruining along the illimitable inane,
Fly on to clash together again,...

The Complete Poetical Works of Tennyson
Lucretius

Thoreau, Henry David
If we knew all the laws of Nature, we should need only one fact, or the description of one actual phenomenon, to infer all the particular results at that point. Now we know only a few laws, and our result is vitiated, not, of course, by any confusion or irregularity in Nature, but by our ignorance of essential elements in the calculation. Our notions of law and harmony are commonly confined to those instances which we detect; but the harmony which results from a far greater number of seemingly conflicting, but really concurring, laws, which we have not detected, is still more wonderful. The particular laws are as our points of view, as to the traveler, a mountain outline varies with every step, and it has an infinite number of profiles, though absolutely but one form. Even when cleft or bored through it is not comprehended in its entireness.

Walden
Chapter XVI (p. 288)

von Schelling, F.W.J.
What then is that secret bond which couples our mind to Nature, or that hidden organ through which Nature speaks to our mind or our mind to Nature?... For what we want is not that Nature should coincide with the laws of our mind by chance (as if through some *third* intermediary), but that *she herself*, necessarily and originally, should not only express, but *even realize*, the laws of our mind, and that she is, and is called, Nature only insofar as she does so.

Nature should be Mind made visible, Mind the invisible Nature. Here then, in the absolute identity of Mind *in us* and Nature *outside us*, the problem of the possibility of a Nature external to us must be resolved.

Ideas for a Philosophy of Nature
Introduction (pp. 41–2)

The purest exercise of man's rightful dominion over dead matter, which was bestowed upon him together with reason and freedom, is that he spontaneously operates upon Nature, determines her according to purpose and intention, lets her act before his eyes, and as it were spies on her at work. But that the exercise of this dominion is possible, he owes yet again to Nature, whom he would strive in vain to dominate, if he would not put her in conflict with herself and set her own forces in motion against her.

Ideas for a Philosophy of Nature
Book I (p. 57)

Whitehead, Alfred North
You cannot talk vaguely about Nature in general.

Nature and Life
Part I (p. 1)

Nature, even in the act of satisfying anticipation, often provides a surprise.
Adventures of Ideas
Part II
Chapter VIII, Section V (p. 130)

Whitman, Walt
The fields of Nature long prepared and fallow,
the silent, cyclic chemistry,
The slow and steady ages plodding, the unoccupied
surface ripening, the rich ores forming beneath...
In James E. Miller, Jr (ed.)
Complete Poetry and Selected Prose
Song of the Redwood Tree

Wordsworth, William
True it is, Nature hides
Her treasures less and less—...

The Complete Poetical Works of Wordsworth
To the Planet Venus

NEUTRINO

Crane, H. Richard

Not everyone would be willing to say that he believes in the existence of the neutrino, but it is safe to say there is hardly one of us who is not served by the neutrino hypothesis as an aid in thinking about beta-decay process.

Review of Modern Physics
The Energy and Momentum Relations in the Beta-Decay
and the Search for the Neutrino (p. 278)
Volume 20, Number 2, March 1948

Eddington, Arthur

I am not much impressed by the neutrino theory. In an ordinary way I might say that I do not believe in neutrinos... Dare I say that experimental physicists will not have sufficient ingenuity to *make* neutrinos?

The Philosophy of Physical Science
Chapter VII (p. 112)

The neutrino is just barely a fact.

Scientific American
The Two-Neutrino Experiment (p. 60)
Volume 208, Number 3, March 1963

Pontecorvo, Bruno

It is difficult to find a case where the word 'intuition' characterises a human achievement better than in the case of the neutrino invention by Pauli.

Journel de Physique
Supplement C8
Volume 48, 1982 (p. 221)

Stenger, Victor J.

Neutrinos are neither rare nor anomalous—just hard to detect.

Physics and Psychics
Chapter 1 (p. 20)

NIGHT

Ackerman, Diane
It is nighttime on the planet Earth. But that is only a whim of nature, a result of our planet rolling in space at 1,000 miles per minute. What we call "night" is the time we spend facing the secret reaches of space, where other solar systems and, perhaps, other planetarians dwell. Don't think of night as the absence of day; think of it as a kind of freedom. Turned away from our sun, we see the dawning of far-flung galaxies. We are no longer sun-blind to the star-coated universe we inhabit.

A Natural History of the Senses
Vision
How to Watch the Sky (p. 245)

Amaldi, Ginestra Giovene
The night sky looks like a giant fistful of glittering diamonds flung carelessly upon a black carpet.

Our World and the Universe Around Us
Volume I
The Universe (p. 13)

Atwood, Margaret
Night falls. Or has fallen. Why is it that night falls, instead of rising, like the dawn? Yet if you look east, at sunset, you can see night rising, not falling; darkness lifting into the sky, up from the horizon, like a black sun behind cloud cover. Like smoke from an unseen fire, a line of fire just below the horizon, brushfire or a burning city. Maybe night falls because it's heavy, a thick curtain pulled up over the eyes. Wool blanket. I wish I could see in the dark.

The Handmaid's Tale
Chapter 30 (p. 201)

Murdin, Paul

Astronomers, literally, and human beings in general, figuratively, *need* the interruption of the night.

In Derek McNally
The Vanishing Universe
The Aims of Astronomy in Science and the Humanities:
Why Astronomy Must Be Protected (p. 19)

Stevenson, Robert Louis

Night is a dead monotonous period under a roof; but in the open world it passes lightly, with its stars and dews and perfumes, and the hours are marked by changes in the face of Nature.

Travels with a Donkey in the Cevennes
A Night Among the Pines (p. 79)

NOTATION

Frayn, Michael
We look at the taciturn, inscrutable universe, and cry, 'Speak to me!'

Construction
Number 7

NOVAE

Gaposchkin, Sergei

When the greatest, the cosmic,
And the most fascinating explosion
Has been probed by the fabulous light
Of the human (but god-like) mind,
It will be remembered
That you shouldered the task
Of exploring the Novae
With intrepid boldness
And richness of thought...

In Arthur Beer
Vistas in Astronomy
Volume 2
Novae Observed (p. 1506)

OBSERVATION

Adams, Douglas

...a scientist must also be absolutely like a child. If he sees a thing, he must say that he sees it, whether it was what he thought he was going to see or not. See first, think later, then test. But always see first. Otherwise you will only see what you were expecting.

So Long & Thanks for All the Fish
Chapter 31 (p. 165)

Ayres, C.E.

When Moses emerged from the cloudy obscurity of Mount Sinai and stood before the people with the stone tablets in his hand, he announced that his laws were based on direct observation. It is not recorded that any one doubted him.

Science: The False Messiah
Chapter III (p. 42)

Bolles, Edmund Blair

...yet there is a difference between scientific and artistic observation. The scientist observes to turn away and generalize; the artist observes to seize and use reality in all its individuality and peculiarity.

A Second Way of Knowing
Chapter 11 (p. 50)

Cohen, Morris Raphael

Accidental discoveries of which popular histories of science make mention never happen except to those who have previously devoted a great deal of thought to the matter. Observation unilluminated by theoretic reason is sterile...Wisdom does not come to those who gape at nature with an empty head. Fruitful observation depends not as Bacon thought upon the absence of bias or anticipatory ideas, but rather on a logical multiplication of them so that having many possibilities in mind we are better prepared to direct our attention to what others have never thought of as within the field of possibility.

Reason and Nature
Chapter I, Section III (p. 17)

Dampier, Sir William Cecil
There was a young man who said, "God
To you it must seem very odd
That a tree as a tree simply ceases to be
When there's no one about in the Quad."
. . .
Young man, your astonishment's odd,
I am always about in the Quad
And that's why the tree continues to be
As observed by, Yours faithfully, God.

In Joseph Needham and Walter Pagel (eds)
Background to Modern Science
from Aristotle to Galileo (pp. 40–1)

Eddington, Sir Arthur
We should be unwise to trust scientific inference very far when it becomes divorced from opportunity for observational test.

The Internal Constitution of the Stars
Chapter I (p. 1)

Faraday, Michael
If in such strivings, we. . . see but imperfectly, still we should endeavor to see, for even an obscure and distorted vision is better than none.

Philosophical Magazine
On the Conservation of Force
Volume 13, Number 4, 1857 (p. 238)

Gay-Lussac, Joseph Louis
In order to draw any conclusion. . . it is prudent to wait until more numerous and exact observations have provided a solid foundation on which we may build a rigorous theory.

In Maurice Grossland
Gay-Lussac: Scientist and Bourgeois
Chapter 4 (p. 71)

Heisenberg, Werner
This again emphasizes a subjective element in the description of atomic events, since the measuring device has been constructed by the observer,

and we have to remember that what we observe is not nature in itself but nature exposed to our method of questioning.

Physics and Philosophy
Chapter III (p. 58)

Herschel, William
Seeing is in some respects an art which must be learnt. To make a person see with such a power is nearly the same as if I were asked to make him play one of Handel's fugues upon the organ. Many a night I have been practicing to see, and it would be strange if one did not acquire a certain dexterity by such constant practice.

In William Hoyt
Planets X and Pluto
Chapter 1 (p. 12)

Holton, Gerald
Roller, Duane H.D.
All intelligent endeavor stands with one foot on observation and the other on contemplation.

Foundations of Modern Physical Science
Chapter 13 (p. 218)

Jeans, Sir James
Each observation destroys the bit of the universe observed, and so supplies knowledge only of a universe which has already become past history...

The New Background of Science
Chapter I (p. 2)

Jonson, Ben
...let mee alone to observe, till I turne my selfe into nothing but observation.

The Poetaster
Act II, Scene I, L. 193

Lee, Oliver Justin
Every bit of knowledge we gain and every conclusion we draw about the universe or about any part or feature of it depends finally upon some observation or measurement. Mankind has had again and again the humiliating experience of trusting to intuitive, apparently logical conclusions without observations, and has seen Nature sail by in her radiant chariot of gold in an entirely different direction.

Measuring Our Universe: From the Inner Atom to Outer Space
Chapter 3 (p. 33)

Lewis, Gilbert N.

I claim that my eye touches a star as truly as my finger touches this table.

The Atlantic Monthly
In George W. Gray
New Eyes on the Universe (p. 608)
Volume 155, Number 5, May 1935

Lubbock, Sir John

What we do see depends mainly on what we look for. When we turn our eyes to the sky, it is in most cases merely to see whether it is likely to rain. In the same field the farmer will notice the crop, geologists the fossils, botanists the flowers, artists the coloring, sportsmen the cover for game. Though we may all look at the same things, it does not at all follow that we should see them.

The Beauties of Nature
Introduction (pp. 3–4)

Meredith, George

Observation is the most enduring of the pleasures of life…

Diana of the Crossways
Chapter XI (p. 104)

Mitchell, Maria

Nothing comes out more clearly in astronomical observations than the immense activity of the universe.

In Phebe Mitchell Kendall
Maria Mitchell: Life, Letters, and Journals
Chapter XI (p. 237)

Orwell, George

To see what is in front of one's nose requires a constant struggle.

In Sonia Orwell and Ian Angus (eds)
The Collected Essays, Journalism and Letters of George Orwell: In Front of Your Nose, 1945–1950
1946
36 (p. 125)

Osler, Sir William

Man can do a great deal by observation and thinking, but with them alone he cannot unravel the mysteries of Nature. Had it been possible the Greeks would have done it; and could Plato and Aristotle have grasped the value of experiment in the progress of human knowledge, the course of European history might have been very different.

Man's Redemption of Man
Address
University of Edinburgh
July 1910 (p. 22)

Shakespeare, William

ARMANDO: How hast thou purchased this experience?
MOTH: By my penny of observation.

Love's Labour's Lost
Act III, scene i, L. 23

Smith, Theobald

...it is the care we bestow on apparently trifling, unattractive and very troublesome minutiae which determines the result.

In W. Bulloch
Journal of Pathology and Bacteriology.
Obituary Notice of Deceased Member (p. 621)
Volume 40, Number 3, May 1935

Thompson, W.R.

The mathematical machine works with unerring precision; but what we get out of it is nothing more than a rearrangement of what we put into it. In the last analysis *observation*—the actual contact with real events—is the only reliable way of securing the data of natural history.

Science and Common Sense
Chapter VI (pp. 114–15)

Unknown

Some say I'm a visionary, others just say I'm seeing things.

Source unknown

The obscure we see eventually, the completely apparent takes longer.

In Tore Frängsmyr (ed.)
Nobel Lectures
Chemistry 1971–80
Nobel Lecture of Peter Mitchell
December 8, 1978 (p. 325)

von Goethe, Johann Wolfgang

Natural objects should be sought and investigated as they are and not to suit observers, but respectfully as if they were divine beings.

In R. Matthaei (ed.)
Goethe's Color Theory
Precautions for the Observer (p. 57)

An extremely odd demand is often set forth but never met, even by those who make it; i.e., that empirical data should be presented without any theoretical context, leaving the reader, the student, to his own devices in judging it. This demand seems odd because it is useless simply to look at something. Every act of looking turns into observation, every act of observation into reflection, every act of reflection into the making of associations; thus it is evident that we theorize every time we look carefully at the world.

In Douglas Miller
Scientific Studies
Volume 12
Chapter VII
Preface (p. 159)

OBSERVATORY

Lowell, Percival

A steady atmosphere is essential to the study of planetary detail; size of instrument being a very secondary matter. A large instrument in poor air will not begin to show what a smaller one in good air will. When this is recognized, as it eventually will be, it will become the fashion to put up observatories where they can see rather than be seen.

Mars
Preface (p. v)

Mitchell, Maria

There is no observatory in this land, nor in any land, probably, of which the question is not asked, 'Are they doing anything? Why don't we hear from them? They should make discoveries, they should publish.'

In Phebe Mitchell Kendall
Maria Mitchell: Life, Letters, and Journals
Chapter XI (p. 223)

Rosseland, S.

...an astronomical observatory of to-day looks more like a factory plant than an abode for philosophers. The poetry of constellations has given way to the lure of plate libraries, and the angel of cosmogenic speculation has been caught in a cobweb of facts insistently clamoring for explanations.

Theoretical Astrophysics
Introduction (p. xi)

OBSERVER

de Chambaud, J.J. Ménuret

The name of *observer* has been given to the physicist who is content to examine the phenomena just as nature presents them to him; he differs from the *experimental* physicist who combines himself and who sees only the result of his own combinations. This latter one never sees nature as it is in fact; he pretends by his labor to render nature more accessible to the senses, to raise the mask which conceals it from our eyes, but often he disfigures it and renders it unintelligible. Nature is always unveiled and bare for him who has eyes—or it is covered only by a slight gauze which the eye and reflection easily pierce—and the pretended mask exists only in the imagination, usually quite limited, of the manipulator of experiments.

In D. Diderot and J.L. d'Alembert (eds)
Encyclopédie, ou Dictionnaire Raisonné des Sciences, des Arts et des Métiers
Observateur
Volume 23 (p. 287D)

Deuteronomy 4:19

And lest thou lift up thine eyes unto heaven, and when thou seest the sun, and the moon, and the stars, even all the host of heaven, shouldest be driven to worship them and serve them.

The Bible

ORDER

Browne, Sir Thomas
All things begin in order, so shall they end, and so shall they begin again; according to the ordainer of order, and the mysticall mathematicks of the City of Heaven.

In John Carter (ed.)
Urne Buriall and The Garden of Cyrus
The Garden of Cyrus
Chapter V (p. 114)

Frankel, Felice
Whitesides, George M.
Order is repetition, regularity, symmetry, simplicity. It forms the spine of our efforts to measure, control, and understand.

On the Surface of Things
Order (p. 63)

Kline, Morris
Is there a law and order in this universe or is its behavior merely the working of chance and caprice? Will the Earth and other planets continue their motions around the sun or will some unknown body, coming from great distances, rush through our planetary system and alter the course of every planet? Cannot the sun some day explode, as other suns are doing daily, and burn us all to a crisp? Was man deliberately planted on a planet especially prepared for his existence or is he merely an insignificant concomitant of accidental cosmic circumstances?

Mathematics In Western Culture
Chapter 24 (p. 276)

Mann, Thomas
...order and simplification are the first steps toward the mastery of a subject—the actual enemy is the unknown.

The Magic Mountain
Encyclopædic (pp. 245–6)

Moulton, Forest Ray

Now we find ourselves a part of a Universal Order of which we did not dream and whose alphabet we are just beginning to learn. Instead of shrinking it to our measure, we contemplate its infinite orderliness and set no limits to the goal our race may hope to attain.

Astronomy
Chapter XVI (p. 533)

Yang, Chen Ning

Nature possesses an order that one may aspire to comprehend.

In Nobel Foundation
Nobel Lecture
Physics 1942–62
Nobel Lecture of Chen Ning Yang
December 11, 1957 (p. 394)

OTHER WORLDS

Hippolytus

There are innumerable cosmos differing in size. In some there is no sun or moon, in others they are larger than with us, in others more numerous. The intervals between the cosmos are unequal: in some places there are more, in others fewer; some are growing, others again are dying; somewhere worlds are coming to be, elsewhere fading. And they are destroyed when they collide with each other. Some cosmos have no living creatures or plants, and no water at all.

Refutations
1.13.2-3

King, Stephen

Go then—there are other worlds than these.

The Dark Tower III
The Waste Lands
Bear and Bone (p. 49)

Magnus, Albertus

Do there exist many worlds, or is there but a single world? This is one of the most noble and exalted questions in the study of Nature.

Annals of Science
Quoted in G. McColley
The Seventeenth-Century Doctrine of a Plurality of Worlds (p. 385)
Volume 1, Number 4, October 15, 1936

PARADOX

Gilbert, William
Sullivan, Arthur
A paradox?
A paradox!
A most ingenious paradox!
We've quips and quibbles heard in flocks,
But none to beat this paradox!

The Complete Plays of Gilbert and Sullivan
Pirates of Penzance
Act II (p. 167)

PARTICLE

Gleick, James
Quantum mechanics taught that a particle was not a particle but a smudge, a traveling cloud of possibilities...

Genius: The Life and Science of Richard Feynman
MIT
Forces in Molecules (p. 89)

Heisenberg, Werner
We can no longer speak of the behaviour of the particle independently of the process of observation. As a final consequence, the natural laws formulated mathematically in quantum theory no longer deal with the elementary particles themselves but with our knowledge of them. Nor is it any longer possible to ask whether or not these particles exist in space and time objectively...

The Physicist's Conception of Nature
Chapter I (p. 15)

Johnson, George
In science's great chain of being, the particle physicists place themselves with the angels, looking down from the heavenly spheres on the chemists, biologists, geologists, meteorologists—those who are applying, not discovering, nature's most fundamental laws. Everything, after all, is made from subatomic particles. Once you have a concise theory explaining how they work, the rest should just be filigree.

The New York Times
New Contenders for a Theory of Everything
F1, Column 1
Tuesday, December 4, 2001

Regnault, Pére
The Imagination is lost here. Rather than the Minds; for if you divide a Particle into the most inconceivably minute Parts, the Mind will always

find therein something that regards the West, and something that regards the East; and what regards the West, is not that which regards the East.

Philosophical Conversations
Volume I
Conversation I (p. 9)

PAST

Barrow, John

Things are as they are because they were as they were.

The Origin of the Universe
Chapter 1 (p. 17)

Whitman, Walt

The past, the infinite greatness of the past!
For what is the present, after all, but a growth out of the past.

In James E. Miller, Jr (ed.)
Complete Poetry and Selected Prose
Passage to India

PATTERNS

Burns, Marilyn

Searching for patterns is a way of thinking that is essential for making generalizations, seeing relationships, and understanding the logic and order of mathematics. Functions evolve from the investigation of patterns and unify the various aspects of mathematics.

About Teaching Mathematics
Patterns and Functions (p. 112)

Gardner, Martin

If the cosmos were suddenly frozen, and all movement ceased, a survey of its structure would not reveal a random distribution of parts. Simple geometrical patterns, for example, would be found in profusion—from the spirals of galaxies to the hexagonal shapes of snow crystals. Set the clockwork going, and its parts move rhythmically to laws that often can be expressed by equations of surprising simplicity. And there is no logical or *a priori* reason why these things should be so.

Order and Surprise
Chapter 4 (p. 57)

Jeffers, Robinson

. . . the old man looked up
At a black eyelet in the white of the Milky Way, and he thought with wonder: "There—or thereabout—
Cloaked in thick darkness in his power's dust-cloud,
There is the hub and heavy nucleus, the ringmaster
Of all this million-shining whirlwind of dancers, the stars of this end of heaven. It is strange, truly,
That great and small, the atoms of a grain of sand and the suns of planets, and all the galactic universes
Are organized on one pattern, the eternal roundabout, the heavy nucleus and whirling electrons, the leashed
And panting runners going nowhere; frustrated flight, unrelieved strain,

endless return—
all—
all—
The eternal fire-wheel."

The Double Axe and Other Poems
Part II of The Double Axe
The Inhumanist
Stanza 22 (p. 67)

Peterson, Ivars

In their search for patterns and logical connections, mathematicians face a vast, mysterious ocean of possibilities. Over the centuries, they have discovered an extensive archipelago of truth and beauty. Much of that accumulated knowledge is passed on to succeeding generations. Even more wonders await future explorers of deep, mathematical waters.

Islands of Truth: A Mathematical Mystery Cruise
Chapter 8 (p. 292)

PHENOMENON

du Noüy, Pierre Lecomte
When we speak of a phenomenon, we speak only of an event, or of a succession of events, arbitrarily isolated from the universe whose evolution they share. By isolating a fact in order to study it, we give it a beginning and an end, which are artificial and relative. In relation to the evolution of the universe, birth is not a beginning, and death is not an end. There are no more isolated phenomena in nature than there are isolated notes in a melody.

The Road to Reason (p. 53)

Jevons, W. Stanley
...every strange phenomenon may be a secret spring which if rightly touched, will open the door to new chambers in the palace of nature.

Principles of Science
Book V
Chapter XXIX (p. 671)

LaPlace, Pierre Simon
The phenomena of nature are most often enveloped by so many strange circumstances, and so great a number of disturbing causes mix their influence, that it is very difficult to recognize them.

A Philosophical Essay on Probabilities
Chapter IX (p. 73)

Wilson, Edward O.
...all tangible phenomena, from the birth of stars to the workings of social institutions, are based on material processes that are ultimately reducible, however long and tortuous the sequences, to the laws of physics.

Consilience: The Unity of Knowledge
Chapter 12 (p. 266)

...There are no more isolated phenomena in nature than there are isolated notes in a melody.
Pierre Lecomte du Noüy – (See p. 239)

PHILOSOPHY

Dennett, Daniel
...there is no such thing as philosophy-free science; there is only science whose philosophical baggage is taken on board without examination.

Darwin's Dangerous Idea
Chapter 1 (p. 21)

Durant, Will
Science gives us knowledge, but only philosophy can give us wisdom.

The Story of Philosophy
Introduction (p. 3)

Faraday, Michael
The philosopher should be a man willing to listen to every suggestion, but determined to judge for himself. He should not be biased by appearances, have no favourite hypotheses; be of no school; and in doctrine have no master. He should not be a respecter of persons, but of things. Truth should be his primary object. If to these qualities he added industry, he may indeed hope to walk within the veil of the temple of nature.

In H. Bence Jones
The Life and Letters of Faraday
Volume I
Chapter IV (p. 198)

Inge, William
...science and philosophy can not be kept in water-tight compartments.

God and the Astronomers
Preface (p. vii)

MacLaurin, Colin
Is it not therefore the business of philosophy, in our present situation in the universe, to attempt to take in at once, in one view, the whole scheme of nature; but to extend, with great care and circumspection, our knowledge,

by just steps, from sensible things, as far as our observations or reasonings from them will carry us, in our enquiries concerning either the greater motions and operations of nature, or her more subtle and hidden works.

An Account of Sir Isaac Newton's Philosophical Discoveries
Chapter I, section 6 (p. 19)

Raether, H.

There are more things between cathode and anode
Than are dreamt of in your philosophy.

Electron Avalanches and Breakdown in Gasses
Introduction (p. 1)

Updike, John

The mad things dreamt up in the sky
Discomfort our philosophy.

Collected Poems 1953–1993
Skyey Developments

Whitehead, Alfred North

Philosophy asks the simple question, What is it all about?

Philosophical Review
Whitehead's Philosophy (p. 178)
Volume XLVI, Number 2, March 1937

Philosophy begins in wonder. And, at the end, when philosophic thought has done its best, the wonder remains. There have been added, however, some grasp of the immensity of things, some purification of emotion by understanding.

Modes of Thought
Chapter III
Lecture VIII (p. 232)

Philosophy is the product of wonder. The effort after the general characterization of the world around us is the romance of human thought.

Nature and Life
Chapter I (p. 1)

Ziman, John M.

One can be zealous for Science, and a splendidly successful research worker, without pretending to a clear and certain notion of what Science really is. In practice it does not seem to matter.

Perhaps this is healthy. A deep interest in theology is not welcome in the average churchgoer, and the ordinary taxpayer should not really concern himself about the nature of sovereignty or the merits of bicameral legislatures. Even though Church and State depend, in the end, upon such abstract matters, we may reasonably leave them to the experts if all goes smoothly. The average scientist will say that he knows from experience and common sense what he is doing, and so long as he is not striking too deeply into the foundation of knowledge he is content to leave the highly technical discussion of the nature of Science to those self-appointed authorities the Philosophers of Science. A rough and ready conventional wisdom will see him through.

Public Knowledge: An Essay Concerning the Social Dimension of Science
Chapter 1 (p. 6)

Philosophy asks the simple question, What is it all about?
Alfred North Whitehead – (See p. 242)

PHOTONS

Einstein, Albert

Every physicist thinks that he knows what a photon is... I spent my life to find out what a photon is and I still don't know it.

In Eugene Hecht
Optics
Chapter 1 (p. 9)

Unknown

If photons have mass, who is their priest?

Source unknown

PHYSICIST

Adams, Douglas
Very strange people, physicists...in my experience the ones who aren't actually dead are in some way very ill.

The Long Dark Tea-Time Of The Soul
Chapter II (p. 111)

Adams, Henry
...the future of thought and therefore of history lies in the hands of physicists, and therefore the future historian must seek his education in the world of mathematical physics.

The Degradation of the Democratic Dogma
The Rule of Phase Applied to History (p. 283)

Einstein, Albert
How wretchedly inadequate is the theoretical physicist as he stands before Nature—and before his students!

Letter dated 15 March 1922
Quoted in Helen Dukas and Banesh Hoffman
Albert Einstein: The Human Side (p. 24)

Green, Celia
If you say to a theoretical physicist that something is inconceivable, he will reply: 'It only *appears* inconceivable because you are naively trying to conceive it. Stop thinking and all will be well.'

The Decline and Fall of Science
Aphorisms (pp. 2–3)

Hanson, N.R.
Physicists do not start from hypotheses; they start from data. By the time a law has been fixed into an H–D [hypothetico–deductive] system, really original physical thinking is over.

Patterns of Discovery
Chapter IV (p. 70)

Johnson, George

Trying to capture the physicists' precise mathematical description of the quantum world with our crude words and mental images is like playing Chopin with a boxing glove on one hand and a catcher's mitt on the other.

The New York Times
On Skinning Schrödinger's Cat (p. 16)
Section 4, Sunday, 2 June 1996

Kush, Polykarp

Our early predecessors observed Nature as she displayed herself to them. As knowledge of the world increased, however, it was not sufficient to observe only the most apparent aspects of Nature to discover her more subtle properties; rather, it was necessary to interrogate Nature and often to compel Nature, by various devices, to yield an answer as to her functioning. It is precisely the role of the experimental physicist to arrange devices and procedures that will compel Nature to make a quantitative statement of her properties and behavior.

In the Nobel Foundation
Nobel Lectures
Physics
1942–62
Nobel Lecture, December 12, 1955 (p. 298)

Michelson, Albert A.

If a poet could at the same time be a physicist, he might convey to others the pleasure, the satisfaction, almost the reverence, which the subject inspires. The aesthetic side of the subject is, I confess, by no means the least attractive to me. Especially is its fascination felt in the branch which deals with light...

Light Waves and Their Uses
Lecture I (p. 1)

Newman, James R.

In this century the professional philosophers have let the physicists get away with murder. It is a safe bet that no other group of scientists could have passed off and gained acceptance for such an extraordinary principle as complementarity, nor succeeded in elevating indeterminacy to a universal law.

Scientific American
Book Review of "Causality and Chance in Modern Physics" (p. 116)
Volume 198, Number 1, January 1958

Rorty, Richard

Here is one way to look at physics: the physicists are men looking for new interpretations of the Book of Nature. After each pedestrian period of normal science, they dream up a new model, a new picture, a new vocabulary, and then they announce that the true meaning of the Book has been discovered. But, of course, it never is, any more than the true meaning of *Coriolanus* or the *Dunciad* or the *Phenomenology of the Spirit* or the *Philosophical Investigations*. What makes them physicists is that their writing are commentaries on the writings of earlier interpreters of Nature, not that they all are somehow "talking about the same thing"...

New Literary History
Philosophy as a Kind of Writing (p. 141)
Volume X, Number 1, Autumn 1978

Toulmin, Stephen
Natural historians... look for regularities of given forms, but physicists seek the form of given regularities.

The Philosophy of Science
Chapter II
Section 2.8 (p. 53)

Unknown
Theoretical physicist—a physicist whose existence is postulated, to make the numbers balance, but who is never actually observed in the laboratory.

Source unknown

PHYSICS

Bohr, Niels
...the new situation in physics is that we are both onlookers and actors in the great drama of existence.

Atomic Theory and the Description of Nature
Chapter IV (p. 119)

Born, Max
Hope is a word one is unlikely to find in the literature of physics.

My Life and My Views
Chapter 6 (p. 190)

Bronowski, Jacob
One aim of the physical sciences has been to give an exact picture of the material world. One achievement of physics in the twentieth century has been to prove that that aim is unattainable.

The Ascent of Man
Chapter 11 (p. 353)

Calvin, Melvin
There is no such thing as pure science. By this I mean that physics impinges on astronomy on the one hand, and chemistry and biology on the other. The synthesis of a really new concept requires some sort of union in one mind of the pertinent aspects of several disciplines...It's no trick to get the right answer when you have all the data. The real creative trick is to get the right answer when you have only half of the data in hand and half of it is wrong and you don't know which half is wrong. When you get the right answer under these circumstances, you are doing something creative.

Following the Trail of Light
Bringing It Together (p. 134)

Condon, E.U.

I take it to be the object of physics so to organize past experience and so to direct the acquisition of new experience that ultimately it will be possible to predict the outcome of any proposed experiment which is capable of being carried out—and to make the prediction in less time than it would have taken actually to carry out the proposed experiment. When this shall have been done I will say that man has a complete understanding of his physical environment. Others may ask more, with this I am satisfied.

Journal of the Franklin Institute
The Philosophical Concepts of Modern Physics
Mathematical models in Modern Physics (p. 257)
Volume 225, Number 3, March 1938

Duhem, Pierre

...physics makes progress because experiment constantly causes new disagreements to break out between laws and facts, and because physicists constantly touch up and modify laws in order that they may more faithfully represent the facts.

The Aim and Structure of Physical Theory
Chapter V (p. 177)

Edelstein, Ludwig

Physics...in antiquity remained closely connected with philosophy, and was predominantly concerned with the philosophical category of the "why," rather than the scientific category of the "how."

In Philip P. Wiener and Aaron Noland
Roots of Scientific Thought
Recent Trends in the Interpretation of Ancient Science (pp. 94–5)

Ehrenfest, Paul

Physics is simple, but subtle.

In Victor F. Weisskopf
Physics in the Twentieth Century: Selected Essays
My Life as a Physicist (p. 3)

Gardner, Martin

In physics and chemistry, like all other branches of science, there is never a sharp line separating pseudo-scientific speculation from the theories of competent men.

Fads and Fallacies
Chapter 7 (p. 80)

Gay-Lussac, Joseph Louis

In the study of physics, we see what are called individual facts but which are by no means isolated and which are not independent of each other;

on the contrary they are related to each other by laws which the physicist devotes all his attention to discovering. It is this which is a measure of the true progress of the science.

In Maurice Grossland
Gay-Lussac: Scientist and Bourgeois
Chapter 3 (p. 70)

Goeppert-Mayer, Maria

Mathematics began to seem too much like puzzle solving. Physics is puzzle solving, too, but of puzzles created by nature, not by the mind of man.

In J. Dash
A Life of One's Own
Maria Goeppert-Mayer (p. 252)

Hasselberg, K.B.

...as for physics, it has developed remarkably as a precision science, in such a way that we can justifiably claim that the majority of all the greatest discoveries in physics are very largely based on the high degree of accuracy which can now be obtained in measurements made during the study of physical phenomena... [Accuracy of measurement] is the very root, the essential condition, of our penetration deeper into the laws of physics—our only way to new discoveries.

In Nobel Foundation
Nobel Lecture
Physics 1901–21
Presentation Speech to Michelson
1907 Nobel Award (pp. 159, 160)

Heidegger, M.

Modern physics is not experimental physics because it applies apparatus to the questioning of nature. Rather the reverse is true. Because physics, indeed already as pure theory, sets nature up to exhibit itself as a coherence of forces calculable in advance, it therefore orders its experiments precisely for the purpose of asking whether and how nature reports itself when set up in this way.

The Question Concerning Technology and other Essays
Part I
The Question Concerning Technology (p. 21)

Heisenberg, Werner

Like all the other natural sciences, Physics advances by two distinct roads. On the one hand, it operates empirically, and thus is enabled to discover and analyse a growing number of phenomena—in this instance, of physical facts; on the other hand, it also operates by theory, which allows it to collect and assemble the known facts in one consistent system, and to predict new ones from the guidance of experimental research.

The Physicist's Conception of Nature
Louis De Broglie (p. 158)

When I was a boy, my grandfather, who was a handicraftsman and knew how to do practical things, once met me when I put a cover on a wooden box with books or so. He saw that I took the cover and I took a nail and I tried to hammer this one nail down to the bottom. 'Oh', he said, 'that is quite wrong what you do there, nobody can do it that way and it is a scandal to look at'. I did not know what the scandal was, but then he said, 'I will show you how you could do it'. He took the cover and he took one nail, put it just a little bit through the cover into the box, and then the next nail a little bit, the third nail a little bit, and so on until all the nails were there. Only when everything was clear, when one could see, that all the nails would fit, then he would start to put the nails really into the box. So, I think this is a good description of how one should proceed in theoretical physics.

In International Centre for Theoretical Physics
From a Life of Physics.
Evening Lectures at the International Centre for Theoretical Physics
Theory, Criticism and a Philosophy
My General Philosophy (p. 46)

Jeans, Sir James

The classical physics seemed to bolt and bar the door leading to any sort of freedom of the will; the new physics hardly does this; it almost seems to suggest that the door may be unlocked—if only we could find the handle. The old physics showed us a universe which looked more like a prison than a dwelling place. The new physics shows us a universe which looks as though it might conceivably form a suitable dwelling place for free men, and not a mere shelter for brutes—a home in which it may at least be possible for us to mould events to our desires and live lives of endeavor and achievement.

Physics and Philosophy
Chapter VII (p. 216)

Koyre, Alexander

Good physics is made *a priori*. Theory precedes fact. Experience is useless because before any experience we are already in possession of the knowledge we are seeking for. Fundamental laws of motion (and of rest), laws that determine the spatio-temporal behavior of material bodies, are laws of a mathematical nature. Of the same nature as those which govern relations and laws of figures and numbers. We find and discover them not in Nature, but in ourselves, in our mind, in our memory, as Plato long ago has taught us.

The Philosophical Review
Galileo and the Scientific Revolution of the Seventeenth Century (p. 347)
Volume 52, Number 3, July 1943

Lindsay, R.B.
...we are essentially viewing the purpose of physics as a scientific discipline as *invention* rather than *discovery*... the term "invention" implies that the physicist uses not only observation but his imaginative powers to construct points of view that identify with experience.

Physics Today
Arbitrariness in Physics (p. 24)
Volume 120, Number 12, December 1967

Mach, Ernst
I only seek to adopt in physics a point of view that need not be changed the moment our glance is carried over into the domain of another science; for ultimately, all must form one whole.

Analysis of Sensations and the Relation of the Physical to the Psychical
Chapter 1 (p. 30, fn 1)

Maritain, J.
Few spectacles are as beautiful and moving for the mind as that of physics thus advancing toward its destiny like a huge throbbing ship.

Distinguish to Unite or The Degrees of Knowledge
Chapter IV (p. 154-5)

Morgan, Thomas H.
Physics has progressed because, in the first place, she accepted the uniformity of nature; because, in the next place, she early discovered the value of exact measurements; because, in the third place, she concentrated her attention on the regularities that underlie the complexities of phenomena as they appear to us; and lastly, and not the least significant, because she emphasized the importance of the experimental method of research. An ideal or crucial experiment is a study of an event, controlled so as to give a definite and measurable answer to a question—*an answer in terms of specific theoretical ideas*, or better still an answer in terms of better understood relations.

Science
The Relation of Biology to Physics (p. 217)
Volume LXV, Number 1679, March 4, 1927

Pines, David

The central task of theoretical physics in our time is no longer to write down the ultimate equations but rather to catalog and understand emergent behavior in its many guises...

The New York Times
In George Johnson
Challenging Particle Physics as Path to Truth
F5, columns 2, 3
Tuesday, December 4, 2001

Rabi, Isidor Isaac

I think that physics should be the central study in all schools. I don't mean physics as it is usually taught—very badly, as a bunch of tricks—but, rather, an appreciation of what it means, and a feeling for it. I don't want to turn everybody into a scientist, but everybody has to be enough of a scientist to see the world in the light of science—to be able to see the world as something that is tremendously important beyond himself, to be able to appreciate the human spirit that could discover these things, that could make instruments to inquire and advance into its own nature. I rate this so highly that with this education people would find something above their religious affiliations, and find a basic unity in the spirit of man.

In Jeremy Bernstein
Experiencing Science
Chapter 2 (p. 126)

Rutherford, Ernest

All science is either physics or stamp collecting.

Quoted by J.B. Birks
Rutherford at Manchester
Memories of Rutherford (p. 108)

Standen, Anthony

Physics is not about the real world, it is about 'abstractions' from the real world, and this is what makes it so scientific.

Science is a Sacred Cow
Chapter III (p. 61)

Tielhard de Chardin, Pierre

The time has come to realise that an interpretation of the universe—even a positive one—remains unsatisfying unless it covers the interior as well as the exterior of things; mind as well as matter. The true physics is that which will, one day, achieve the inclusion of man in his wholeness in a coherent picture of the world.

The Phenomenon of Man
Forward (pp. 35–6)

Truesdell, Clifford
Noll, Walter

Pedantry and sectarianism aside, the aim of theoretical physics is to construct mathematical models such as to enable us, from the use of knowledge gathered in a few observations, to predict by logical processes the outcomes in many other circumstances. Any logically sound theory satisfying this condition is a good theory, whether or not it be derived from "ultimate" or "fundamental" truth. It is as ridiculous to deride continuum physics because it is not obtained from nuclear physics as it would be to reproach it with lack of foundation in the Bible.

The Non-Linear Field Theories of Mechanics (pp. 2–3)

Unknown

Introductory physics courses are taught at three levels: physics with calculus, physics without calculus, and physics without physics.

Source unknown

We apologise for the fact that in the title of a recent talk in the last newsletter, the words 'theoretical physics' came out as 'impossible ideas'.

Punch
Country Life
Volume 290, Issue 7575, 26 February 1986

von Goethe, Johann Wolfgang

Physics must be sharply distinguished from mathematics. The former must stand in clear independence, penetrating into the sacred life of nature in common with all the forces of love, veneration and devotion. The latter, on the other hand, must declare its independence of all externality, go its own grand spiritual way, and develop itself more purely than is possible so long as it tries to deal with actuality and seeks to adapt itself to things as they really are.

Werke
Schriften zur Naturwissenschaft
XXXIX (p. 92)

Weinburg, Steven

Physics is not a finished logical system. Rather, at any moment it spans a great confusion of ideas, some that survive like folk epics from the heroic periods of the past, and others that arise like utopian novels from our dim premonitions of a future grand synthesis.

Gravitation and Cosmology:
Principles and Applications of the General Theory of Relativity
Part I
Chapter I (p. 3)

Weiszaecker, von Karl Friedrich

Physics begins by facing a mystery. It transforms the mystery into a puzzle. It solves the puzzle. And it finds itself facing a new mystery.

In Pekka Lahti and Peter Mittelstaedt
Symposium on the Foundations of Modern Physics:
50 Years of the Einstein–Podolsky–Rosen Gedankenexperiment
Quantum Theory and Space–Time (p. 237)

Wigner, Eugene P.

We have ceased to expect from physics an explanation of all events, even of the gross structure of the universe, and we aim only at the discovery of the laws of nature, that is the regularities, of the events.

In the Nobel Foundation
Nobel Lectures
Physics
1963–70
Nobel Lecture, December 12, 1963 (p. 9)

Zukav, Gary

Unfortunately, when most people think of "physics", they think of chalkboards covered with undecipherable symbols of an unknown mathematics. The fact is that physics is not mathematics. Physics, in essence, is simple wonder at the way things are and a divine (some call it compulsive) interest in how that is so. Mathematics is the tool of physics. Stripped of mathematics, physics becomes pure enchantment.

The Dancing Wu Li Masters
Chapter I (p. 31)

PION

Marshak, Robert E.

The cement that holds the universe together is the force of gravity. The glue holding the atom together is electromagnetic attraction. But the glue that holds the nucleus of the atom together is a mystery that defies all our experience and knowledge of the physical world.

Scientific American
Pions (p. 84)
Volume 196, Number 1, January 1957

PLANET

Aristotle
. . . that which does not twinkle is near—we must take this truth as having been reached by induction or sense-perception.

Posterior Analytics
Book I, Chapter 13, 78[a] [34–5]

Burroughs, William
After one look at this planet any visitor from outer space would say, "I WANT TO SEE THE MANAGER."

The Adding Machine
Women: A Biological Mistake (p. 125)

Chapman, Clark R.
Planets are like living creatures. They are born, full of life and activity. They mature, consume energy, and settle into established ways. Finally, they run down, become dormant, and die. On a human time scale planetary lives are virtually eternal. We see only a snapshot of each planet and can only surmise its evolution.

The Inner Planets
Chapter 6 (pp. 88–9)

Chaucer, Geoffrey
The seven bodies I'll describe anon:
Sol, gold is, Luna's silver, as we see,
Mars iron, and quicksilver's Mercury,
Saturn is lead, and Jupiter is tin,
And Venus copper, by my father's kin!

Canterbury Tales
Canon Yeoman's Tale, L. 16,293–7

Emerson, Ralph Waldo

He who knows what sweets and virtues are in the ground, the waters—the planets, the heavens, and how to come at these enchantments, is the rich and royal man.

Essays and Lectures
Essays
Second Series
Nature (p. 543)

Hammond, Allen L.

With the beginning of direct exploration of the solar system, planetary science has revived to become not only respectable but one of the active, forefront areas of research. How active can be gauged by the assessment, widely agreed on, that the rate of new discoveries and the rate of obsolescence of old ideas have never been so rapid as at present. Investigators are now confronted with such an overwhelming array of new observations and theories that what amounts to a revolution in understanding the solar system is in progress.

Science
Exploring the Solar System (I): An Emerging New Perspective (p. 720)
Volume 186, Number 4165, 22 November 1974

Joyce, James

Gasballs spinning about, crossing each other, passing. Same old dingdong always. Gas, then solid, then world, then cold, then dead shell drifting around, frozen rock like that pineapple rock. The moon.

Ulysses (p. 155)

Kahn, Fritz

This is the universe: infinity. Space without beginning, without end, dark, empty, cold. Through the silent darkness of this space move gleaming spheres, separated from each other by inconceivable distances. Around them, again inconceivably far away, like bits of dust lost in immensity, circle smaller dark spheres, receiving light and life from their "mother suns."

Design of the Universe
Chapter 1 (p. 2)

Marlowe, Christopher

. . . whose faculties can comprehend
The wondrous architecture of the world,
And measure every wand'ring planet's course,. . .

Tamberlaine the Great
Scene VIII

Miller, Hugh
The planet which we inhabit is but one vessel in the midst of a fleet sailing on through the vast ocean of space, under convoy of the sun.

Geology Versus Astronomy
Chapter II (p. 14)

Moliére (Jean-Baptiste Poquelin)
We had a narrow escape, Madame, While asleep;
A neighboring planet did pass us close by,
Cutting a swathe right through our whirlpool;
Had its path led to a collision with mother earth,
She would have shattered in pieces like glass.

Femmes Savantes
Act IV, scene iii

Shapley, Harlow
Millions of planetary systems must exist, and *billions* is the better word. Whatever the methods of origin, and doubtless more than one type of genesis has operated, planets may be the common heritage of all stars except those so situated that planetary materials would be swallowed up by greater masses or cast off through gravitational action.

Of Stars and Men: The Human Response to the Expanding Universe
Chapter 8
The Fourth Adjustment (p. 90)

Siegel, Eli
The planets show grandeur and nicety in their operations; the question is, how did they learn this?

Damned Welcome
Aesthetic Realism Maxims
Part 1, No. 50 (p. 26)

Standage, Tom
A planet is, by definition, an unruly object.

The Neptune Files
Chapter 2 (p. 17)

Teilhard de Chardin, Pierre
Despite their vastness and splendor the stars cannot carry the evolution of matter much beyond the atomic series: it is only on the very humble planets, on them alone, that the mysterious ascent of the world into the sphere of higher complexity has a chance to take place. However inconsiderable they may be in the history of sidereal bodies, however accidental their coming into existence, the planets are finally nothing less

than the key-points of the Universe. It is through them that the axis of life now passes; it is upon them that the energies of an Evolution principally concerned with the building of large molecules is now concentrated.

The Future of Man
Chapter VI, Section I (p. 109)

Tombaugh, Clyde

Behold the heavens and the great vastness thereof, for a planet could be anywhere therein.

Thou shalt dedicate thy whole being to the search project with infinite patience and perseverance.

Thou shalt set no other work before thee, for the search shall keep thee busy enough.

Thou shalt take the plates at opposition time lest thou be deceived by asteroids near their stationary positions.

Thou shalt duplicate the plates of a pair at the same hour angle lest refraction distortions overtake thee.

Thou shalt give adequate overlap of adjacent plate regions lest the planet play hide and seek with thee.

Thou must not become ill at the dark of the moon lest thou fall behind the opposition point.

Thou shalt have no dates except at full moon when long-exposure plates cannot be taken at the telescope.

Many false planets shall appear before thee, hundreds of them, and thou shalt check every one with a third plate.

Thou shalt not engage in any dissipation, that thy years may be many, for thou shalt need them to finish the job.

In David H. Levy
Clyde Tombaugh: Discoverer of Planet Pluto
Chapter 12
Ten Special Commandments for a Would-Be Planet Hunter (p. 180)

MERCURY

Ackerman, Diane

A prowling holocaust keeling low in the sky heads westward for another milk run. The Sun never sets on the Mercurian empire: it only idles on each horizon and lurches back, broiling the same arc across the sky.

The Planets
Mercury (p. 15)

VENUS

Ball, Robert S.

The lover of nature turns to admire the sunset, as every lover of nature will. In the golden glory of the west a beauteous gem is seen to glitter; it is the evening star—the planet Venus... All the heavenly host—even Sirus and Jupiter—must pale before the splendid lustre of Venus, the unrivalled queen of the firmament.

The Story of the Heavens
Venus (p. 168)

Tennyson, Alfred

For a breeze of morning moves,
And the planet of love is on high,
Beginning to faint in the light that she loves
On a bed of daffodil sky,
To faint in the light of the sun she loves,
To faint in his light, and to die.

The Complete Poetical Works of Tennyson
Maude

EARTH

Ackerman, Diane

Long ago, Earth bunched its granite to form the continents, ground molar Alps and Himalayas, rammed Africa and Italy into Europe, gnashing its teeth, till mountain ranges buckled and churned, and oceans (salty once rivers bled flavor from the seasoned earth) gouged their kelpy graves. And the rest is history...

The Planets
Earth
Cosmogony (p. 36)

Albran, Kehlog

The Earth is like a grain of sand, only much heavier.

The Profit (p. 88)

Cloos, Hans

Earth: beautiful, round, colorful planet. You carry us safely through the emptiness and deadness of space. Graciously you cover the black abyss with air and water. You turn as towards the sun, that we may be warm and content, that we may wander, with open eyes, through your meadows, and look upon your splendor. And then you turn us away from the too fiercely burning sun, that we may rest in the coolness of the night from life's heat and the struggle of the day.

Conversation with the Earth
Prologue (p. 3)

Coleridge, Samuel Taylor

Earth! Thou mother of numberless children, the nurse and the mother,
Sister thou of the stars, and beloved by the Sun, the rejoicer!
Guardian and friend of the moon, O Earth, whom the comets forget not,
Yea, in the measureless distance wheel round and again they behold thee!

The Poems of Samuel Taylor Coleridge
Hymn to the Earth (p. 328)

Eddington, Sir Arthur Stanley

My subject disperses the galaxies, but it unites the Earth.

In Arthur Beer (ed.)
Vistas in Astronomy
Volume II
Meeting of the International Astronomical Union
Cambridge, MA, USA
September, 1932 (p. i)

Guiterman, Arthur

We dwell within the Milky Way,
Our Earth, a paltry little mommet,
Suspended in a grand array
Of constellation, moon and comet.

Gaily the Troubadour
Outline of the Universe (p. 70)

Irwin, James

The Earth reminded us of a Christmas tree ornament hanging in the blackness of space. As we got farther and farther away it diminished in size. Finally it shrank to the size of a marble, the most beautiful marble you can imagine. That beautiful, warm, living object looked so fragile, so delicate, that if you touched it with a finger it would crumble and fall apart. Seeing this has to change a man, has to make a man appreciate the creation of God and the love of God.

In Kevin W. Kelley
The Home Planet
With Plate 38

Johnson, Lyndon B.

Think of our world as it looks from that rocket that is heading toward Mars. It is like a child's globe, hanging in space, the continents stuck to its side like colored maps. We are all fellow passengers on a dot of earth. And each of us, in the span of time, has really only a moment among his companions.

Inaugural Address
January 20, 1965

MacLeish, Archibald
To see the earth as we now see it, small and beautiful in that eternal silence where it floats, is to see ourselves as riders on the earth together, brothers on that bright loveliness in the unending night—brothers who see now they are truly brothers.

Riders on the Earth
Bubble of Blue Air (p. xiv)

Sagan, Carl
...if we are to understand the Earth, we must have a comprehensive knowledge of the other planets.

Scientific American
The Solar System (p. 27)
Volume 233, Number 3, 1975

The surface of the Earth is the shore of the cosmic ocean. From it we have learned most of what we know. Recently, we have waded a little out to sea, enough to dampen our toes or, at most, wet our ankles. The water seems inviting. The ocean calls. Some part of our being knows this is from where we came. We long to return. These aspirations are not, I think, irreverent, although they may trouble whatever gods may be.

Cosmos
The Shores of the Cosmic Ocean (p. 5)

Teilhard de Chardin, Pierre
We can only bow before this universal law, whereby, so strangely to our minds, the play of large numbers is mingled and confounded with a final purpose. Without being overawed by the Improbable, let us now concentrate our attention on the planet we call Earth. Enveloped in the blue mist of oxygen which its life breathes, it floats at exactly the right distance from the sun to enable the higher chemisms to take place on its surface. We do well to look at it with emotion. Tiny and isolated though it is, it bears clinging to its flanks the destiny and future of the Universe.

The Future of Man
Chapter VI, Section I (p. 110)

Thomas, Lewis
The overwhelming astonishment, the queerest structure we know about so far in the whole universe, the greatest of all cosmological scientific puzzles, confounding all our efforts to comprehend it, is the earth.

Late Night Thoughts on Listening to Mahler's Ninth Symphony
The Corner of the Eye (p. 16)

Vizinczey, Stephen
Is it possible that I am not alone in believing that in the dispute between Galileo and the Church, the Church was right and the centre of man's universe is the earth?

Truth and Lies in Literature
Rules of the Game (p. 269)

Whipple, Fred L.
Our Earth seems so large, so substantial, and so much with us that we tend to forget the minor position it occupies in the solar family of planets. Only by a small margin is it the largest of the other terrestrial planets. True, it does possess a moderately thick atmosphere that overlies a thin patchy layer of water and it does have a noble satellite, about $\frac{1}{4}$ its diameter. These qualifications of the Earth, however, are hardly sufficient to bolster our cosmic egotism. But, small as is the Earth astronomically, it is our best-known planet and therefore deserves and has received careful study.

Earth, Moon and Planets
The Earth (p. 55)

MARS

Bradbury, Ray
We are all...children of this universe. Not just Earth, or Mars, or this System, but the whole grand fireworks. And if we are interested in Mars at all, it is only because we wonder over our past and worry terribly about our possible future.

In Ray Bradbury, Arthur C. Clarke, Bruce Murray, Carl Sagan and Walter Sullivan
Mars and the Mind of Man
Forward (p. x)

Longfellow, Henry Wadsworth
There is no light in earth or heaven
  But the cold light of stars;
And the first watch of night is given

To the red planet Mars.

The Poetical Works of Henry Wadsworth Longfellow
The Light of Stars
Stanza 2

Lowell, Percival
There are celestial sights more dazzling, spectacles that inspire more awe, but to the thoughtful observer who is privileged to see them well, there is nothing in the sky so profoundly impressive as the canals of Mars.

Mars (p. 228)

JUPITER

Ackerman, Diane
Vibrant as an African trade-bead with bone
chips in orbit round it, Jupiter floods the night's
black scullery, all those whirlpools and burbling
aerosols little changed since the solar-system began.

The Planets
Jupiter (p. 81)

SATURN

Huygens, Christiaan
a a a a a a a c c c c c d e e e e e g h i i i i i i i l l l l m m n n n n n n n n n o o o o p p r r s t t t t u u u u u
Annulo cingitur, tenui, plano, nusquam cohaerente, ad eclipticam inclinato
[It is surrounded by a thin flat ring, inclined to the ecliptic, and nowhere touches the body of the planet]

De Saturni luna observatio nova

Melville, Herman
Seat thyself sultanically among the moons of Saturn.

Moby Dick
Chapter 107

Pallister, William
The planet Saturn, to the naked eye
Appears an oval star; in seeking why
The telescope shows us a startling sight
Which seems some lovely vision on a night
Of dreams. A giant, wide, sunlit, tilted ring,

More strange than any other heavenly thing.

Poems of Science
Other Worlds and Ours
Saturn (p. 205)

Thayer, John H.
If you want to see a picture painted as only the hand of God can paint it, go with me to Saturn...

Popular Astronomy
Saturn. The Wonder of the Worlds (p. 175)
Volume XXVII, Number 263, March 1919

URANUS

Herschel, William
In the fabulous ages of ancient times the appellations of Mercury, Venus, Mars, Jupiter, and Saturn were given to the planets as being the names of their principal heroes and divinities. In the present more philosophical era, it would hardly be allowable to have recourse to the same method, and call on Juno, Pallas, Apollo, or Minerva for a name to our new heavenly body...I cannot but wish to take this opportunity of expressing my sense of gratitude, by giving the name Georgium Sidus, to a star, which (with respect to us) first began to shine under His auspicious reign.

In James Sime
William Herschel and His Work
Chapter V
Letter to Sir Joseph Banks (p. 74)

NEPTUNE

Clerke, Agnes M.
Forever invisible to the unaided eye of man, a sister-globe to our earth was shown to circulate, in frozen exile, at 30 times its distance from the sun. Nay, the possibility was made apparent that the limits of our system were not even thus reached, but that yet profounder abysses of space might shelter obedient, though little favoured members of the solar family, by future astronomers to be recognized through the sympathetic thrillings of Neptune, even as Neptune himself was recognized through the tell-tale deviations of Uranus.

A Popular History of Astronomy during the Nineteenth Century
Part I
Chapter IV (p. 82)

PLUTO

Hoyt, William Graves

The planet was named Pluto, of course, the first two letters of the name as well as its planetary symbol of the superimposed letters "P" and "L" standing for the initials of Pecival Lowell's name.

Lowell and Mars
Chapter 14 (p. 280)

POSITRON

Eddington, Sir Arthur
A positron is a hole from which an electron has been removed; it is a bung-hole which would be evened up with its surroundings if an electron were inserted...

You will see that the physicist allows himself even greater liberty than the sculptor. The sculptor removes material to obtain the form he desires. The physicist goes further and adds material if necessary—an operation which he describes as removing negative material. He fills up a bung-hole, saying he is removing a positron.

The Philosophy of Physical Science
Chapter VIII
Section II (pp. 120–1)

PROBLEM

Douglas, A. Vibert

It is a solemn thought that no man liveth unto himself. It is equally true that no star, no atom, no electron, no ripple of radiant energy, exists unto itself. All the problems of the physical universe are inextricably bound up with one another in the relations of space and time.

The Atlantic Monthly
From Atoms to Stars (p. 165)
Volume 144, August 1929

Einstein, Albert

There are so many unsolved problems in physics. There is so much that we do not know; our theories are far from adequate.

Scientific American
In I. Bernard Cohen
An Interview with Einstein (p. 69)
Volume 193, Number 1, July 1955

Halmos, Paul R.

A teacher who is not always thinking about solving problems—ones he does not know the answer to—is psychologically simply not prepared to teach problem solving to his students.

I Want to Be a Mathematician
Chapter 14 (p. 322)

Hawkins, David

There are many things you can do with problems beside solving them. First you must define them, pose them. But then of course you can also *refine* them, *depose* them, or expose them, even *dissolve* them! A given problem may send you looking for analogies, and some of these may lead you astray, suggesting new and different problems, related or not to the original. Ends and means can get reversed. You had a goal, but the means you found didn't lead to it, so you found a new goal they did lead to.

It's called play. Creative mathematicians play a lot; around any problem really interesting they develop a whole cluster of analogies, of playthings.

In Necia Grant Cooper (ed.)
From Cardinals to Chaos
The Spirit of Play (p. 44)

Hilbert, David

As long as a branch of science offers an abundance of problems, so long is it alive; a lack of problems foreshadows extinction of the cessation of independent development. Just as every human undertaking pursues certain objects, so also mathematical research requires its problems. It is by the solution of problems that the investigator tests the temper of his steel; he finds new methods and new outlooks, and gains a wider and freer horizon.

Bulletin of the American Mathematical Society
Mathematical Problems (p. 438)
Volume 8, July 1902

Kiepenheuer, Karl

For the astronomer, the inexhaustible store of problems in the world he has set out to conquer remains the real mainspring of all his arduous researches.

The Sun
Conclusion (p. 158)

Unknown

What is the best you can do for this problem? Leave it alone and invent another problem.

In George Polya
Mathematical Discovery
Volume II
The Traditional Mathematical Professor (p. 36)

PROGRESS

Clerke, Agnes M.

Progress is the result, not so much of sudden flights of genius, as of sustained patient, often commonplace endeavor; and the true lesson of scientific history lies in the close connection which it discloses between the most brilliant developments of knowledge and the faithful accomplishment of his daily task by each individual thinker and worker.

A Popular History of Astronomy during the Nineteenth Century
Part I
Chapter VI (p. 108)

Duhem, Pierre

Scientific progress has often been compared to a mounting tide; applied to the evolution of physical theories, this comparison seems to us very appropriate, and it may be pursued in further detail.

Whoever casts a brief glance at the waves striking a beach does not see the tide mount; he sees a wave rise, run, uncurl itself, and cover a narrow strip of sand, then withdraw by leaving dry the terrain which it had seemed to conquer; a new wave follows, sometimes going a little farther than the preceding one, but also sometimes not even reaching the sea shell made wet by the former wave. But under this superficial to-and-fro motion, another movement is produced, deeper, slower, imperceptible to the casual observer; it is a progressive movement continuing steadily in the same direction and by virtue of it the sea continually rises. The going and coming of the waves is the faithful image of those attempts at explanation which arise only to be crumbled, which advance only to retreat; underneath there continues the slow and constant progress whose flow steadily conquers new lands, and guarantees to physical doctrines the continuity of a tradition.

The Aim and Structure of Physical Theory
Part I
Chapter III (pp. 38–9)

Feyerabend, Paul

The only principle that does not inhibit progress is: Anything goes.

Against Method: Outline of an Anarchistic Theory of Knowledge
Chapter 1 (p. 23)

Lee, Tsung Dao

The progress of science has always been the result of a close interplay between our concepts of the universe and our observations on nature. The former can only evolve out of the latter and yet the latter is also conditioned greatly by the former. Thus, in our exploration of nature, the interplay between our concepts and our observations may sometimes lead to totally unexpected aspects among already familiar phenomena.

In Nobel Foundation
Nobel Lecture
Physics 1942–62
Nobel Lecture of Tsung Dao Lee
December 11, 1957 (p. 417)

Poincaré, Lucien

There are no limits to progress, and the field of our investigations has no boundaries. Evolution will continue with invincible force. What we to-day call the unknowable, will retreat further and further before science, which will never stay her onward march. Thus physics will give greater and increasing satisfaction to the mind by furnishing new interpretations of phenomena; but it will accomplish, for the whole of society, more valuable work still, by rendering, by the improvements it suggests, life every day more easy and more agreeable, and by providing mankind with weapons against the hostile forces of Nature.

The New Physics and Its Evolution
Chapter XI (p. 328)

Whitehead, Alfred North

The progress of Science consists in observing interconnections and in showing with a patient ingenuity that the events of this ever-shifting world are but examples of a few general relations, called laws. To see what is general in what is particular, and what is permanent in what is transitory, is the aim of scientific thought.

An Introduction to Mathematics
Introduction (p. 4)

PROOF

Aristotle
...to prove what is obvious by what is not is the mark of a man who is unable to distinguish what is self-evident from what is not.

Physics
Book II, Chapter I, 193[a] [5–7]

Cabell, James Branch
"But I can prove it by mathematics, quite irrefutably. I can prove anything you require of me by whatever means you may prefer," said Jurgen, modestly, "for the simple reason that I am a monstrous clever fellow."

Jurgen
Chapter 32 (p. 236)

Davis, Philip
Hersh, Reuben
He rests his faith on rigorous proof; he believes that the difference between a correct proof and an incorrect one is an unmistakable and decisive difference. He can think of no condemnation more damning than to say of a student, "He doesn't even know what a proof is." Yet he is able to give no coherent explanation of what is meant by rigor, or what is required to make a proof rigorous. In his own work, the line between complete and incomplete proof is always somewhat fuzzy, and often controversial.

The Mathematical Experience
The Ideal Mathematician (p. 34)

Gleason, Andrew
...proofs really aren't there to convince you that something is true—they're there to show why it is true.

In D. Albers, G. Alexanderson and C. Reid (eds)
More Mathematical People
(p. 86)

Unknown

Proof by obviousness
"The proof is so clear that it need not be mentioned."

Proof by general agreement
"All in favor?..."

Proof by imagination
"Well, we'll pretend it's true..."

Proof by convenience
"It would be very nice if it were true, so..."

Proof by necessity
"It had better be true, or the entire structure of mathematics would crumble to the ground."

Proof by plausibility
"It sounds good, so it must be true."

Proof by intimidation
"Don't be stupid; of course it's true."

Proof by lack of sufficient time
"Because of the time constraint, I'll leave the proof to you."

Proof by postponement
"The proof for this is long and arduous, so it is given in the appendix."

Proof by accident
"Hey, what have we here?"

Proof by insignificance
"Who really cares, anyway?"

Proof by mumbo-jumbo
$\forall(B \subset \Pi), \exists(X \in \Omega)$

Proof by profanity
(example omitted)

Proof by definition
"We define it to be true."

Proof by tautology
"It's true because it's true."

Proof by plagiarism
"As we see on page 289..."

Proof by lost reference
"I know I saw it somewhere..."

Proof by intimidation
"Don't be stupid; of course it's true."
Unknown – (See p. 274)

Proof by calculus
"This proof requires calculus, so we'll skip it."

Proof by terror
"When intimidation fails..."

Proof by lack of interest
"Does anyone really want to see this?"

Proof by illegibility
(scribble, scribble) QED

Proof by logic
"If it is on the problem sheet, then it must be true!"

Proof by majority rule
Only to be used if general agreement is impossible

Proof by clever variable choice
"Let A be the number such that this proof works..."

Proof by tessellation
"This proof is the same as the last."

Proof by divine word
"And the Lord said, 'Let it be true,' and it was true."

Proof by stubbornness
"I don't care what you say—it is true!"

Proof by simplification
"This proof reduces to the statement $1 + 1 = 2$."

Proof by hasty generalization
"Well, it works for 17, so it works for all reals."

Proof by deception
"Now everyone turn their backs..."

Proof by supplication
"Oh please, let it be true."

Proof by poor analogy
"Well, it's just like..."

Proof by avoidance
Limit of proof by postponement as it approaches infinity.

Proof by design
If it's not true in today's math, invent a new system in which it is.

Proof by authority
"Well, Don Knuth says it's true, so it must be!"

Proof by intuition
"I just have this gut feeling..."

Source unknown

QUANTUM

Bohr, Niels
There is no quantum world. There is only an abstract quantum physical description. It is wrong to think that the task of physics is to find out how nature *is*. Physics concerns what we can say about nature.

Bulletin of the Atomic Scientists
The Philosophy of Niels Bohr (p. 12)
Volume 19, Number 7, September 1963

Carroll, Lewis
"Curiouser and curiouser," cried Alice.

Alice's Adventures in Wonderland
The Pool of Tears
Chapter II (p. 13)

Einstein, Albert
The more one chases after quanta, the better they hide themselves.

Letter to Paul Ehrenfest
dated 12 July 1924
Quoted in Helen Dukas and Banesh Hoffman
Albert Einstein: The Human Side (p. 69)

Feynman, Richard P.
The theory of quantum electrodynamics describes Nature as absurd from the point of view of common sense. And it agrees with experiment. So I hope you can accept Nature as She is—absurd.

QED: The Strange Theory of Light and Matter
Chapter 1 (p. 10)

Gell-Mann, Murray
All of modern physics is governed by that magnificent and thoroughly confusing discipline called quantum mechanics, invented more than fifty years ago. It has survived all tests and there is no reason to believe that

there is any flaw in it. We suppose that it is exactly correct. Nobody understands it, but we all know how to use it and how to apply it to problems; and so we have learned to live with the fact that nobody can understand it.

In Frank Durham and Robert D. Purrington (eds)
Some Truer Method
Chapter 2 (p. 51)

Heisenberg, Werner

If anything like mechanics were true then one would never understand the existence of atoms. Evidently there exists another "quantum mechanics."

In Keith Hannabuss
An Introduction to Quantum Theory
Letter to Wolfgang Pauli
June 21, 1925 (p. 21)

...quantum theory reminds us, as Bohr has put it, of the old wisdom that when searching for harmony in life one must never forget that in the drama of existence we are ourselves both players and spectators. It is understandable that in our scientific relation to nature our own activity becomes very important when we have to deal with parts of nature into which we can penetrate only by using the most elaborate tools.

Physics and Philosophy
The Copenhagen Interpretation of Quantum Theory

Lawrence, D.H.

I like relativity and quantum theories
because I don't understand them
and they make me feel as if space shifted
about like a swan that can't settle,
refusing to sit still and be measured;
and as if the atom were an impulsive thing
Always changing its mind.

In Vivian de Sola Pinto and Warren Roberts (eds)
The Complete Poems of D.H. Lawrence
Volume 1
Relativity (p. 524)

Polkinghorne, J.C.

Quantum theory is both stupendously successful as an account of the small-scale structure of the world and it is also the subject of unresolved debate and dispute about its interpretation. That sounds rather like being shown an impressively beautiful palace and being told that no one is quite sure whether its foundations rest on bedrock or shifting sand.

The Quantum World
Chapter 1 (p. 1)

Zee, A.
Welcome to the strange world of the quantum, where one cannot determine how a particle gets from here to there. Physicists are reduced to bookies, posting odds on the various possibilities.

Fearful Symmetry
Chapter 10 (p. 141)

QUASAR

Mundell, Carole

. . . observing quasars is like observing the exhaust fumes of a car from a great distance and then trying to figure out what is going on under the hood.

Attributed
Scientific American
In Michael Disney
A New Look At Quasars (p. 57)
Volume 278, Number 6, June 1998

QUESTION

Kundera, Milan
...the only truly serious questions are the ones that even a child can formulate. Only the most naive of questions are truly serious. They are questions with no answers. A question with no answer is a barrier that cannot be breached. In other words, it is questions with no answers that set the limits of human possibilities, describe the boundaries of human existence.

The Unbearable Lightness of Being
Part 4
Section 6 (p. 139)

Landau, Lev
Physicists have learned that certain questions cannot be asked, not because the level of our knowledge does not yet permit us to find the answer, but because such an answer simply isn't stored in nature.

In Alexandre Dorozynski
The Man They Wouldn't Let Die
Chapter 7 (p. 108)

Payne-Goposchkin, Cecila H.
Whenever we look in nature we can see spiral forms in the uncurling fern, the snail, the nautilus shell, the hurricane, the stirred cup of coffee, the water that swirls out of a wash bowl. Perhaps we shouldn't be surprised to see spirals in the great star systems whirling in space. Yet they remain a great, intriguing question.

Scientific American
Why Do Galaxies Have a Spiral Form? (p. 89)
Volume 189, Number 3, September 1953

Popper, Karl
...a young scientist who hopes to make discoveries is badly advised if his teacher tells him, "Go round and observe", and that he is well advised if

his teacher tells him: "Try to learn what people are discussing nowadays in science. Find out where the difficulties arise, and take an interest in disagreements. These are the questions which you should take up."

Conjectures and Refutations
Chapter 4 (p. 129)

RADIO ASTRONOMY

Christiansen, Chris
Radio astronomy was not born with a silver spoon in its mouth. Its parents were workers. One parent was the radio-telescope, the other was radar.

The Daily Telegraph (Sydney)
August 25, 1952

Gingerich, Owen
But even if radio astronomy has not so much destroyed our older astronomical viewpoint, it has enormously enlarged and enriched it. It is like that magical moment in the old Cinerama, when the curtains suddenly opened still further, unveiling the grandeur of the wide screen. Optical astronomy in the 1950s, on that narrow, central screen, offered a quiescent view of a slowly burning universe, the visible radiations from thermal disorder. But then the curtains abruptly parted, adding a grand and breathtaking vista, a panorama of swift and orderly motions that revealed themselves through the synchrotron radiation they generated—the so-called violent universe.

In W.T. Sullivan
The Early Years of Radio Astronomy
Radio Astronomy and the Nature of Science (p. 404)

Kraus, John
The radio sky is no carbon copy of the visible sky; it is a new and different firmament, one where the edge of the universe stands in full view and one which bears the tell-tale marks of a violent past.

Big Ear
Chapter 21 (p. 166)

Mitton, Simon
During the last 20 years radio astronomers have led a revolution in our knowledge of the Universe that is paralleled only by the historic contributions of Galileo and Copernicus. In particular, the poetic picture

of a serene Cosmos populated by beautiful wheeling galaxies has been replaced by a catalogue of events of astonishing violence: a primeval fireball, black holes, neutron stars, variable quasars and exploding galaxies.

New Scientist
Newest Probe of the Radio Universe (p. 138)
Volume 56, Number 816, 19 October 1972

Unsold, Albrecht

The old dream of wireless communication through space has now been realized in an entirely different manner than many had expected. The cosmos' short waves bring us neither the stock market nor jazz from distant worlds. With soft noises they rather tell the physicist of the endless love play between electrons and protons.

In W.T. Sullivan, III
Classics in Radio Astronomy
Preface (p. xiii)

REALITY

Dürrenmatt, Friedrich
Our researches are perilous, our discoveries are lethal. For us physicists there is nothing left but to surrender to reality.

The Physicists
Act II (p. 81)

Egler, Frank E.
Reality is not what *is*; it is what the layman *wishes* it to be.

The Way of Science
Science Concepts (p. 22)

Einstein, Albert
Physics is an attempt conceptually to grasp reality as it is thought independently of its being observed. In this sense one speaks of "physical reality."

In Paul Schlipp
Albert Einstein, Philosopher–Scientist
Autobiographical Notes (p. 81)

Einstein, Albert
Infeld, Leopold
In our endeavor to understand reality we are somewhat like a man trying to understand the mechanism of a closed watch. He sees the face and moving hands, and even hears the ticking, but he has no way of opening the case. If he is ingenious enough he may form some picture of a mechanism which could be responsible for all the things he observes, but he may never be quite sure his picture is the only one which could explain his observations.

The Evolution of Physics
Chapter I (p. 31)

Frankel, Felice
Whitesides, George M.
Our reality is illusion: We don't know for sure what's out there.

On the Surface of Things
Illusion (p. 121)

Frost, Robert
You're searching, Joe
For things that don't exist.
I mean beginnings
Ends and beginnings
Ends and beginnings—there are no such things
There are only middles.

Complete Poems of Robert Frost
Mountain Interval

Goodwin, Brian
There is no truth beyond magic... reality is strange. Many people think reality is prosaic. I don't. We don't explain things away in science. We get closer to the mystery.

In Roger Lewin
Complexity
Chapter 2 (p. 32)

Jeans, Sir James
...the most outstanding achievement of twentieth-century physics is not the theory of relativity with its welding together of space and time, or the theory of quanta with its present apparent negation of the laws of causation, or the dissection of the atom with the resultant discovery that things are not what they seem; it is the general recognition that we are not yet in contact with ultimate reality.

The Mysterious Universe
Chapter V (p. 127)

Mead, George H.
...the ultimate touchstone of reality is a piece of experience found in an unanalyzed world. The approach to the crucial experiment may be a piece of torturing analysis, in which things are physically and mentally torn to shreds, so that we seem to be viewing the dissected tissues of objects in ghostly dance before us, but the actual objects in the experimental experience are the common things of which we say that seeing is believing, and of whose reality we convince ourselves by handling. We extravagantly advertise the photograph of the path of an electron, but in fact we could never have given as much reality to the

electrical particle as does now inhabit it, if the photograph had been of naught else than glistening water vapour.

The Philosophy of the Act
Chapter II (p. 32)

Raymo, C.
Science is a map of reality.

The Virgin and the Mousetrap
Chapter 16 (p. 147)

Reeve, F.D.
Because all things balance—as on a wheel—and we cannot see nine-tenths of what is real, our claims of self-reliance are pieced together by unpanned gold. The whole system is a game: the planets are the shells; our earth, the pea. *May there be no moaning of the bar*. Like ships at sunset in a reverie, We are shadows of what we are.

The American Poetry Review
Coasting (p. 38)
Volume 24, Number 4, July–August 1995

Weinberg, Steven
When we say that a thing is real we are simply expressing a sort of respect.

Dreams of a Final Theory
Chapter 2 (p. 46)

Weyl, Hermann
A picture of reality drawn in a few sharp lines cannot be expected to be adequate to the variety of all its shades. Yet even so the draftsman must have the courage to draw the lines firm.

Philosophy of Mathematics and Natural Science
Appendix D (p. 274)

Wilde, Oscar
Well, the way of paradoxes is the way of truth. To test Reality we must see it on the tight rope. When the Verities become acrobats we can judge them.

The Picture of Dorian Gray
Chapter 3 (p. 45)

REASON

Gore, George
That which is beyond reason at present may not be so in the future; but it has now no place in science for want of a basis of verified truth.

The Art of Scientific Discovery
Chapter III (p. 25)

John of Salisbury
Reason, therefore, is a mirror in which all things are seen...

In John van Laarhoven (ed.)
Entheticus Maior and Minor
Volume 1
Part II
Section I, Notes from Epicuris, L. 567–9

Kline, Morris
What mathematics accomplishes with its reasoning may be more evident from an analogy. Suppose a farmer takes over a tract of land in a wilderness with a view to farming it. He clears a piece of ground but notices wild beasts lurking in a wooded area surrounding the clearing who may attack him at any time. He decides therefore to clear that area. He does so but the beasts move to still another area. He therefore clears this one. And the beasts move to still another spot just outside the new clearing. The process goes on indefinitely. The farmer clears away more and more land but the beasts remain on the fringe. What has the farmer gained? As the cleared area gets larger the beasts are compelled to move farther back and the farmer becomes more and more secure as long as he works in the interior of his cleared area. The beasts are always there and one day they may surprise and destroy him but the farmer's relative security increases as he clears more land. So, too, the security with which

we use the central body of mathematics increases as logic is applied to clear up one or another of the foundational problems.

Mathematics: The Loss of Certainty
Chapter XIV (pp. 317–18)

Shakespeare, William
Fool: "The reason why the seven stars are no more than seven is a pretty reason."
Lear: "Because they are not eight?"
Fool: "Yes, indeed.: Thou wouldst make a good fool."

King Lear
Act I, scene v, L. 38–40

von Goethe, Johann Wolfgang
The texture of this world is made up of necessity and chance. Human reason holds the balance between them, treating necessity as the basis of existence, but manipulating and directing chance, and using it.

In Eric A. Blackall (ed.)
Wilhelm Meister's Apprenticeship
Book 1
Chapter 17 (p. 38)

RED SHIFT

Gamow, George
The discovery of the red shift in the spectra of distant stellar galaxies revealed the important fact that our universe is in the state of uniform expansion, and raised an interesting question as to whether the present features of the universe could be understood as the result of its evolutionary development... We conclude first of all that the relative abundance of various atomic species (which were found to be essentially the same all over the observed region of the universe) must represent the most ancient archaeological document pertaining to the history of the universe.

Nature
The Evolution of the Universe (p. 680)
Volume 162, Number 4122, October 1948

Gray, George
...just as the shifting of bookkeeping accounts into the red measures disintegrating, scattering, dissipating financial resources, so the shifting of starlight into the red indicates disintegrating, scattering, dissipating physical resources. It says that the universe is running down,... To entertain this preposterous idea of all these massive star systems racing outward was to accept a radically new picture of the cosmos—a universe in expansion, a vast bubble blowing, distending, scattering, thinning out into gossamer, losing itself. The snug, tight, stable world of Einstein had room for no such flights.

The Atlantic
Universe in the Red (pp. 233, 236)
Volume 1151, Number 2, February 1933

Jeans, Sir James
Another possibility... is that the universe retains its size, while we and all material bodies shrink uniformly. The red shift we observe in the spectra of the nebulae is then due to the fact that the atoms which emitted the light

millions of years ago were larger than the present-day atoms with which we measured the light—the shift is, of course, proportional to distance.

Supplement to *Nature*
Contributions to a British Association Discussion
on the Evolution of the Universe (pp. 703–4)
Volume 128, Number 3234, November 1931

Stapledon, Olaf

I noticed that the sun and all the stars in his neighbourhood were ruddy. Those at the opposite pole of the heaven were of an icy blue. The explanation of this strange phenomenon flashed upon me. I was still traveling, and traveling so fast that light itself was not wholly indifferent to my passage. The overtaking undulations took long to catch me. They therefore affected me as slower pulsations than they normally were, and I saw them therefore as red. Those that met me on my headlong flight were congested and shortened, and were seen as blue.

Last and First Men and Star Maker
Star Maker
Chapter II (p. 262)

RELATIVITY

Eddington, Sir Arthur Stanley

Results of measurements are the subject-matter of physics; and the moral of the theory of relativity is that we can only comprehend what the physical quantities *stand for* if we first comprehend what they *are*.

The Mathematical Theory of Relativity
Conclusion (p. 240)

Einstein, Albert

There is something attractive in presenting the evolution of a sequence of ideas in as brief a form as possible, and yet with a completeness sufficient to preserve throughout the continuity of development. We shall endeavor to do this for the Theory of Relativity, and to show that the whole ascent is composed of small, almost self-evident steps of thought.

Nature
A Brief Outline of the Development of the Theory of Relativity (p. 782)
Volume 106, Number 2677, February 1921

Page, Leigh

The rotating armatures of every generator and motor in this age of electricity are steadily proclaiming the truth of the relativity theory to all who have ears to hear.

American Journal of Physics
Volume 43, Number 4, April 1975 (p. 330)

RESEARCH

Bachrach, Arthur J.
...people don't usually do research the way people who write books about research say that people do research.

Psychological Research
Introduction (pp. 19–20)

Bush, Vannevar
In these circumstances it is not at all strange that the workers sometimes proceed in erratic ways. There are those who are quite content, given a few tools, to dig away, unearthing odd blocks, piling them up in the view of fellow workers and apparently not caring whether they fit anywhere or not...Some groups do not dig at all, but spend all their time arguing as to the exact arrangement of a cornice or an abutment. Some spend all their days trying to pull down a block or two that a rival has put in place. Some, indeed, neither dig nor argue, but go along with the crowd, scratch here and there, and enjoy the scenery. Some sit by and give advice, and some just sit.

Endless Horizons
Chapter 17 (p. 180)

Veblen, Thorstein
...the outcome of any serious research can only be to make two questions grow where only one grew before.

The Place of Science in Modern Civilization and Other Essays
The Evolution of the Scientific Point of View (p. 33)

SCATTERING

Rutherford, Ernst

It was quite the most incredible event that has ever happened to me in my life. It was almost as incredible as if you fired a 15-inch shell at a piece of tissue paper and it came back and hit you. On consideration, I realized that this scattering backward must be the result of a single collision, and when I made calculations I saw that it was impossible to get anything of that order of magnitude unless you took a system in which the greater part of the mass of the atom was concentrated in a minute nucleus. It was then that I had the idea of an atom with a minute massive center carrying a charge.

In Joseph Needham and W. Pagel (eds)
Background to Modern Science
The Development of the Theory of Atomic Structure (p. 68)

SCIENCE

Astbury, W.T.

...science is truly one of the highest expressions of human culture—dignified and intellectually honest, and withal a never-ending adventure. Personally, I feel much the same with regard to the more ecstatic moments in science as I do with regard to music. I see little difference between the thrill of scientific discovery and what one experiences when listening to the opening bars of the Ninth Symphony.

School Science Review
Science in Relation to the Community (p. 279)
Number 109, 1948

Black, Hugh

The limits of science are not limits of its methods, but limits of its spheres.

Everybody's Magazine
Our Made-Over World (p. 710)
November 1914

Bloom, Allan

Science, in freeing men, destroys the natural condition that makes them human. Hence, for the first time in history, there is the possibility of a tyranny grounded not on ignorance, but on science.

The Closing of the American Mind
Part 3
Swift's Doubts (p. 295)

Born, Max

...I believe that there is no philosophical highroad in science, with epistemological signposts. No, we are in a jungle and find our way by trial and error, building our road *behind* us as we proceed. We do not *find* signposts at crossroads, but our own scouts *erect* them, to help the rest.

Experiment and Theory in Physics (p. 44)

Buffon, Georges
The only good science is the knowledge of facts, and mathematical truths are only truths of definition, and completely arbitrary, quite unlike physical truths.

In L. Ducros
Les Encyclopedistes (p. 326)

Bury, J.B.
Science has been advancing without interruption during the last three or four hundred years; every new discovery has led to new problems and new methods of solution, and opened up new fields for exploration. Hitherto men of science have not been compelled to halt, they have always found means to advance further. But what assurance have we that they will not come up against impassable barriers?

The Idea of Progress
Introduction (p. 3)

Buzzati-Traverso, Adriano
Science is a game: it can be exhilarating, it can be useful, it can be frightfully dangerous. It is a play prompted by man's irrepressible curiosity to discover the universe and himself, and to increase his awareness of the world in which he lives and operates.

The Scientific Enterprise, Today and Tomorrow
Part I
Chapter 1 (p. 3)

Campbell, Norman
First, science is a body of useful and practical knowledge and a method of obtaining it. It is science of this form which played so large a part in the destruction of war, and, it is claimed, should play an equally large part in the beneficent restoration of peace... In its second form or aspect, science has nothing to do with practical life, and cannot affect it, except in the most indirect manner, for good or for ill. Science of this form is a pure intellectual study... its aim is to satisfy the needs of the mind and not those of the body; it appeals to nothing but the disinterested curiosity of mankind.

What is Science
Chapter I (p. 1)

Camus, Albert
At the final stage you teach me that this wondrous and multicolored universe can be reduced to the atom and that the atom itself can be reduced to the electron. All this is good and I wait for you to continue. But you tell me of an invisible planetary system in which electrons gravitate

around a nucleus. You explain this world to me with an image. I realize then that you have been reduced to poetry: I shall never know. Have I the time to become indignant? You have already changed theories. So that science that was to teach me everything ends up in a hypothesis, that lucidity founders in metaphor, that uncertainty is resolved in a work of art.

The Myth of Sisyphus and Other Essays
The Myth of Sisyphus
An Absurd Reasoning (pp. 19–20)

Carrel, Alexis

There is a strange disparity between the sciences of inert matter and those of life. Astronomy, mechanics, and physics are based on concepts which can be expressed, tersely and elegantly, in mathematical language. They have built up a universe as harmonious as the monuments of ancient Greece. They weave about it a magnificent texture of calculations and hypotheses. They search for reality beyond the realm of common thought up to unutterable abstractions consisting only of equations of symbols.

Man the Unknown
Chapter 1, section 1 (p. 1)

Chargaff, Erwin

Science cannot be a mass occupation, any more than the composing of music or the painting of pictures.

Perspectives in Biological Medicine
In Praise of Smallness—How Can We Return to Small Science (p. 373)
Volume 23, Number 3, Spring 1980

. . . in most sciences the question 'Why?' is forbidden and the answer is actually to the question, 'How?' Science is much better in explaining than in understanding, but it likes to mistake one for the other.

Perspectives in Biological Medicine
Voices in the Labyrinth (p. 322)
Volume 18, Spring 1975

The sciences have started to swell. Their philosophical basis has never been very strong. Starting as modest probing operations to unravel the works of God in the world, to follow its traces in nature, they were driven gradually to ever more gigantic generalizations. Since the pieces of the giant puzzle never seemed to fit together perfectly, subsets of smaller, more homogeneous puzzles had to be constructed, in each of which the fit was better.

Perspectives in Biological Medicine
Voices in the Labyrinth
VII (p. 323)
Volume 18, Spring 1975

To be a pioneer in science has lost much of its attraction: significant scientific facts and, even more, fruitful scientific concepts pale into oblivion long before their potential value has been utilized. New facts, new concepts keep crowding in and are in turn, within a year or two, displaced by even newer ones... Now, however, in our miserable scientific mass society, nearly all discoveries are born dead; papers are tokens in a power game, evanescent reflections on the screen of a spectator sport, news items that do not outlive the day on which they appeared.

Heraclitean Fire
More Foolish and More Wise (pp. 78, 81)

In science, there is always one more Gordian knot than there are Alexanders. One could almost say that science, as it is practiced today, is an arrangement through which each Gordian knot, once cut, gives rise to two new knots, and so on. Out of one problem considered as solved, a hundred new ones arise; and this has created the myth of the limitlessness of the natural sciences. Actually, many sciences now look as feeble and emaciated as do mothers who have undergone too many deliveries.

Heraclitean Fire
More Foolish and More Wise (p. 116)

Davis, W.M.
Science is therefore not final any more than it is infallible.

In H. Shapley, H. Wright, and S. Rapport (eds)
Readings in the Physical Sciences
The Reasonableness of Science (p. 25)

del Rio, A.M.
It is impossible that he who has once imbibed a taste for science can ever abandon it.

Annals of Philosophy
Analysis of an Alloy of Gold and Rhodium from the Parting House at Mexico
Volume 10, Number 2, October 1825

Dobshansky, Theodosius
Science is cumulative knowledge. Each generation of scientists works to add to the treasury assembled by its predecessors. A discovery made today may not be significant or even comprehensible by itself, but it will make sense in conjunction with what was known before. Indeed this will usually have been necessary to its achievement.

In Robert M. Hutchins and Mortimer J. Adler (eds)
The Great Ideas Today 1974
Advancement and Obsolescence in Science (p. 52)

Science does more than collect facts; it makes sense of them. Great scientists are virtuosi of the art of discovering the meaning of what otherwise might seem barren observations.

In Robert M. Hutchins and Mortimer J. Adler (eds)
The Great Ideas Today 1974
Advancement and Obsolescence in Science (p. 56)

Science has been called "the endless frontier." The more we know, the better we realize that our knowledge is a little island in the midst of an ocean of ignorance.

In Robert M. Hutchins and Mortimer J. Adler (eds)
The Great Ideas Today 1974
Advancement and Obsolescence in Science (p. 61)

Egler, Frank E.

...science...ever reflecting a faith in the intelligibility of nature.

The Way of Science
The Nature of Science (p. 2)

Science is a product of man, of his mind; and science creates the real world in its own image.

The Way of Science
Science Concepts (p. 22)

Einstein, Albert

Although it is true that it is the goal of science to discover rules which permit the association and foretelling of facts, this is not its only aim. It also seeks to reduce the connections discovered to the smallest possible number of mutually independent conceptual elements. It is in this striving after the rational unification of the manifold that it encounters its greatest successes, even though it is precisely this attempt which causes it to run the greatest risk of falling a prey to illusion. But whoever has undergone the intense experience of successful advances made in this domain, is moved by profound reverence for the rationality made manifest in existence.

Ideas and Opinions
Science and Religion (p. 49)

Feyerabend, Paul

The idea that science can, and should, be run according to fixed and universal rules, is both unrealistic and pernicious.

Against Method
Chapter 18 (p. 295)

Feynman, Richard

Science is a way to teach how something gets to be known, what is not known, to what extent things *are* known (for nothing is known absolutely), how to handle doubt and uncertainty, what the rules of evidence are, how to think about things so that judgments can be made, how to distinguish truth from fraud, and from show.

Engineering and Science
The Problem of Teaching Physics in Latin America
November 1963

Fulbright, J. William

Science has radically changed the conditions of human life on earth. It has expanded our knowledge and our power but not our capacity to use them with wisdom.

Old Myths and New Realities
Conclusion (p. 142)

Gould, Stephen Jay

The net of science covers the empirical universe: What is it made of (fact) and why does it work this way (theory)?

Natural History
Non Overlapping Magisteria (p. 61)
Volume 106, Number 2, March 1997

Heisenberg, Werner

Almost every progress in science has been paid by a sacrifice, for almost every new intellectual achievement previous positions and conceptions had to be given up. Thus, in a way, the increase of knowledge and insight diminishes continually the scientist's claim to 'understand' nature.

In A. Sarlemijn and M.J. Sparnaay (eds)
Physics in the Making
Chapter I (p. 9)

Science no longer confronts nature as an objective observer, but sees itself as an actor in this interplay between man and nature. The scientific method of analysing, explaining, and classifying has become conscious of its limitations... Method and object can no longer be separated.

The Physicist's Conception of Nature
Chapter I (p. 29)

Hippocrates

There are, indeed, two things, knowledge and opinion, of which the one makes its possessor really to know, the other to be ignorant...

The Law
Paragraph 4 (p. 144)

Hodgson, Leonard

...science can only deal with what is, and can say nothing about what ought to be, which is the concern of ethics; science can tell us about means to ends, but not about what the ends should be.

Theology in an Age of Science
An Inaugural Lecture
November 3, 1944 (p. 9)

Holton, Gerald
Roller, Duane H.D.

Science is an ever-unfinished quest to discover facts and establish relationships between them.

Foundations of Modern Physical Science
Chapter 13 (p. 214)

...science has grown almost more by what it has learned to ignore than by what it has had to take into account.

Foundations of Modern Physical Science
Chapter 2 (p. 25)

Hoyle, Sir Fred

The fragmentation of science is a source of difficulty to all teachers and to all students—the connection of one research area to another is not always apparent. This is because science is rather like a vast and subtle jig-saw puzzle, and the usual way to attack a jig-saw puzzle is to work simultaneously on several parts of it. Only at the end do we seek to fit the different parts of it together into a coherent whole.

In Eugene H. Kone and Helene J. Jordan (eds)
The Greatest Adventure
Cosmology and its Relation to the Earth (p. 22)

Huxley, Aldous

Science is dangerous; we have to keep it most carefully chained and muzzled.

Brave New World
Chapter 16 (p. 270)

Jeffers, Robinson

"...Science is not to serve but to know. Science is for itself its own value,
 it is not for man,
His little good and big evil: it is a noble thing, which to use
Is to degrade..."
"Therefore astronomy is the most noble science; is the
 most useless..."

"...Science is not a chambermaid-woman." "Brother," the old man said, "you are right. Science is an adoration; a kind of worship."

The Double Axe and Other Poems
The Inhumanist
Part II of The Double Axe
Stanza 36 (p. 92)

Science is dangerous; we have to keep it most carefully chained and muzzled.
Aldous Huxley – (See p. 301)

Jourdain, Philip E.B.
Our ideal in natural science is to build up a working model of the universe out of the sort of ideas that all people carry with them everywhere "in their heads," as we say, and to which ideas we appeal when we try to teach mathematics. These ideas are those of *number*, *order*, the numerical measures of times and distances, and so on.

The Nature of Mathematics
Chapter IV (p. 54)

Keyser, Cassius Jackson
Science is destined to appear as the child and the parent of freedom blessing the earth without design. Not in the ground of need, not in bent and painful toil, but in the deep-centered play-instinct of the world, in

the joyous mood of the eternal Being, her spirit, which is always young, Science has her origin and root; and her spirit, which is the spirit of genius in moments of elevation, is but a sublimated form of play, the austere and lofty analogue of the kitten playing with the entangled skein or of the eaglet sporting with the mountain winds.

Mathematics (p. 44)

Kuhn, Thomas

To understand why science develops as it does, one need not unravel the details of biography and personality that lead each individual to a particular choice, though that topic has vast fascination. What one must understand, however, is the manner in which a particular set of shared values interacts with the particular experiences shared by a community of specialists to ensure that most members of the group will ultimately find one set of arguments rather than another decisive.

The Structure of Scientific Revolutions
Postscript—1969 (p. 200)

Lowell, Percival

Now in science there exists two classes of workers. There are men who spend their days in amassing material, in gathering facts. They are the collectors of specimens in natural history, the industrious takers of routine measurements in physics and astronomy or the mechanical accumulators of photographic plates. Very valuable such collections are. They may not require much brains to get, but they enable other brains to get a great deal out of them later... The rawer they are the better. For the less mind enters into them the more they are worth. When destitute altogether of informing intelligence, they become priceless, as they then convey nature's meaning unmeddled of man...

The second class of scientists are the architects of the profession. They are the men to whom the building up of science is due. In their hands, the acquired facts are put together to that synthesizing of knowledge from which new conceptions spring... Though the gathering of material is good, without the informing mind to combine the facts they had forever remained barren of fruit.

In William Graves Hoyt
Lowell and Mars
Chapter 2 (p. 22)

Mach, Ernst

Where neither confirmation nor refutation is possible, science is not concerned. Science acts and only acts in the domain of *uncompleted* experience.

Science of Mechanics
Chapter IV
Section IV (pp. 587–8)

Maffei, Paolo

We are now moving beyond those concepts and the knowledge familiar to us in the first half of this century, and we are entering a world in which science and fantasy intertwine...

Beyond the Moon
Chapter 10 (p. 301)

Margulis, Lynn
Sagan, Dorian

...Science has become a social method of inquiring into natural phenomena, making intuitive and systematic explorations of laws which are formulated by observing nature, and then rigorously testing their accuracy in the form of predictions. The results are then stored as written or mathematical records which are copied and disseminated to others, both within and beyond any given generation. As a sort of synergetic, rigorously regulated group perception, the collective enterprise of science far transcends the activity within an individual brain.

Microcosmos
Chapter 12 (p. 233)

Mendeléeff, D.

Knowing how contented, free and joyful is life in the realms of science, one fervently wishes that many would enter their portals.

Principles of Chemistry
Volume 1
Preface (p. ix)

While science is pursuing a steady onward movement, it is convenient from time to time to cast a glance back on the route already traversed, and especially to consider new conceptions which aim at discovering the general meaning of the stock of facts accumulated from day to day in our laboratories.

Journal of the Chemical Society
Volume 55, 1889 (p. 634)

Millikan, Robert A.

. . . Science walks forward on two feet, namely theory and experiment. . .

In the Nobel Foundation
Nobel Lectures
Physics
1922–41
Nobel Lecture, May 23, 1924 (p. 54)

Mitchell, Maria

The phrase 'popular science' has in itself a touch of absurdity. That knowledge which is popular is not scientific.

In Phebe Mitchell Kendall
Maria Mitchell: Life, Letters, and Journals
Chapter VII (p. 138)

Oppenheimer, J. Robert

Both the man of science and the man of art live always at the edge of mystery, surrounded by it; both always as to the measure of their creation, have had to do with the harmonization of what is new with what is familiar, with the balance between novelty and synthesis, with the struggle to make partial order in total chaos.

Prospects in the Arts and Sciences
Speech 26 December 1954
Columbia University Bicentennial

Peacock, Thomas Love
Science is one thing and wisdom is another. Science is an edged tool with which men play like children and cut their own fingers. If you look at the results which science has brought in its train, you will find them to consist almost wholly in elements of mischief... The day would fail, if I should attempt to enumerate the evils which science has inflicted on mankind. I almost think it is the ultimate destiny of science to exterminate the human race.

Gryll Grange
Chapter 19 (p. 127)

Popper, Karl
...science is most significant as one of the greatest spiritual adventures that man has yet known...

Poverty of Historicism
Chapter III, Section 19 (p. 56)

Russell, Bertrand
A life devoted to science is therefore a happy life, and its happiness is derived from the very best sources that are open to dwellers on this troubled and passionate planet.

Mysticism and Logic
Chapter II (p. 45)

Smolin, Lee
Science is, above everything else, a search for an understanding of our relationship with the rest of the universe.

The Life of the Cosmos
Part 1
Chapter 1 (p. 23)

Soddy, Frederick
As science advances and most of the more accessible fields of knowledge have been gleaned of their harvest, the need for more and more powerful and elaborate appliances and more and more costly materials ever grows.

Science and Life
Science and the State (p. 60)

Thomson, Sir George

[The method of science is] a collection of pieces of advice, some general, some rather special, which may help to guide the explorer in his passage through the jungle of apparently arbitrary facts... In fact, the sciences differ so greatly that it is not easy to find any sort of rule which applies to all without exception.

The Inspiration of Science
Chapter II (p. 7)

Whitehead, Alfred North

Science can find no individual enjoyment in Nature; science can find no aim in Nature; science can find no creativity in Nature.

Modes of Thought
Chapter III
Lecture VIII (p. 211)

SCIENTIFIC

Bridgman, P.W.
The scientific method, as far as it is a method, is nothing more than doing one's damnedest with one's mind.

Reflections of a Physicist
Chapter 21 (p. 351)

Medawar, Peter
If the purpose of scientific methodology is to prescribe or expound a system of enquiry or even a code of practice for scientific behavior, then scientists seem able to get on very well without it.

Pluto's Republic
Induction and Intuition in Scientific Thought (p. 78)

Rabi, I.I.
The essence of the scientific spirit is to use the past only as a springboard to the future.

In A.A. Warner, Dean Morse and T.E. Cooney (eds)
The Environment of Change
The Revolution in Science (p. 47)

Renan, Ernest
Orthodox people have as a rule very little *scientific honesty*. They do not *investigate*, they try to *prove* and this must necessarily be so. The result has been given to them beforehand; this result is true, undoubtedly true. Science has no business with it, science which starts from doubt without knowing whither it is going, and gives itself up bound hand and foot to criticism which leads it wheresoever it lists.

The Future of Science
Chapter III (p. 33)

Verne, Jules

There is no more envious race of men than scientific discoverers.

A Journey to the Center of the Earth
Chapter 3 (p. 22)

Wells, H.G.

Scientific truth is the remotest of mistresses. She hides in strange places, she is attained by tortuous and laborious roads, but *she is always there*! Win to her and she will not fail you; she is yours and mankind's forever. She is reality, the one reality I have found in this strange disorder of existence...

Tono-Bungay
Book III, Chapter 3, I (p. 278)

SCIENTIST

Calder, Alexander

Scientists leave their discoveries, like foundlings, on the doorstep of society, while the stepparents do not know how to bring them up.

In Alan J. Friedman and Carol C. Donley
Einstein as Myth and Muse
Chapter 1 (p. 7)

Erasmus, Desiderius

And next these come our Philosophers, so much reverenc'd for their Fur'd Gowns and Starcht Beards, that they look upon themselves as the onely Wise Men, and all others as Shadows. And yet how pleasantly do they dote while they frame in their heads innumerable worlds; measure out the Sun, the Moon, the Stars, nay and Heaven it self, as it were with a pair of Compasses...

In P.S. Allen (ed.)
The Praise of Folly (p. 112)

Gardner, Martin

When reputable scientists correct flaws in an experiment that produced fantastic results, then fail to get those results when they repeat the test with flaws corrected, they withdraw their original claims. They do not defend them by arguing irrelevantly that the failed replication was successful in some other way, or by making intemperate attacks on whomever dares to criticize their competence.

The New York Review of Books
Claims for ESP
Reply by Martin Gardner
February 19, 1981

Gray, George W.

The modern scientist is like a detective who finds clues, but never gets a glimpse of the fugitive he seeks.

The Atlantic Monthly
New Eyes of the Universe (p. 607)
Volume 155, Number 5, May 1935

Mitchell, Maria

The true scientist must be self-forgetting. He knows that under the best circumstances he is sowing what others must reap—or rather he is striking the mine which others must open up—for human life at longest has not the measure of a single breath in the long life of science.

In Helen Wright
Sweeper in the Sky
Chapter 9 (p. 168)

It is the highest joy of the true scientist...that he can reap no lasting harvest—that whatever he may bring into the storehouse today will be surpassed by the gleaners tomorrow—he studies Nature because he loves her and rejoices to 'look through Nature up to Nature's God.'

In Helen Wright
Sweeper in the Sky
Chapter 9 (p. 168)

Ting, Samuel C.C.

...scientists must go beyond what is taught in the textbook, and they must think independently. Also, they cannot hesitate to ask questions, even when their view may be unpopular.

In Janet Nomura Morey and Wendy Dunn
Famous Asian Americans
Samuel C.C. Ting (p. 143)

Unknown

The poet, according to Gilbert K. Chesterton, is content to walk along with his head in the heavens, while the scientist must ever seek in vain to cram the heavens into his head!

Journal of Chemical Education
March 1945 (p. 106)

Young, J.Z.

One of the characteristics of scientists and their work, curiously enough, is a certain confusion, almost a muddle. This may seem strange if you have come to think of science with a big S as being all clearness and light.

Doubt and Certainty in Science
First Lecture (p. 1)

SENSES

Born, Max

The problem of physics is how the actual phenomena, as observed with the help of our sense organs aided by instruments, can be reduced to simple notions which are suited for precise measurement and used for the formulation of quantitative laws.

Experiment and Theory in Physics (p. 9)

Einstein, Albert

We can only see the universe by the impressions of our senses reflecting indirectly the things of reality.

Cosmic Religion
On Science (p. 101)

SHADOW

Aesop

Beware that you do not lose the substance by grasping at the shadow.

Fables
The Dog and the Shadow

Eddington, Sir Arthur Stanley

In the world of physics we watch a shadowgraph performance of familiar life. The shadow of my elbow rests on the shadow-tables as the shadow-ink flows over the shadow-paper...The frank realisation that physical science is concerned with a world of shadows is one of the most significant of recent advances.

The Nature of the Physical World
Introduction (p. xiv, xv)

In the world of physics we watch a shadowgraph performance of familiar life. The shadow of my elbow rests on the shadow-tables as the shadow-ink flows over the shadow-paper... The frank realisation that physical science is concerned with a world of shadows is one of the most significant of recent advances.
Sir Arthur Stanley Eddington – (See p. 313)

SIMPLICITY

Bailey, Janet
It is an article of faith in physics that the world's bewildering mask of complexity hides an ultimate simplicity.

The Good Servant
Chapter 4 (p. 110)

Chandrasekhar, Subrahmanyan
The simple is the seal of the true and beauty is the splendour of truth.

In Tore Frängsmyr (ed.)
Nobel Lectures
Physics 1981–90
Nobel Acceptance Speech
8 December 1983

Einstein, Albert
In every important advance the physicist finds that the fundamental laws are simplified more and more as experimental research advances. He is astonished to notice how sublime order emerges from what appeared to be chaos. And this cannot be traced back to the workings of his own mind but is due to a quality that is inherent in the world of perception. Leibniz well expressed this quality by calling it a pre-established harmony.

In Max Planck
Where is Science Going
Introduction (p. 13)

Gore, George
Simplicity, whether truthful or not, is often attractive to unphilosophical minds, because it requires less intellectual exertion.

The Art of Scientific Discovery
Chapter IV (p. 29)

Haldane, J.B.S.

In scientific thought we adopt the simplest theory which will explain all the facts under consideration and enable us to predict new facts of the same kind. The catch in this criterion lies in the word "simplest." It is really an aesthetic canon such as we find implicit in our criticisms of poetry or painting. The layman finds such a law as $\partial x/\partial t = \kappa(\partial^2 x/\partial y^2)$ less simple than "it oozes," of which it is the mathematical statement. The physicist reverses this judgment.

On Being the Right Size and Other Essays
Science and Theology as Art-Forms (pp. 33–4)

Heisenberg, Werner

You may object that by speaking of simplicity and beauty I am introducing aesthetic criteria of truth, and I frankly admit that I am strongly attracted by the simplicity and beauty of the mathematical schemes which nature presents us. You must have felt this too: the almost frightening simplicity and wholeness of the relationships, which nature suddenly spreads out before us...

Physics and Beyond
Chapter 5 (pp. 68–9)

Poincaré, Henri

...it is because simplicity, because grandeur, is beautiful that we preferably seek simple facts, sublime facts, that we delight now to follow the majestic courses of the stars, now to examine with the microscope that prodigious littleness which is also grandeur, now to seek in geologic time the traces of a past which attracts because it is far away.

The Foundations of Science
Science and Method
Book I
Chapter I (p. 367)

Reid, Thomas

Men are often led into errors by the love of simplicity, which disposes us to reduce things to few principles, and to conceive a greater simplicity in nature than there really is.

Essays on the Intellectual Powers of Man
Essay VI
Chapter VIII (p. 696)

Wheeler, John

Some day a door will surely open and expose the glittering central mechanism of the world in all its beauty and simplicity.

In Charles W, Misner, Kip W. Thorne and John Wheeler
Gravitation (p. 1197)

SKY

Astronomy Survey Committee

Nature offers no greater splendor than the starry sky on a clear, dark night. Silent, timeless, jeweled with the constellations of ancient myth and legend, the night sky has inspired wonder throughout the ages.

Astronomy and Astrophysics for the 1980's
Volume 1
Report to the Astronomy Survey Committee (p. 3)

Brandt, John C.
Chapman, Robert D.

...each step forward in unraveling the mystery of comets (or any other natural phenomenon) brings great pleasure to all who look to the sky as a source of beauty and intellectual challenge.

Introduction to Comets
Chapter 10 (p. 226)

Browning, Robert

Sky—what a scowl of cloud
Till, near and far,
Ray on ray split the shroud
Splendid, a star!

The Complete Poetical Works of Browning
The Two Poets of Croisic

de Saint-Exupéry, Antoine

A sky as pure as water bathed the stars and brought them out.

Southern Mail
Chapter I (p. 9)

Emerson, Ralph Waldo

The sky is the daily bread of the eyes.

In Edward Waldo Emerson (ed.)
Journals of Ralph Waldo Emerson
1841–44
25 May 1843 (p. 410)

Friedman, Herbert

It is impossible for any sensitive person to look at a star-filled sky without being stirred by thoughts of creation and eternity. The mystery of the origin and destiny of the universe haunts us throughout our lives.

The Amazing Universe
Chapter 7 (p. 166)

Kremyborg, Alfred

The sky is that beautiful old parchment in which the sun and the moon keep their diary.

Old manuscript

Lockwood, Marion

He who has again and again consciously watched the sky passing through its changing gamut of color and light, has experienced one of the most exciting adventures which can come to the human spirit.

The Sky
Winter Stars (p. 16)
January 1939

Lowell, Amy

A wise man,
Watching the stars pass across the sky,
Remarked:
In the upper air the fireflies move more slowly.

The Complete Poetical Works of Amy Lowell
Meditation

Manilus

It is my delight to traverse the very air and spend my life touring the boundless skies, learning of the constellations and the contrary motions of the planets.

Astronomica
Book I

Maunder, E. Walter

The oldest picture book in our possession is the Midnight Sky.

Nineteenth Century
The Oldest Picture-Book of All (p. 451)
Volume XLVIII, Number CCLXXXIII, September 1900

Moulton, Forest Ray

It is doubtful whether there is in the whole range of human experience any more awe-inspiring spectacle than that presented by the sky on a clear and moonless night. Under the vault of the sparkling heavens one is raised, if ever, to an actual realization of the fact that the earth beneath his feet is a relatively tiny mass in comparison with the infinite cosmos spread out above.

Astronomy
Chapter II (p. 14)

Schaefer, Bradley E.

The sky is beautiful and vast and harbors many secrets.

Sky and Telescope
Inventory of Cosmic Mysteries (p. 68)
Volume 94, Number 4, October 1994

Shakespeare, William

My soul is in the sky.

A Midsummer Night's Dream
Act V, scene i

Shore, Jane

Each night the sky splits open like a melon
its starry filaments
the astronomer examines with great intensity.

Eye Level
An Astronomer's Journal (p. 31)

Smoot, George

There is something about looking at the night sky that makes a person wonder.

Wrinkles in Time
Chapter 1 (p. 1)

Swings, Pol

The sky belongs to everyone, with stars to spare for all.

In Henry Margenau and David Bergamini (eds)
The Scientist (p. 116)

Whitman, Walt

Over all the sky—the sky! Far, far out of reach, studded, breaking out, the eternal stars.

In James E. Miller, Jr (ed.)
Complete Poetry and Selected Prose
Bivouac on a Mountain Side

SOLAR SYSTEM

Carlyle, Thomas

Did not the Boy Alexander weep because he had not two Planets to conquer; or a whole Solar System; or after that, a whole universe?

Sartor Resartus, On Heroes, Hero-Worship and the Heroic in History
Sartor Resartus
Book II, Chapter VIII (p. 137)

Davies, Paul

The secret of our success on planet Earth is space. Lots of it. Our solar system is a tiny island of activity in an ocean of emptiness.

The Last Three Minutes
Chapter I (p. 4)

Horowitz, Norman H.

If the exploration of the solar system in our time brings home to us a realization of the uniqueness of our small planet and thereby increase our resolve to avoid self-destruction, they will have contributed more than just science to the human future.

To Utopia and Back: The Search for Life in the Solar System
Chapter 8 (p. 146)

Lambert, Johann Heinrich

Nothing is more simple than the plan of the Solar System...

The System of the World
Part I
Chapter I (p. 1)

Lowell, Percival

Now when we think that each of these stars is probably the center of a solar system grander than our own, we cannot seriously take ourselves to be the only minds in it all.

Mars
Chapter I (p. 5)

MacLennan, Hugh

We have just reached the outer fringes of the Solar System. Can any sane man possibly argue that we should stop there?

Scotchman's Return and Other Essays
Remembrance Day, 2010 AD (p. 89)

SPACE

Archytas of Tarentum
If I am at the extremity of the heaven of the fixed stars, can I stretch outwards my hand or staff? It is absurd to suppose that I could not; and if I can, what is outside must be either body or space. We may then in the same way get to the outside of that again, and so on; and if there is always a new place to which the staff may be held out, this clearly involves extension without limit.

In H.A.L. Fisher and Others (eds)
Essays in Honor of Gilbert Murray
The Invention of Space (p. 233)

Bradley, John Hodgdon Jr
A sea whose shores no eyes have ever seen, whose depth no instrument can fathom, whose waters no scientist can analyze—such is the sea of space. Nothing can be as empty and cold as the gulf wherein our destinies are immersed. Star worlds, like fish in schools, drift through the void, star worlds as large as our sun and many times larger, in schools of hundreds of millions. Unlike a school of fish, whose direction may be changed by a whim of the leader, whose organization may be destroyed by the rush of an enemy, whose fate is in the hands of a shifty environment, the stars in their galaxy move with the majesty of perfect orderliness. From the smallest satellite slave of the smallest star to the largest super-galaxy of worlds in space, everything bows to the first law of nature. Chaos and caprice do not exist.

Parade of the Living
Chapter I (p. 3)

Carlyle, Thomas
...has not a deeper meditation taught certain of every climate and age, that the WHERE and WHEN, so mysteriously inseparable from all our thoughts, are but superficial terrestrial adhesions to thought; that the Seer may discern them where they mount up out of the celestial

EVERYWHERE and FOREVER: have not all nations conceived their GOD as Omnipresent and Eternal; as existing in a universal HERE, an everlasting NOW? Think well, thou too wilt find that Space is but a mode of our human Sense, so likewise Time; there is no Space and no Time: We are—we know not what—light-sparkles floating in the aether of Deity!...

Sartor Resartus, On Heroes, Hero-Worship and the Heroic in History
Sartor Resartus
Book I, Chapter VIII (p. 40)

Deudney, Daniel
Space is only 80 miles from every person on earth—far closer than most people are to their own national capitals...

Space: The High Frontier in Perspective
Introduction (p. 6)

Eddington, Sir Arthur Stanley

...space is not a lot of points close together; it is a lot of distances interlocked.

The Mathematical Theory of Relativity
Chapter I, section 1 (p. 10)

Empson, William

Space is like earth, rounded, a padded cell;
Plumb the stars depth, your lead bumps you behind...

Collected Poems
The Worlds End

Gibran, Kahlil

Space is not space between the earth and the sun to one who looks down from the windows of the Milky Way.

Sand and Foam (p. 7)

Glenn, John Jr

In space one has the inescapable impression that here is a virgin area of the universe in which civilized man, for the first time, has the opportunity to learn and grow without the influence of ancient pressures. Like the mind of a child, it is yet untainted with acquired fears, hate, greed, or prejudice.

In Kevin W. Kelley
The Home Planet
With Plate 136

Hale, George Edward

Like buried treasures, the outposts of the universe have beckoned to the adventurous from immemorial times. Princes and potentates, political or industrial, equally with men of science, have felt the lure of the uncharted seas of space, and through their provision of instrumental means the sphere of exploration has rapidly widened...

Harper's Magazine
Possibilities of Large Telescopes (p. 639)
April 1928

Joubert, Joseph

There is something divine about the ideas of space and eternity which is wanting in those of pure duration and simple extension.

Pensées and Letters of Joseph Joubert
XII (p. 90)

Murray, Bruce

Space... is a colorful thread intimately woven into the enormous tapestry of human existence and experience.

In Ray Bradbury, Arthur C. Clarke, Bruce Murray, Carl Sagan and Walter Sullivan
Mars and the Mind of Man
Bruce Murray (p. 47)

Ockels, Wubbo

Space is so close: It took only eight minutes to get there and twenty to get back.

In Kevin W. Kelley
The Home Planet
With Plate 126

Reade, Winwood

And then, the earth being small, mankind will migrate into space, and will cross the airless Saharas which separate planet from planet and sun from sun. The earth will become a Holy Land which will be visited by pilgrims from all the quarters of the universe. Finally, men will master the forces of nature; they will become themselves architects of systems, manufacturers of worlds.

The Martyrdom of Man
Chapter IV (p. 515)

Russell, Bertrand

How can a certain line, or a certain surface, form an impassable barrier to space, or have any mobility different in kind from that of all other lines or surfaces? The notion cannot, in philosophy, be permitted for a moment, since it destroys that most fundamental of all the axioms, the homogeneity of space.

An Essay on the Foundations of Geometry
Chapter I, section 45 (p. 49)

Siegel, Eli

Space won't keep still, and it won't budge either: so give up trying.

Damned Welcome
Aesthetic Realism Maxims
Part 2, No. 396 (p. 153)

Smith, Logan Pearsall

So gazing up on hot summer nights at the London stars, I cool my thoughts with a vision of the giddy, infinite, meaningless waste of

Creation, the blazing Suns, the Planets and frozen Moons, all crashing blindly forever across the void of space.

Trivia
Book II
Mental Vice (p. 97)

Valéry, Paul
Space is an imaginary body, as time is fictive movement.
When we say "in space" or "space is filled with..." we are positing a body.

The Collected Works of Paul Valéry
Volume 14, Analects, CIX (p. 321)

vas Dias, Robert
The premise... is that outer space is as much a territory of the mind as it is a physical concept.

Inside Outer Space: New Poems of the Space Age
Introduction (p. xxxix)

von Braun, Wernher
...don't tell me that man doesn't belong out there. Man belongs wherever he wants to go—and he'll do plenty well when he gets there.

Time
Reach for the Stars (p. 25)
Volume 71, 17 February 1958

Weyl, H.
Nowhere do mathematics, natural science, and philosophy permeate one another so intimately as in the problem of space.

Philosophy of Mathematics and Natural Science
Chapter III (p. 67)

SPACETIME

Berlinski, David
Yet everything has a beginning, everything comes to an end, and if the universe actually began in some dense explosion, thus creating time and space, so time and space are themselves destined to disappear, the measure vanishing with the measured, until with another ripple running through the primordial quantum field, something new arises from nothingness once again.

A Tour of the Calculus
Chapter 26 (p. 309)

MacLeish, Archibald
Space–time has no beginning and no end.
It has no door where anything can enter.
How break and enter what will only bend.

Songs for Eve
Reply to Mr. Wordsworth

Minkowski, Hermann
The views of space and time which I wish to lay before you have sprung from the soil of experimental physics, and therein lies their strength. Henceforth space by itself and time by itself, are doomed to fade away into mere shadows, and only a kind of union of the two will preserve an independent reality.

Space and Time
80th Assembly of German Natural Scientists and Physicians
September 21, 1908

Taylor, Edwin F.
Wheeler, John A.
Never make a calculation until you know the answer: Make an estimate before every calculation, try a simple physical argument (symmetry! invariance! conservation!) before every derivation, guess the answer to every puzzle.

Courage: no one else needs to know what the guess is. Therefore make it quickly, by instinct. A right guess reinforces this instinct. A wrong guess brings the refreshment of surprise. In either case life as a spacetime expert, however long, is more fun!

Spacetime Physics
Chapter 1 (p. 60)

Thorne, Kip S.
...spacetime is like a piece of wood impregnated with water. In this analogy, the wood represents space, the water represents time, and the two (wood and water; space and time) are tightly interwoven, unified. The singularity and the laws of quantum gravity that rule it are like a fire into which the water impregnated wood is thrown. The fire boils the water out of the wood, leaving the wood alone and vulnerable; in the singularity, the laws of quantum gravity destroy time, leaving space alone and vulnerable. The fire then converts the wood into a froth of flakes and ashes; the laws of quantum gravity then convert space into a random, probabilistic froth.

Black Holes and Time Warps
Chapter 13 (p. 477)

Wheeler, John A.
...Space and time, unified as spacetime, do not merely witness great masses struggling to bend the motion of other masses. Like the gods of ancient Greece, spacetime helps guide the battle and itself participates...The scope and power of this century's new view of gravity and spacetime is seen nowhere more dramatically than in its prediction of the expansion of the universe. To have predicted, and predicted against all expectation, a phenomenon so fantastic is the greatest token yet of our power to understand this strange and beautiful universe.

A Journey into Gravity and Spacetime
Chapter 1 (p. 2)

SPACE TRAVEL

Arnold, James R.

Space is the empty place next to the full place where we live. I believe we will be true to our nature and go there.

American Scientist
The Frontier in Space. Will one be true to our nature and accept the challenge of the next frontier? (p. 304)
Volume 68, Number 3, May–June 1980

Blagonravov, Anatoly A.

The exploration of the cosmos—the moon and the planets—is a noble aim. Our generation has the right to be proud of the fact that it has opened the space era of mankind.

Bulletin of the Atomic Scientists
In Mose L. Harvey
The Lunar Landing and the US-Soviet Equation (p. 29)
Volume XXV, Number 7, September 1969

Burroughs, William

...man is an artifact designed for space travel. He is not designed to remain in his present biologic state any more than a tadpole is designed to remain a tadpole.

The Adding Machine
Civilian Defense (p. 82)

Dyson, Freeman J.

When we are a million species spreading through the galaxy, the question "Can man play God and still stay sane?" will lose some of its terrors. We shall be playing God, but only as local deities and not as lords of the universe. There is safety in numbers. Some of us will become insane, and rule over empires as crazy as Doctor Moreau's island. Some of us will shit on the morning star. There will be conflicts and tragedies. But in the long run, the sane will adapt and survive better than the insane. Nature's pruning of the unfit will limit the spread of insanity among species in the galaxy, as it does among individuals on earth. Sanity is, in its essence, nothing more than the ability to live in harmony with nature's laws.

Disturbing the Universe
Chapter 21 (pp. 236–7)

Firsoff, V.A.
Yet if we go into space, let us do so humbly, in the spirit of cosmic piety. We know very little. We are face to face with the great unknown and have no right to assume that we are alone in the Solar System.

Exploring the Planets
Chapter XV (p. 160)

Jung, C.G.
Space flights are merely an escape, a fleeing away from oneself, because it is easier to go to Mars or to the moon than it is to penetrate one's own being.

In Miguel Serrano
C.G. Jung and Hermann Hesse
The Farewell (p. 102)

Lowell, Percival
From time immemorial travel and discovery have called with strange insistence to him who, wandering on the world, felt adventure in his veins. The leaving familiar sights and faces to push forth into the unknown has with magnetic force drawn the bold to great endeavor and fired the thought of those who stayed at home.

Mars and Its Canals
Chapter I (p. 3)

Lucretius
...he passed far beyond the flaming walls of the world and traversed throughout in mind and spirit the immeasurable universe...

The Nature of the Universe
Book I, 62 (p. 2)

Murray, Bruce
I think space exploration is as important as music, as art, as literature. It's one of the things that we can do very well because of the way we're constructed as a society. It is one of the most important long-term endeavors of this generation, one upon which our grandchildren and great-grandchildren will look back and say, "That was good."

In Ray Bradbury, Arthur C. Clarke, Bruce Murray, Carl Sagan and Walter Sullivan
Mars and the Mind of Man
We Want Mars to be Like the Earth (p. 25)

Thoreau, Henry David

Perchance, coming generations will not abide the dissolution of the globe, but, availing themselves of future inventions in aerial locomotion, and the navigation of space, the entire race may migrate from the earth, to settle some vacant and more western planet... It took but little art, a simple application of natural laws, a canoe, a paddle, and a sail of matting, to people the isles of the Pacific, and a little more will people the shining isles of space. Do we not see in the firmament the lights carried along the shore by night, as Columbus did? Let us not despair or mutiny.

The Writings of Henry David Thoreau
Volume 4
Paradise (To Be) Regained (p. 292)

Tsiolkovsky, Konstantin Eduardovich

Man will not stay on earth forever, but in the pursuit of light and space will first emerge timidly from the bounds of the atmosphere and then advance until he has conquered the whole of circumsolar space.

Inscription on Tombstone
In John Noble Wilford
We Reach the Moon
Chapter 4 (p. 60)

Unknown

Living on Earth may be expensive, but it includes an annual free trip around the Sun.

Source unknown

Verne, Jules

...I repeat that the distance between the earth and her satellite is a mere trifle, and undeserving of serious consideration. I am convinced that before twenty years are over one-half of our earth will have paid a visit to the moon.

From Earth to the Moon
Chapter XIX (p. 99)

von Braun, Wernher

[Space travel] will free man from his remaining chains, the chains of gravity which still tie him to this planet. It will open to him the gates of heaven.

Time
The Jupiter People (p. 18)
February 10, 1958

Wells, H.G.

All this world is heavy with the promise of greater things, and a day will come, one day in the unending succession of days, when beings, beings who are now latent in our thoughts and hidden in our loins, shall stand upon this earth as one stands upon a footstool and laugh and reach out their hands amidst the stars.

Nature
The Discovery of the Future
Volume 65, 1902 (pp. 326–31)

SPECTRA

Huggins, William

I looked into the spectroscope. No spectrum such as I expected! A single bright line only!...The riddle of the nebulae was solved. The answer, which had come to us in the light itself, read: Not an aggregation of stars, but a luminous gas. Stars after the order of our own sun, and of the brighter stars, would give a different spectrum; the light of this nebula had clearly been emitted by a luminous gas.

The Scientific Papers of Sir William Huggins
Historical Statement (p. 106)

Maxwell, James Clerk

The vast interplanetary and interstellar regions will no longer be regarded as waste places in the universe, which the Creator has not seen fit to fill with the symbols of the manifold order of His kingdom. We shall find them to be already full of this wonderful medium; so full, that no human power can remove it from the smallest portion of space, or produce the slightest flaw in its infinite continuity. It extends unbroken, from star to star; and when a molecule of hydrogen vibrates in the dog-star, the medium receives the impulses of these vibrations; and after carrying them in its immense bosom for three years, delivers them in due course, regular order, and full tale into the spectroscope...

The Scientific Papers of James Clerk Maxwell
Volume II
Action at a Distance (p. 322)

Stedman, Edmund

White orbs like angels pass
Before the triple glass
That men may scan the record of each flame,—
Of spectral line and line
The legendary divine
Finding their mould the same, and aye the same,

The atoms that we knew before
Of which ourselves are made,—dust, and no more.

Journal of the Royal Astronomical Society of Canada
Volume 27, 1933 (p. 375)

Twain, Mark

Spectrum analysis enabled the astronomer to tell when a star was advancing head on, and when it was going the other way. This was regarded as very precious. Why the astronomer wanted to know, is not stated; nor what he could sell out for, when he did know. An astronomer's notions about preciousness were loose. They were not much regarded by practical men, and seldom excited a broker.

The Secret History of Eddypus

SPIN

Goudsmit, Samuel A.
It was a little over fifty years ago that George Uhlenbeck and I introduced the concept of spin... It is therefore not surprising that most young physicists do not know that spin had to be introduced. They think that it was revealed in Genesis or perhaps postulated by Sir Isaac Newton, which most young physicists consider to be about simultaneous.

In A.P. French and Edwin F. Taylor
An Introduction to Quantum Physics (p. 424)

SPIRAL ARMS

van de Hulst, H.C.

The discovery of spiral arms and—later—of molecular clouds in our Galaxy, combined with a rapidly growing understanding of the birth and decay process of stars, changed interstellar space from a stationary 'medium' into an 'environment' with great variations in space and time.

In A. Bonetti, J.M. Greenberg and S. Aiello (eds)
Evolution in Interstellar Dust and Related Topics
The Dim Past of Dusty Space (p. 5)

STAR

Acton, Loren
When you look out the other way toward the stars you realize it's an awful long way to the next watering hole.

In Kevin W. Kelley
The Home Planet
With Plate 84

Aeschylus
This waste of year-long vigil I have prayed
God for some respite, watching elbow-stayed,
As sleuthhounds watch, above the Atreidae's hall,
Till well I know yon midnight festival
Of swarming stars, and them that lonely go,
Bearers to man of summer and of snow...

The Agamemnon
Watchman (p. 1)

Alighieri, Dante
...and thence we issued forth again to see the stars.

The Divine Comedy of Dante Alighieri
Hell
Canto XXXIV, L. 138

Aurelius, Marcus
Look round at the courses of the stars, as if thou were going along with them; and constantly consider the changes of the elements into one another; for such thoughts purge away the filth of the terrene life..

Meditations
Book VII.47

Brown, Fredric

Overhead and in the far distance are the lights in the sky that are stars. The stars they tell us we can never reach because they are too far away. They lie; we'll get there. If rockets won't take us, *something* will.

The Lights in the Sky are Stars
1997 (p. 20)

Bryant, William Cullen

The sad and solemn night
Hath yet her multitude of cheerful fires;
The glorious host of light
Walk the dark hemisphere till she retires;
All through her silent watches, gliding slow,
Her constellations come and climb the heavens and go.

In Parke Godwin (ed.)
The Poetical Works of William Cullen Bryant
Volume I
Hymn to the North Star

Bunting, Basil

Furthest, fairest thing, stars, free of our humbug,
each his own, the longer known, the more alone,
wrapt in emphatic fire roaring out to a black flue...
Then is Now. The star you steer by is gone.

Collected Poems
Briggflats
V (p. 58)

Burke, Edmund

The starry heaven, though it occurs so very frequently to our view, never fails to excite an idea of grandeur. This cannot be owing to the stars themselves, separately considered. The number is certainly the cause. The apparent disorder augments the grandeur, for the appearance of care is highly contrary to our ideas of magnificence. Besides, the stars lie in such apparent confusion, as makes it impossible on ordinary occasions to reckon them. This gives them the advantage of a sort of infinity.

A Philosophical Enquiry into the Origin of Our Ideas of the Sublime and Beautiful
Magnificence (p. 139)

Burnet, Thomas

They lie carelessly scatter'd, as if they had been sown in the Heaven, like Seed, by handfuls; and not by a skilful hand neither. What a beautiful Hemisphere they would have made, if they had been plac'd in rank and order, if they had been all dispos'd into regular figures, and the little ones set with due regard to the greater. Then all finisht and made up into one fair piece or great Composition, according to the rules of Art and Symmetry.

The Sacred Theory of the Earth
Book II
Chapter XI (p. 220)

Campbell, Thomas

And the sentinel stars set their watch in the sky.

The Poetical Works of Thomas Campbell
The Soldier's Dream

In yonder pensile orb, and every sphere
That gems the starry girdle of the year.

The Poetical Works of Thomas Campbell
Pleasures of Hope
Part II

Carlyle, Thomas

...when I gaze into these Stars, have they not looked down on me as if with pity, from their serene spaces: like Eyes glistening with heavenly tears over the little lot of man! Thousands of human generations, all as noisy as our own, have been swallowed-up by Time, and there remains no wreck of them any more; and Arcturus and Orion and Sirius and the Pleiades are still shining in their courses, clear and young, as when the Shepard first noted them in the plain of Shinar.

Sartor Resartus, On Heroes, Hero-Worship and the Heroic in History
Sartor Resartus
Book II, Chapter VIII (p. 138)

Canopus shining-down over the desert, with its blue diamond brightness (that wild, blue, spirit-like brightness far brighter than we ever witness here), would pierce into the heart of the wild Ishmaelitish man, whom it was guiding through the solitary waste there.

Sartor Resartus, On Heroes, Hero-Worship and the Heroic in History
On Heroes, Hero-Worship and the Heroic in History
Lecture I

Clarke, Arthur C.

The thing's hollow—it goes on forever—an—oh my God!—it's full of stars.

2001: A Space Odyssey
Chapter XXXIV (p. 119)

Cole, Thomas

How lovely are the portals of

the night,
When stars come out to
watch the daylight die.

Twilight

Coleridge, Samuel Taylor
...the stars hang bright above her dwelling,
Silent as though they watched the sleeping Earth!

The Poems of Samuel Taylor Coleridge
Dejection : An Ode
Stanza VIII

Crane, Hart
Stars scribble on our eyes the frosty sagas,
The gleaming cantos of unvanquished space.

In Brom Weber (ed.)
The Complete Poems and Selected Letters and Prose of Hart Crane
Cape Hatteras

Darwin, Erasmus
Roll on, ye stars! Exult in youthful prime,
Mark with bright curves the printless steps of time...
Flowers of the sky! Ye too to age must yield,
Frail as your silken sisters of the field.

The Botanic Garden
Part I, Canto IV, X, L. 379

de Saint-Exupéry, Antoine
All men have the stars...but they are not the same things for different people. For some, who are travelers, the stars are guides. For others they are no more than little lights in the sky. For others, who are scholars, they are problems. For my businessman they were wealth. But all these stars are silent. You—you alone—will have the stars as no one else has them...

The Little Prince
XXVI (p. 85)

de Tabley, Lord
The May-fly lives an hour,
The star a million years;
But as a summer flower,
Or as a maiden's fears,
They pass, and heaven is bare

As tho' they never were.

The Collected Poems of Lord de Tabley
Hymn to Astarte

Eddington, Sir Arthur
...it is reasonable to hope that in a not too distant future we shall be competent to understand so simple a thing as a star.

The Internal Constitution of the Stars
Chapter XIII (p. 393)

Eddington, Sir Arthur Stanley
I am aware that many critics consider the conditions in the stars not sufficiently extreme...the stars are not hot enough. The critics lay themselves open to an obvious retort: we tell them to go and find a hotter place.

Stars and Atoms
Lecture 3

Eliot, George
The stars are golden fruit
Upon a tree
All out of reach

The Spanish Gypsy
The World is Great

Emerson, Ralph Waldo
The stars awaken a certain reverence, because though always present, they are inaccessible.

Essays and Lectures
Nature: Addresses and Lectures
Chapter I (p. 9)

If a man would be alone, let him look at the stars. The rays that come from those heavenly worlds, will separate between him and what he touches. One might think the atmosphere was made transparent with this design, to give man, in the heavenly bodies, the perpetual presence of the sublime. Seen in the streets of cities, how great they are!

Essays and Lectures
Nature: Addresses and Lectures
Chapter I (p. 9)

...these delicately emerging stars, with their private and ineffable glances.

Essays and Lectures
Essays
Second Series
Nature (p. 543)

Hitch your wagon to a star.

Society and Solitude
Civilization (p. 27)

Flecker, James Elroy

West of these out to seas colder than the Hebrides I must go
Where the fleet of stars is anchored and the young Star-captains glow.

The Collected Poems of James Elroy Flecker
The Dying Patriot

Goodenough, Ursula

I lie on my back under the stars and the unseen galaxies and I let their enormity wash over me. I assimilate the vastness of the distances, the impermanence, the *fact* of it all. I go all the way out and then I go all the way down, to the fact of photons without mass and gauge bosons that become massless at high temperatures. I take in the abstractions about forces and symmetries and they caress me, like Gregorian chants because the words are so haunting.

The Sacred Depths of Nature
Chapter I (pp. 12–13)

Grondal, Florence Armstrong

...if all the wondrous phenomena of visible stars could be seen on but one of the nights of our long ride about the sun, the civilized world would spend its last cent on glasses and sit up until dawn to feast its eyes on the sublimity of the spectacle.

The Music of the Spheres
Chapter II (p. 16)

If all the diamonds in the world were melted into one huge magical jewel, its sparkling brilliance would pale beside Sirus, the diamond of the heavens.

The Music of the Spheres
Chapter VIII (p. 159)

Guiterman, Arthur

When the bat's on the wing and the bird's in the tree,
Comes the starlighter, whom none may see.

First in the West where the low hills are,
He touches his wand to the Evening Star.

Then swiftly he runs on his rounds on high,
Till he's lit every lamp in the dark blue sky.

Gaily the Troubadour
The Starlighter (p. 190)

While poets feign that, passing earthly bars,
We Fireflies shall someday shine as Stars,
Our scientists, more plausibly surmise
That Stars are underdeveloped Fireflies.

Gaily the Troubadour
My Firefly Stars (p. 187)

Habington, William

The starres, bright cent'nels of the skies.

The Poems of William Habington
Dialogue between Night and Araphil, L. 3

When I survey the bright
Celestial sphere,
So rich with jewels hung, that night
Doth like an Ethiop bride appear:
My soul her wings doth spread,
And heavenward flies,
The Almighty's mysteries to read
In the large volumes of the skies.

Source unknown

Hardy, Thomas

The sky was clear—remarkably clear—and the twinkling of all the stars seemed to be but throbs of one body, timed by a common pulse.

Far From the Madding Crowd
Chapter 2 (p. 7)

The sovereign brilliancy of Sirius pierced the eye with a steely glitter, the star called Capella was yellow, Albebaran and Betelgueux shone with a fiery red.

To persons standing alone on a hill during a clear midnight such as this, the roll of the world eastward is almost a palpable movement.

Far From the Madding Crowd
Chapter 2 (p. 7)

Hearn, Lafcadio

The infinite gulf of blue above seems a shoreless sea, whose foam is stars, a myriad million lights are throbbing and flickering and palpitating...

In Elizabeth Bisland
The Life and Letters of Lafcadio Hearn
Volume I
Letter to H.E. Krehbiel
1877 (p. 170)

Heine, Heinrich

Perhaps the stars in the sky only appear to us to be so beautiful and pure because we are so far away from them and do not know their intimate lives. Up above there are certainly a few stars that lie and beg; stars that put on airs; stars that are forced to commit all possible transgressions; stars that kiss and betray each other; stars that flatter their enemies and, what is even more painful, their friends, just as we do here below.

The Romantic School and Other Essays
The Romantic School
Book 2
Chapter III (p. 73)

Herrick, Robert

The starres of the night
Will lend thee their light
Like Tapers cleare without number.

In J. Max Patrick (ed.)
The Complete Poetry of Robert Herrick
The Night-piece, to Julia
Stanza 3

Hodgson, Ralph

I stood and stared, the sky was lit,
The sky was stars all over it,
I stood, I knew not why,
Without a wish, without a will,
I stood upon that silent hill
And stared into the sky until
My eyes were blind with stars and still
I stared into the sky.

Collected Poems
The Song of Honour

Homer

And on the windless night the stars shine clear
Around the moon, as if the veiling sky
Has broken open to reveal its lamps...

Iliad
Book VIII, L. 555–7

Hoyle, Fred

The stars are best seen as a spectacle, not from everyday surroundings where trees and buildings, to say nothing of street lighting, distract the attention too much, but from a steep mountainside on a clear night, or from a ship at sea. Then the vault of heaven appears incredibly large and seems to be covered by an uncountable number of fiery points of light.

The Nature of the Universe
Chapter 3 (p. 51)

Huxley, Julian
And all about the cosmic sky,
The black that lies beyond our blue,
Dead stars innumerable lie,
And stars of red and angry hue
Not dead but doomed to die.

The Captive Shrew
Cosmic Death

Isaiah 40:26
Lift up your eyes on high and see: who created these? He who brings out their host by number, calling them all by name; by the greatness of his might, and because he is strong in power, *not one is missing*.

The Bible

Jacobson, Ethel
Crystal fish
Caught in the seine
Of the trawler, Night.

Nature Magazine
Stars (p. 260)
May 1958

Jeffers, Robinson
I seem to have stood a long time and watched the stars pass.
They also shall perish, I believe.

The Selected Poetry of Robinson Jeffers
Margrave

Job 38:7
The morning stars sang together,
and all the sons of God shouted for joy.

The Bible

Job 38:32
Canst thou guide Arcturus with his sons?

The Bible

Keats, John

Bright star, would I were steadfast as thou art—
Not in lone splendor hung aloft the night
And watching, with eternal lids apart,
Like Nature's patient, sleepless Ermite,
The moving waters at their priest-like task
Of pure ablution round earth's human shores.

The Complete Poetical Works of Keats
Bright Star

Krutch, Joseph Wood

The stars are little twinkling rogues who light us home sometimes when we are drunk but care for neither you nor me nor any man.

The Twelve Seasons
June (p. 46)

Longfellow, Henry Wadsworth

The stars arise, and the night is holy.

The Poetical Works of Henry Wadsworth Longfellow
Hyperion
Book I, Chapter 1

Lowell, Percival

Bright points in the sky or a blow on the head will equally cause one to see stars.

Mars
Chapter IV (p. 159)

Mandino, Og

I will love the light for it shows me the way; yet I will love the darkness for it shows me the stars.

The Greatest Salesman in the World
Chapter 9 (p. 64)

Meredith, George

He reached a middle height, and at the stars,
Which are the brain of heaven, he looked, and sank.
Around the ancient track marched, rank on rank,
The army of unalterable law.

Poems of George Meredith
Lucifer in Starlight

Bright points in the sky or a blow on the head will equally cause one to see stars.
Percival Lowell – (See p. 347)

Milton, John
And all the spangled hosts keep watch in squadrons bright.

Miscellaneous Poems
Hymn on the Morning of Christ's Nativity, L. 21

Witness this new-made World, another Heav'n
from Heaven Gate not farr, founded in view
On the clear *Hyaline*, the Glassie Sea;
Of amplitude almost immense, with starr's
Numerous, and every Starr perhaps a World
Of destined habitation...

Paradise Lost
Book VII, L. 617–22

Mitchell, Maria
When we are chaffed and fretted by small cares, a look at the stars will show us the littleness of our own interests.

In Phebe Mitchell Kendall
Maria Mitchell: Life, Letters, and Journals
Chapter VII (p. 138)

We call the stars garnet and sapphire; but these are, at best, vague terms. Our language has not terms enough to signify the different delicate shades; our factories have not the stuff whose hues might make a chromatic scale for them.

In Phebe Mitchell Kendall
Maria Mitchell: Life, Letters, and Journals
Chapter XI (p. 235)

Pagels, Heinz R.
Stars are born, they live and they die. Filling the night sky like beacons in an ocean of darkness, they have guided our thoughts over the millennia to the secure harbor of reason.

Perfect Symmetry
Part 1
Chapter 2 (p. 30)

Plato
Vain would be the attempt of telling all the figures of them circling as in a dance, and their juxtapositions, and the return of them in their revolutions upon themselves, and their approximations...

Timaeus
Section 40

Poe, Edgar Allan
Look down into the abysmal distances!—attempt to force the gaze down the multitudinous vistas of the stars, as we sweep slowly through them thus—and thus—and thus! Even the spiritual vision, is it not all points arrested by the continuous golden walls of the universe?—the walls of the myriads of the shining bodies that mere number has appeared to blend into unity?

In H. Beaver (ed.)
The Science Fiction of Edgar Allan Poe
The Power of Words (p. 171)

Ptolemy
I know that I am mortal and ephemeral. But when I search for the close-knit encompassing convolutions of the stars, my feet no longer touch the

earth, but in the presence of Zeus himself I take my fill of ambrosia which the gods produce.

Attributed
In Johannes Kepler
Mysterium Cosmographicum
Title page

Raymo, Chet
I weigh out nebulas. I dam up the Milky Way and use it to grind my grain. I put up summer stars like vegetables in jars for my delectation in winter. I have winter stars folded in boxes in the attic for cloudy summer nights.

Sky and Telescope
Night Brought to Numbers (p. 555)
Volume 71, Number 6, June 1966

Service, Robert
The waves have a story to tell me,
As I lie on the lonely beach;
Chanting aloft in the pine-tops,
The wind has a lesson to teach;
But the stars sing an anthem of glory
I cannot put into speech.

Collected Poems of Robert Service
The Three Voices

Shakespeare, William
. . . these blessed candles of the night. . .

The Merchant of Venice
Act V, scene I, L. 220

Smythe, Daniel
The years of sky are now galactic,
So deep that we have little trace.
Our spectrographs, cool and emphatic,
Betray the depths of stars and space.

What do we seek on dizzying borders
Or groups of systems we have classified?
We cannot search in these huge orders
And find the answers they have passed.

Nature Magazine
Strange Horizons (p. 101)
February 1958

Spenser, Edmund
He that strives to touch the stars
Oft stumbles at a straw.

The Complete Poetical Works of Edmund Spenser
The Shepherdess Calendar

Taylor, Bayard
Each separate star
Seems nothing, but a myriad scattered stars
Break up the night, and make it beautiful.

Lars
Book III
Conclusion

Teasdale, Sara
Stars over snow,
And in the west a planet
Swinging below a star—
Look for a lovely thing and you will find it
It is not far—It will never be far.

The Collected Poems of Sara Teasdale
Night

Tennyson, Alfred
. . . the fiery Sirius alters hue
And bickers into red and emerald.

The Complete Poetical Works of Tennyson
The Princess

Thompson, Francis
Thou canst not stir a flower
Without troubling of a star.

Complete Poetical Works of Francis Thompson
The Mistress of Vision
Stanza XXII

Thoreau, Henry David
The stars are the apexes of what wonderful triangles! What distant and different beings in the various mansions of the universe are contemplating the same one at the same moment!

Walden
Economy (p. 8)

Truly the stars were given for a consolation to man.

The Writings of Henry David Thoreau
Volume 5
A Walk to Wachusett (p. 146)

Travers, P.L.

Jane was watching Mrs Corry splashing the glue on the sky and Mary Poppins sticking on the stars…

"What I want to know," said Jane, "is this: Are the stars gold paper or is the gold paper stars?"

There was no reply to her question and she did not expect one. She knew that only someone very much wiser than Michael could give her the right answer…

Mary Poppins
Chapter 8 (p. 88)

Trevelyan, G.M.

The stars out there rule the sky more than in England, big and lustrous with the honour of having shone upon the ancients and been named by them.

Clio, a Muse
Walking (p. 65)

Twain, Mark

There's another trouble about theories: there's always a hole in them somewheres, sure, if you look close enough. It's just so with this one of Jim's. Look what billions and billions of stars there is. How does it come that there was just exactly enough star-stuff, and none left over? How does it come there ain't no sand-pile up there?

The Complete Works of Mark Twain
Volume 14
Tom Sawyer Abroad (pp. 78–9)

There are too many stars in some places and not enough in others, but that can be remedied presently, no doubt.

The Diaries of Adam and Eve
Eve's Diary
Sunday (p. 7)

Unknown

The meek shall inherit the Earth, the rest of us will go to the stars.

Source unknown

LECTURER: "Fundamentally, a star is a pretty simple structure..."
VOICE FROM THE AUDIENCE: "You would look pretty simple, too, at a distance of ten parsecs."

In Arthur Beer
Vistas in Astronomy
Volume 1
Colloquium in Cambridge University, 1954 (p. 247)

Vaughan, Henry
The Jewel of the Just,
Shining nowhere but in the dark;
What mysteries do lie beyond thy dust,
Could man outlook that mark!

Poetry and Selected Prose
Accession Hymn

Webster, John
We are merely the stars' tennis-balls, struck and bandied
Which way please them.

In J.M. Morrell
Four English Tragedies
The Duchess of Malfy
Act V, Scene 4

Whitman, Walt
I was thinking the day most splendid till I
saw what the not-day exhibited;
I was thinking of this globe enough till there
sprang out so noiseless around me myriads of other globes.

In James E. Miller, Jr (ed.)
Complete Poetry and Selected Prose
Night on the Prairies

Wordsworth, William
Look for the stars, you'll say that there are none;
Look up a second time, and, one by one,
You mark them twinkling out with silvery light,
And wonder how they could elude the sight!

The Complete Poetical Works of Wordsworth
Calm is the Fragrant Air

STUDY

Rowland, Henry Augustus

The whole universe is before us to study. The greatest labor of the greatest minds has only given us a few pearls; and yet the limitless ocean, with its hidden depths filled with diamonds and precious stones, is before us. The problem of the universe is yet unsolved, and the mystery involved in one single atom yet eludes us. The field of research only opens wider and wider as we advance, and our minds are lost in wonder and astonishment at the grandeur and beauty unfolded before us.

The Physical Papers of Henry Augustus Rowland
A Plea for Pure Science (p. 613)

Skinner, B.F.

When you run into something interesting, drop everything else and study it.

Cumulative Record
A Case History in Scientific Method (p. 81)

Tsiolkovsky, K.E.

To place one's feet on the soil of asteroids, to lift a stone from the moon with your hand, to construct moving stations in ether space, to organize inhabited rings around Earth, moon and sun, to observe Mars at the distance of several tens of miles, to descend to its satellites or even to its own surface—what could be more insane! However, only at such a time when reactive devices are applied, will a great new era begin in astronomy: the era of more intensive study of the heavens.

In M.K. Tikhonravov (ed.)
Works on Rocket Technology by K.E. Tsiolkovsky
The Investigation of Universal Space by Means of Reactive Devices (p. 95)

STUPIDITY

Einstein, Albert
Everyone has to sacrifice at the altar of stupidity from time to time, to please the Deity and the human race.

In Max Born
The Born–Einstein Letters
Letter 21
9 September, 1920 (p. 35)

SUN

Bourdillon, Francis William
The night has a thousand eyes,
And the day but one;
Yet the light of the bright world dies,
With the dying Sun.

Among the Flowers, and Other Poems
The Night Has a Thousand Eyes

Deutsch, Armin J.
The face of the sun is not without expression, but it tells us precious little of what is in its heart.

Scientific American
The Sun (p. 38)
Volume 179, Number 5, November 1948

Dryden, John
The glorious lamp of heaven, the radiant sun,
Is Nature's eye;...

The Poetical Works of Dryden
The Fable of Acis, Polyphemus, and Galatea from the Thirteenth Book of Ovid's Metamorphoses (p. 405)

Ecclesiastes 11:7
Truly the light is sweet,
And it is pleasant for the eyes to behold the sun...

The Bible

Falconer, William
High in his chariot glow'd the lamp of day.

The Poetical Works of Beattie, Blair, and Falconer
The Shipwreck
Canto I, L. 334

Gilbert, William
Sullivan, Arthur
The Sun, whose rays
Are all ablaze
With ever-living glory,
Does not deny
His majesty—
He scorns to tell a story!

The Complete Plays of Gilbert and Sullivan
The Mikado
Act II (p. 373)

Heraclitus
The sun. . . is new each day.

In G.S Kirk and J.E. Raven
The Presocratic Philosophers
Fragment 228 (p. 202)

Kelvin, Lord
Now, if the sun is not created a miraculous body, to shine on and give out heat forever, we must suppose it to be a body subject to the laws of matter (I do not say there may not be laws which we have not discovered) but, at all events, not violating any laws we have discovered or believe we have discovered. We should deal with the sun as we should with any large mass of molten iron, or silicon, or sodium.

On Geological Time (p. 18)

Longfellow, Henry Wadsworth
Down sank the great red sun, and in golden glimmering vapours
Veiled the light of his face, like the Prophet descending from Sinai.

The Poetical Works of Henry Wadsworth Longfellow
Evangeline
Part I, section 4

Macpherson, James
Whence are thy beams, O sun! thy everlasting light! Thou comest forth in thy awful beauty; the stars hide themselves in the sky; the moon, cold and pale, sinks in the western wave; but thou thyself movest alone.

The Poems of Ossian
Carthon (p. 233)

Muir, John
The sun, looking down on the tranquil landscape, seems conscious of the presence of every living thing on which he is pouring his blessings, while

they in turn, with perhaps the exception of man, seem conscious of the presence of the sun as a benevolent father and stand hushed and waiting.

Steep Trails
Chapter XVII (p. 226)

Nietzsche, Friedrich

. . . the Moon's love affair has come to an end!
Just look! There it stands; pale and dejected—before the dawn!
For already it is coming, the glowing Sun—*its* love of the Earth is coming!
All sun-love is innocence and creative desire!
Just look how it comes impatiently over the sea! Do you not feel the thirst and hot breath of its love?

Thus Spoke Zarathustra
Of Immaculate Perception (p. 146)

Parker, E.N.

The riddles the sun presents are signposts to new horizons.

Scientific American
The Sun (p. 50)
Volume 233, Number 3, September 1975

Pascal, Blaise

Let man contemplate the whole of nature in her full and grand mystery, and turn his vision from the low objects which surround him. Let him gaze on that brilliant light, set like and eternal lamp to illumine the universe.

Pensées
Aphorism 72

Starr, Victor P.
Gilman, Peter A.

It has always been easier to record and describe solar events than to provide theoretical explanations for them.

Scientific American
The Circulation of the Sun's Atmosphere (p. 100)
Volume 218, Number 1, January 1968

Swift, Jonathan

These people are under continual disquietudes, never enjoying a minute's peace of mind; and their disturbances proceed from causes which very little affect the rest of mortals. Their apprehension arises from several changes they dread in the celestial bodies. For instance. . . that the sun, daily spending its rays without any nutriment to supply them, will at last be wholly consumed and annihilated; which must be attended with the destruction of this earth, and all the planets that receive their light from it.

Gulliver's Travels
A Voyage to Laputa
Chapter II (p. 98)

Thoreau, Henry D.

It is true, I never assisted the sun materially in his rising; but doubt not, it was of the last importance only to be present at it.

Walden
Chapter 1 (p. 15)

Wells, H.G.

...the sun, red and very large, halted motionless upon the horizon, a vast dome glowing with a dull heat, and now and then suffering a momentary extinction...[it] grew larger and duller in the westward sky, and the life of the old earth ebbed away. At last, more than thirty million years hence, the huge red-hot dome of the sun had come to obscure nearly a tenth part of the darkling heavens.

Seven Famous Novels By H.G. Wells
The Time Machine
Chapter 11 (p. 59, 61)

Xenophanes

The sun comes into being each day from little pieces of fire that are collected...

In G.S Kirk and J.E. Raven
The Presocratic Philosophers
Fragment 178 (p. 172)

SUNSPOT

Galilei, Galileo
Neither the satellites of Jupiter nor any other stars are spots or shadows, nor are the sunspots stars. It is indeed true that I am quibbling over names, while I know that anyone may impose them to suit himself. So long as a man does not think that by names he can confer inherent and essential properties on things, it would make little difference whether he calls these "stars."

Discoveries and Opinions of Galileo
Letters on Sunspots
Third Letter on Sunspots, From Galileo Galilei to Mark Welser (p. 139)

Zirin, Harold
Just like the green fields and virgin forests, the granules, the sunspots, the elegant prominences reflect the pure beauty of nature. They offer aesthetic pleasure, as well as scientific challenge, to those who study them.

Astrophysics of the Sun
Preface (p. ix)

SUPERNOVA

Crowley, Abraham

So when by various Turns of the Celestial Dance,
In many thousand years,
A Star, so long unknown, appears,
Though Heaven it self more beauteous by it grow,
It troubles and alarms the World below,
Does to the Wise a Star, to Fools a Meteor show.

In Thomas Sprat
The History of the Royal-Society of London for Improving of Natural Knowledge
To the Royal Society

Herschel, J.F.W.

An immense impulse was now given to science and it seemed as if the genius of mankind, long pent up, had at length rushed eagerly upon Nature, and commenced, with one accord, the great work of turning up her hitherto unbroken soil, and exposing her treasures so long concealed. A general sense now prevailed of the poverty and insufficiency of existing knowledge in *matters of fact*; and, as information flowed fast in, an era of excitement and wonder commenced to which the annals of mankind had furnished nothing similar. It seemed, too, as if Nature herself seconded the impulse; and while she supplied new and extraordinary aids to those senses which were henceforth to be exercised in her investigation—while the telescope and the microscope laid open *the infinite* in both directions—as if to call attention to her wonders, she signalised the epoch, she displayed the rarest, the most splendid any mysterious, of all astronomical phenomena, the appearance and subsequent total extinction of a new and brilliant fixed star twice within the life of Galileo himself.

The Cabinet of Natural Philosophy
Section 106 (pp. 114–15)

Schaaf, Fred

...a star gone to seed—a star spectacularly sowing space with heavy elements and the promise of new stars, worlds, life, and eyes.

The Starry Room: Naked Eye Astronomy in the Intimate Universe
Chapter 11 (p. 194)

Woosley, Stan
Weaver, Tom

The collapse and explosion of a massive star is one of nature's grandest spectacles. For sheer power nothing can match it. During the supernova's first 10 seconds...it radiates as much energy from a central region 20 miles across as all the other stars and galaxies in the rest of the visible universe combined...It is a feat that stretches even the well-stretched minds of astronomers.

Scientific American
The Great Supernova of 1987 (p. 32)
Volume 261, Number 2, August 1989

SYMMETRY

Ferris, Timothy
...let us pause to slake our thirst one last time at symmetry's bubbling spring.

Coming of Age in the Milky Way
Chapter 20 (p. 385)

Weyl, Hermann
Symmetry is a vast subject, significant in art and nature. Mathematics lies at its root, and it would be hard to find a better one on which to demonstrate the working of the mathematical intellect.

Symmetry
Crystals: The General Mathematical Idea of Symmetry (p. 145)

Wickham, Anna
God, Thou great symmetry,
Who put a biting lust in me
From whence my sorrows spring
For all the frittered days
That I have spent in shapeless ways
Give me one perfect thing.

The Contemplative Quarry
Envoi

Yang, C.N.
Nature seems to take advantage of the simple mathematical representations of the symmetry laws. When one pauses to consider the elegance and the beautiful perfection of the mathematical reasoning involved and contrast it with the complex and far-reaching physical consequences, a

deep sense of respect for the power of the symmetry laws never fails to develop.

In Nobel Foundation
Nobel Lecture
Physics 1942–62
Nobel Lecture of Chen Ning Yang
December 11, 1957 (pp. 394–5)

TEACH

Gauss, Carl Friedrich
I am giving this winter two courses of lectures to three students, of which one is only moderately prepared, the other less than moderately, and the third lacks both preparation and ability. Such are the onera of a mathematical profession.

Briefwechsel zwischen Gauss und Bessel
Letter 46
Letter to Bessel
January 7, 1810 (p.107)

Regnault, Pére
Will you discover to me...those Secrets which Nature has imparted to you?

Philosophical Conversations
Volume I
Conversation XII (p. 154)

Stoppard, Tom
THOMASINA: If *you* do not teach me the true meaning of things, who will?
SEPTIMUS: Ah. Yes, I am ashamed. Carnal embrace is sexual congress, which the insertion of the male genital organ into the female genital organ for purposes of procreation and pleasure. Fermat's last theorem, by contrast, asserts that when x, y, and z are whole numbers each raised to power of n, the sum of the first two can never equal the third when n is greater than 2. (*Pause.*)
THOMASINA: Eurghhh!
SEPTIMUS: Nevertheless, that is the theorem.
THOMASINA: It is disgusting and incomprehensible. Now when I am grown to practice it myself I shall never do so without thinking of you.

Arcadia
Act I, Scene I (p. 3)

TELESCOPE

Bierce, Ambrose
Telescope (n): A device having a relation to the eye similar to that of a telephone to the ear, enabling distant objects to plague us with a multitude of needless details.

The Devil's Dictionary

Emerson, Ralph Waldo

The sight of a planet through a telescope is worth all the course on astronomy...

Essays
Second Series
New England Reformers

Holmes, Oliver Wendell

I love all sights of earth and skies,
From flowers that grow to stars that shine;
The comet and the penny show,
All curious things above, below
...
But most I love the tube that spies
The orbs celestial in their march;
That shows the comet as it whisks
Its tail across the planet's disk,
Or wheels so close against the sun
We tremble at the thought of risks
Our little spinning ball may run.

The Flaneur

Hubble, Edwin

With increasing distance our knowledge fades and fades rapidly. Eventually we reach the dim boundary, the utmost limits of our telescope. There we measure shadows, and we search among ghostly errors of measurement for landmarks that are scarcely more substantial. The search will continue. Not until the empirical resources are exhausted need we pass on to the dreamy realm of speculation.

The Realm of the Nebulae
Chapter VIII (p. 202)

Kepler, Johannes

What now, dear reader, shall we make out of our telescope? Shall we make a Mercury's magic-wand to cross the liquid ether with, and like Lucian lead a colony to the uninhabited evening star, allured by the sweetness of the place? Or shall we make it a Cupid's arrow which, entering by our eyes, has pierced our inmost mind, and fired us with love of Venus?... O telescope, instrument of much knowledge, more precious than any scepter! Is not he who holds thee in his hand made king and lord of the works of God?

Dioptrice
Preface (pp. 86, 103)

Lovell, A.C.B.

Astronomy has marched forward with the growth in size of its telescopes.

Scientific American
Radio Stars (p. 21)
Volume 188, Number 1, January 1953

Mitchell, Maria

The tube of Newton's first telescope... was made from the cover of an old book—a little glass at one end of the tube and a large brain at the other...

In Helen Wright
Sweeper in the Sky
Chapter 9 (p. 168)

Mullaney, James

The telescope in particular needs to be regarded as not just another gadget or material possession but a wonderful, magical gift to humankind—a window on creation, a time machine, a spaceship of the mind that enables us to roam the universe in a way that is surely the next best thing to being out there.

Sky and Telescope
Focal Point (p. 244)
March 1990

Panek, Richard

The relationship between the telescope and our understanding of the dimensions of the universe is in many ways the story of modernity. It's the story of how the development of one piece of technology has changed the way we see ourselves and of how the way we see ourselves has changed this piece of technology, each set of changes reinforcing the other over the course of centuries until, in time, we've been able to look back and say with some certainty that the pivotal division between the world we inhabit today and the world of our ancestors was the invention of this instrument.

Seeing and Believing: How the Telescope Opened our Eyes and Minds to the Heavens
Prologue (p. 4)

Peltier, Leslie C.

Old telescopes never die, they are just laid away.

Starlight Nights
Chapter 28 (p. 232)

Rowan-Robinson, Michael

Once it was the navigators crossing the oceans to find new continents and new creatures, the globe opening up before their eyes, and at the same time the unknown areas, white on the map, shrinking.

Now it is the astronomers' telescopes penetrating the void to find new worlds, voyages of discoveries made with giant metal eyes, seeing light we cannot see.

Our Universe: An Armchair Guide (p. x)

Ryder-Smith, Roland

All night he watches roving worlds go by
Through tempered glass, his window on the sky
Feels in his own beat
Of some far mightier heart, and hears
The mystic concert of the spheres.

The Scientific Monthly
Astronomer (p. 253)
Volume 67, Number 4, October 1948

Toogood, Hector B.

The telescope, an instrument which, if held the right way up, enables us to examine the stars and constellations at close quarters. If held the wrong way up, however, the telescope is of little or no use.

The Outline of Everything
Chapter VIII (p. 96)

Tsiolkovsky, Konstantin

All that which is marvelous, and which we anticipate with such thrill, already exists but we cannot see it because of the remote distances and the limited power of our telescopes...

In Adam Starchild (ed.)
The Science Fiction of Konstantin Tsiolkovsky
Dreams of the Earth and Sky (p. 154)

Vehrenberg, Hans

It is a fundamental human instinct to collect, whether berries and roots in the prehistoric past or knowledge of the universe today. For several decades, my favorite pastime has been to collect celestial objects in photographs. I will never forget the many thousands of hours I have spent with my instruments, working peacefully in my telescope shelter as I listened to good music and dreamed about the infinity of the universe.

Atlas of Deep Sky Splendors
Preface

Vezzoli, Dante

Cyclopean eye that sweeps the sky,
Whose silvered iris gathers light
From galaxies that unseen pierce

The silent blanket of the night.

The Sky
Eye of Palomar (p. 8)
January 1940

Wordsworth, William

WHAT crowd is this? what have we here! we must not pass it by;
  A Telescope upon its frame, and pointed to the sky:
  Long is it as a barber's pole, or mast of little boat,
  Some little pleasure-skiff, that doth on Thames's waters float.

The Complete Poetical Works of Wordsworth
Star-Gazers

THEORY

Bethe, Hans
Scientific theories are not overthrown; they are expanded, refined, and generalized.

In Victor Weisskopf
Physics in the Twentieth Century
Forward (p. x)

d'Abro, A.
...a theory of mathematical physics is not one of pure mathematics. Its aim and its *raison d'être* are not solely to construct the rational scheme of some possible world, but to construct that particular rational scheme of the particular real world in which we live and breathe. It is for this reason that a theory of mathematical physics, in contradistinction to one of pure mathematics, is constantly subjected to the control of experiment.

The Evolution of Scientific Thought
Chapter XXI (p. 215)

Duhem, Pierre
Contemplation of a set of experimental laws does not, therefore, suffice to suggest to the physicist what hypotheses he should choose in order to give a theoretical representation of these laws; it is also necessary that the thoughts habitual with those among whom he lives and the tendencies impressed on his own mind by his previous studies come and guide him, and restrict the excessively great latitude left to this day a merely empirical form until circumstances prepare the genius of a physicist to conceive the hypothesis which will organize them into a theory!

The Aim and Structure of Physical Theory
Chapter VII (p. 255)

Einstein, Albert
For the creation of a theory the mere collection of recorded phenomena never suffices—there must always be added a free invention of the human

mind that attacks the heart of the matter. And: the physicist must not be content with the purely phenomenological considerations that pertain to the phenomenon. Indeed, he should press on to the speculative method, which looks for the underlying pattern.

Lecture at the Berlin Planetarium
4 October 1931
Quoted in Helen Dukas and Banesh Hoffman
Albert Einstein: The Human Side (pp. 29–30)

Creating a new thory is not like destroying a barn or erecting a new skyscraper in its place. It is rather like climbing a mountain, gaining new and wider views, discovering unexpected connections between our starting point and its rich environment. But the point from which we started out still exists and can be seen, although it appears smaller and forms a tiny part of out broad view gained by the mastery of the obstacles on our adventurous way up.

The Evolution of Physics

Einstein, Albert
Infeld, Leopold
Physical theories try to form a picture of reality and to establish its connection with the wide world of sense impressions. Thus the only justification for our mental structures is whether and in what way our theories form such a link.

The Evolution of Physics
Chapter IV (p. 294)

Goldhaber, Maurice
Antaeus was the strongest person alive, invincible as long as he was in contact with his mother, the earth. Once he lost contact with the earth, he grew weak and was vanquished. Theories in physics are like that. They have to touch the ground for their strength.

The Atlantic Monthly
In Robert P. Crease and Charles C. Mann
How the Universe Works (p. 91)
August 1984

Koestler, Arthur
The history of cosmic theories...may without exaggeration be called a history of collective obsessions and controlled schizophrenias.

The Sleepwalkers

Laszlo, E.
Ours is a complex world. But human knowledge is finite and circumscribed. 'Nature does not come as clean as you can think it,'

warned Alfred North Whitehead, and went on to propound an extremely clean and elegant cosmology. Since theories, like window panes, are clear only when they are clean, and the world does not come as cleanly as all that, we must know where we perform a clean-up operation. Scientific theories while simpler than reality, must nevertheless reflect its essential structure. Science then must beware of rejecting the structure for the sake of simplicity; that would be to throw out the baby with the bath water.

The Systems View of the World:
The Natural Philosophy of the New Developments in the Sciences
Chapter 1, section 2 (p. 13)

Libes, Antoine

Let us add a word in favor of theories, which certain physicists still dare to present as invincible obstacles to the discovery of truth. It is incontestable that experience and observation ought to serve as the basis of our physical knowledge. But without the help of theory the most well-certified experiments, the most numerous observations will be only isolated facts in the hands of the physicist, isolated facts which cannot serve for the advancement of physics. The man of genius must seize upon these scattered links and bring them together skillfully to form a continuous chain. This continuity constitutes the theory, which alone can give us a glimpse of the relations which bind the facts to one another and of their dependence on the causes which have produced them.

In Russell McCormmach (ed.)
Historical Studies in the Physical Sciences
Volume 4
In Robert H. Stilliman
Fresnel and the Emergence of Physics as a Discipline (p. 143)

Lowell, Percival

All deductions rests ultimately upon the data derived from experience. This is the tortoise that supports our conception of the cosmos. For us, therefore, the point at issue in any theory is not whether there is a possibility of its being false, but whether there is a probability of its being true. This...is too often lost sight of in discussing theories on their way to recognition. Negative evidence is no evidence at all, and the possibility that a thing might be otherwise, no proof whatever that it is not so. The test of a theory is, first, that it shall not be directly contradicted by any facts, and secondly, that the probabilities in its favor shall be sufficiently great.

In William Graves Hoyt
Lowell and Mars
Chapter 2 (p. 25)

Popper, Karl

We have no reason to regard the new theory as better than the old theory—to believe that it is nearer to the truth—until we have derived from the new theory *new predictions* which were unobtainable from the old theory (the phases of Venus...) and until we have found that these new predictions were successful.

Conjectures and Refutations
Chapter 10 (p. 246)

The more a theory forbids, the better it is.

Conjectures and Refutations
Chapter I, Section I (p. 41)

Rothman, Tony

It is not difficult to calculate that if one inflated the world to keep up with the current rate of population growth, then after 2598 years the earth would be expanding at the speed of light. The growth of science is proceeding even faster. Several years ago, in physics at least, we crossed the point at which the expected lifetime of a theory became less than the lead time for publication in the average scientific journal. Consequently, most theories are born dead on arrival and journals have become useless, except as historical documents.

A Physicist on Madison Avenue
Chapter 8 (p. 118)

Slater, John C.

A theoretical physicist in these days asks just one thing of his theories: if he uses them to calculate the outcome of an experiment, the theoretical prediction must agree, within limits, with the result of the experiment. He does not ordinarily argue about the philosophical implications of his theory. Almost his only recent contribution to philosophy has been the operational idea, which is essentially only a different way of phrasing the statement I have just made, that the one and only thing to be done with a theory is to predict the outcome of an experiment. As a physicist, I find myself very well satisfied with this attitude. Questions about a theory which do not affect its ability to predict experimental results correctly seem to me quibbles about words, rather than anything more substantial, and I am quite content to leave such questions to those who derive some satisfaction from them.

Journal of the Franklin Institute
Electrodynamics of Ponderable Bodies
Volume 225, Number. 3, March 1938 (pp. 277–87)

Synge, J.L.

A well built theory has three merits: (i) it has an aesthetic appeal, (ii) it is comparatively easy to understand, and (iii), if its postulates are clearly stated, it may be taken out of its original physical context and applied in another.

Proceedings of the Irish Academy
The Hamiltonian Method and its Application to Water Waves (p. 1)
Volume 63, Section A, Number 1, May 1962

von Goethe, Johann Wolfgang

An extremely odd demand is often set forth but never met, even by those who make it: i.e., that empirical data should be presented without any theoretical context, leaving the reader, the student, to his own devices in judging it. This demand seems odd because it is useless simply to look at something. Every act of looking turns into observation, every act of observation into reflection, every act of reflection into the making of associations; thus it is evident that we theorize every time we look carefully at the world.

In Douglas Miller
Scientific Studies
Volume 12
Theory of Color
Preface (p. 159)

The highest is to understand that all fact is really theory. The blue of the sky reveals to us the basic law of color. Search nothing beyond the phenomena, they themselves are the theory.

In Rupprecht Matthaei (ed.)
Goethe's Color Theory (p. 76, note)

Wheeler, John A.

To hate is to study, to study is to understand, to understand is to appreciate, to appreciate is to love. So maybe I'll end up loving your theory.

Scientific American
June 1991

Wisdom, J.O.

[Theory] Sometimes it is used for a hypothesis, sometimes for a confirmed hypothesis; sometimes for a train of thought; sometimes for a wild guess at some fact or for a reasoned claim about what some fact is—or even for a philosophical speculation.

Foundations of Inference in Natural Science
Chapter III (p. 33)

THERMODYNAMICS

Atkins, P.W.

Everything is driven by motiveless, purposeless decay.

The Creation
Chapter 2 (p. 23)

Cardenal, Ernesto

The second law of thermodynamics!:
energy is indestructible in quantity
but continually changes in form.
And it always runs down like water.

Cosmic Canticle
Cantiga 3
Autumn Fugue

Dickerson, Richard E.

It is possible to know thermodynamics without understanding it...

Molecular Thermodynamics
Chapter 7 (p. 387)

Hoffmann, Roald

My second law, your second law, ordains
that local order, structure in space
and time, be crafted in ever-so-losing
contention with proximal disorder in
this neat but getting messier universe.

The Metamict State
The Devil Teaches Thermodynamics (p. 3)

Pippard, A.B.

It may be objected by some that I have concentrated too much on the dry bones, and too little on the flesh which clothes them, but I would ask such critics to concede at least that the bones have an austere beauty of their own.

Classical Thermodynamics for Advanced Students of Physics
Preface (p. vii)

Reiss, H.

...the almost certain truth that nobody (authors included) understands thermodynamics completely. The writing of a book therefore becomes a kind of catharsis in which the author exorcises his own demon of incomprehension and prevents it from occupying the soul of another.

Methods of Thermodynamics
Preface (p. vii)

Almost all books on thermodynamics contain some errors which are not purely typographical.

Methods of Thermodynamics
Preface (p. ix)

Truesdell, Clifford

...thermodynamics is the kingdom of deltas.

The Tragicomical History of Thermodynamics
Chapter 1 (p. 1)

Every physicist knows exactly what the first and the second law mean, but...no two physicists agree about them.

In Mario Bunge (ed.)
Delaware Seminar in the Foundations of Physics
Foundations of Continuum Mechanics (p. 37)

Unknown

If you think things are mixed up now, just wait a while.

Source unknown

TIME

Barnett, Lincoln

Time itself will come to an end. For entropy points the direction of time. Entropy is the measure of randomness. When all system and order in the universe have vanished, when randomness is at its maximum, and entropy cannot be increased, when there is no longer any sequence of cause and effect, in short when the universe has run down, there will be no direction to time—there will be no time.

The Universe and Dr Einstein
Chapter 14 (p. 103)

Carlyle, Thomas

...no hammer in the horologue of Time peals through the universe when there is a change from Era to Era. Men understand not what is among their hands...

On History

Eddington, Sir Arthur

Whatever may be time *de jure*, the Astronomer Royal's time is time *de facto*. His time permeates every corner of physics.

The Nature of the Physical World
Chapter III (p. 36)

Lightman, Alan

There is a place where time stands still...illuminated by only the most feeble red light, for light is diminished to almost nothing at the center of time, its vibrations slowed to echoes in vast canyons, its intensity reduced to the faint glow of fireflies.

Einstein's Dreams
14 May 1905 (pp. 70, 72–3)

Lyell, Charles

We aspire in vain to assign limits to the works of creation in *space*, whether we examine the starry heavens, or that world of minute animalcules which is revealed to us by the microscope. We are prepared, therefore, to find that in *time* also the confines of the universe lie beyond the reach of mortal ken.

Principles of Geology
Concluding Remarks

Mann, Thomas
Time has no division to mark its passage, there is never a thunder-storm or blare of trumpets to announce the beginning of a new month or year. Even when a new century begins it is only we mortals who ring bells and fire off pistols.

The Magic Mountain
Whims of Mercurius (p. 225)

McLuhan, Marshall
Fiore, Quentin
Our time is a time for crossing barriers, for erasing old categories—for probing around.

The Medium is the Message (p. 10)

Poinsot, Louis
If anyone asked me to define *time*, I should reply: 'Do you know what it is that you speak of?' If he said 'Yes,' I should answer, 'Very well, let us talk about it.' If he said 'No,' I should answer, 'Very well, let us talk about something else.'

In William Maddock Bayliss
Principles of General Physiology
Preface (p. xvii)

Saint Augustine
For what is time? Who can easily and briefly explain it? Who even in thought can comprehend it, even to the pronouncing of a word concerning it? But what in speaking do we refer to more familiarly and knowingly than time? And certainly we understand when we speak of it; we understand also when we hear it spoken of by another. What, then, is time? If no one ask of me, I know; if I wish to explain to him who asks, I know not.

Confessions
Book XI, XIV, 17

Shakespeare, William
There are many events in the womb of time which will be delivered.

Othello, The Moor of Venice
Act I, scene iii, L. 376

Unless hours were cups of sack, and minutes capons, and clocks the tongues of bawds, and dials the signs of leaping-houses, and the blessed sun himself a fair hot wench in flame-colour'd taffeta, I see no reason why thou shouldst be so superfluous to inquire the nature of time.

The First Part of King Henry the Fourth
Act I, scene ii, L. 7–10

Silesius, Angelus

Do not compute eternity
as light-year after year
One step across
that line called Time
Eternity is here.

The Book of Angelus Silesius
Of Time and Eternity (p. 42)

Swinburne, Richard

It would be an error to suppose that if the universe is infinitely old, and each state of the universe at each instant of time has a complete explanation which is a scientific explanation in terms of a previous state of the universe and natural laws (and so God is not invoked), that the existence of the universe throughout infinite time has a complete explanation, or even a full explanation. It has not. It has neither. It is totally inexplicable.

The Existence of God
Chapter 7 (p. 122)

Unknown

In astronomy, we are concerned, not with defining time, but only with measuring it.

Explanatory Supplement to the Astronomical Ephemeris and the American Ephemeris and Nautical Almanac

Weil, Simone

Time is an image of eternity, but it is also a substitute for eternity.

Gravity and Grace
Renunciation of Time (p. 65)

Wells, H.G.

"Can an *instantaneous* cube exist?"
"Don't follow you," said Filby.
"Can a cube that does not last for any time at all, have a real existence?"
Filby became pensive. "Clearly," the Time Traveler proceeded, "any real body must have extension in *four* directions: it must have Length,

Breadth, Thickness, and—Duration... There are really four dimensions, three which we call the three planes of Space and a fourth, Time. There is, however, a tendency to draw an unreal distinction between the former three dimensions and the latter, because it happens that our consciousness moves intermittently in one direction along the latter from the beginning to the end of our lives."

Seven Famous Novels by H.G. Wells
The Time Machine
Chapter 1 (pp. 3–4)

TIME TRAVEL

Allen, Elizabeth Akers
Backward, turn backward, O Time, in your flight.
Make me a child again just for to-night.

Rock Me to Sleep, Mother
Rock Me to Sleep, Mother (p. 11)

TRUTH

Balfour, Arthur James

It is not by mere accumulation of material, nor even by a plant-like development, that our beliefs grow less inadequate to the truths which they strive to represent. Rather we are like one who is perpetually engaged in altering some ancient dwelling in order to satisfy new-born needs. The ground-plan of it is being perpetually modified. We build here; we pull down there. One part is kept in repair, another part is suffered to decay. And even those portions of the structure which may in themselves appear quite unchanged, stand in such new relations to the rest, and are put to such different uses, that they would scarce be recognized by their original designer.

The Foundations of Belief
Appendix
Section I (p. 350)

Black, Max

Scientists can never hope to be in a position to know *the* truth, nor would they have any means of recognizing it if it came into their possession.

Critical Thinking
Chapter 19 (p. 396)

Cole, William

Whoever attempts to erect a building should take care that the foundation be securely laid; so also in our inquiries after truth, all our proceedings should be founded upon just and incontrovertible grounds.

Philosophical Remarks on the Theory of Comets
Introduction (p. xi)

Faraday, Michael

Nothing is too wonderful to be true.

Faraday's Diary
March 19, 1849

Feynman, Richard
It is possible to know when you are right way ahead of checking all the consequences. You can recognize truth by its beauty and simplicity.

The Character of Physical Law
Chapter 7 (p. 171)

Gore, George
The deepest truths require still deeper truths to explain them.

The Art of Scientific Discovery
Chapter III (p. 26)

Halmos, Paul
The joy of suddenly learning a former secret and the joy of suddenly discovering a hitherto unknown truth are the same to me—both have the flash of enlightenment, the almost incredibly enhanced vision, and the ecstasy and euphoria of released tension.

I Want to be a Mathematician
Chapter 1 (p. 3)

Hawkins, Michael
"Scientific truths" is simply another way of saying "the fittest, most beautiful, and most elegant survivors of scientific debate and testing."

Hunting Down the Universe
Chapter 1 (p. 6)

Teilhard de Chardin, Pierre
We are given to boasting of our age being an age of science... Yet though we may exalt research and derive enormous benefits from it, with what pettiness of spirit, poverty of means and general haphazardness do we pursue truth in the world today!... [W]e leave it to grow as best it can, hardly tending it, like those wild plants whose fruits are plucked by primitive peoples in their forests.

The Phenomenon of Man
Book 4
Chapter III
Section 2A (pp. 278, 279)

UFO

Sagan, Carl

After I give lectures—on almost any subject—I am often asked, "Do you believe in UFOs?". I'm always struck by how the question is phrased, the suggestion that this is a matter of belief and not evidence. I'm almost never asked, "How good is the evidence that UFOs are alien spaceships?"

The Demon Haunted World
Chapter 3 (p. 82)

UNCERTAINTY

Hoyle, Fred
If matters still seem very uncertain it must always be remembered that clearly sign-posted roads are not to be expected at a pioneering frontier.

Frontiers of Astronomy
Chapter 19 (p. 341)

Unknown
Heisenberg might have slept here.

Source unknown

HEISENBERG MIGHT HAVE SLEPT HERE ALSO!
HEISENBERG MIGHT HAVE SLEPT HERE

UNDERSTAND

Atiyah, Michael

. . . it is hard to communicate understanding because that is something you get by living with a problem for a long time. You study it, perhaps for years, you get the feel of it and it is in your bones. You can't convey that to anybody else. Having studied the problem for five years you may be able to present it in such a way that it would take somebody else less time to get to that point than it took you but if they haven't struggled with the problem and seen all the pitfalls, then they haven't really understood it.

Mathematical Intelligencer
An Interview with Michael Atiya (p. 17)
Volume 6, Number 1, 1984

Ferris, Timothy

We might eventually obtain some sort of bedrock understanding of cosmic structure, but we will never understand the universe in detail; it is just too big and varied for that. If we possessed an atlas of our galaxy that devoted but a single page to each star system in the Milky Way (so that the sun and all its planets were crammed in on one page), that atlas would run to more than ten million volumes of ten thousand pages each. It would take a library the size of Harvard's to house the atlas, and merely to flip through it, at the rate of a page per second, would require over ten thousand years.

Coming of Age in the Milky Way
Chapter 20 (p. 383)

Galilei, Galileo

And finally I ask you, foolish man, can your mind grasp that magnitude of the universe, which you consider to be too vast? If you can grasp it would you consider that your comprehension extends beyond that of the Divine Power? Do you mean to say that you can imagine greater things than those that God can create? But if you do not grasp it then why do you wish to give an opinion on things that you do not understand?

Dialogues on the Two Chief Systems of the World
Third Day

Heisenberg, Werner
Even for a physicist the description in plain language will be a criterion of the degree of understanding that has been reached.

Physics and Philosophy
Chapter X (p. 168)

Pagels, Heinz R.
The attempt to understand the origin of the universe is the greatest challenge confronting the physical sciences. Armed with the new concepts, scientists are rising to meet that challenge, although they know that success may be far away. Yet when the origin of the universe is understood, it will open a new vision that is beautiful, wonderful and filled with the mystery of existence. It will be our intellectual gift to our progeny and our tribute to the scientific heroes who began this great adventure of the human mind, never to see it completed.

Perfect Symmetry
Part 1
Chapter 7 (p. 156)

Rabi, Isidor
Scientific understanding...is an essential step to our finding a home for ourselves in the universe. Through understanding the universe, we become at home in it. In a certain sense we have made this universe out of human concepts and human discoveries. It ceases to be a lonely place, because we can to some extent actually navigate in it.

In A.A. Warner, Dean Morse and T.E. Cooney (eds)
The Environment of Change
The Revolution in Science (p. 49)

Walker, Kenneth
It may be said that all understanding of the universe comes from the combined action of two faculties in us, the power to register impressions and the capacity to reason and reflect on them.

Meaning and Purpose
Chapter II (p. 18)

UNIVERSE

Adams, Douglas
The Universe, as has been observed before, is an unsettlingly big place, a fact which for the sake of a quiet life most people tend to ignore.

The Restaurant at the End of the Universe
Chapter 10 (p. 71)

Amaldi, Ginestra Giovene
Our imagination has roamed far and wide through distant reaches of the Universe. Understandably, we may have become dazed by the immense dimensions of space and the enormous sizes of some of its occupants.

Our World and the Universe Around Us
Volume I
First Steps into Space (p. 124)

Bacon, Francis
The Universe is not to be narrowed down to the limits of the Understanding, which has been men's practice up to now, but the Understanding must be stretched and enlarged to take in the image of the Universe as it is discovered.

Parasceve
Aphorism 4

For the fabric of this universe is like a labyrinth to the contemplative mind, where doubtful paths, deceitful imitations of things and their signs, winding and intricate folds and knots of nature everywhere present themselves, and a way must constantly be made through the forests of experience and particular natures, with the aid of the uncertain light of the senses, shining and disappearing by fits.

In Basil Montague
The Works
Volume 3
The Great Instauration
Preface (p. 336)

Barth, John

All the scientists hope to do is describe the universe mathematically, predict it, and maybe control it. The philosopher, by contrast, seems unbecomingly ambitious: He wants to *understand* the universe; to get behind phenomena and operation and solve the logically prior riddles of being, knowledge, and value. But the artist, and in particular the novelist, in his essence wishes neither to explain nor to control nor to understand the universe. He wants to make one of his own, and may even aspire to make it more orderly, meaningful, beautiful, and interesting than the one God turned out. What's more, in the opinion of many readers of literature, he sometimes succeeds.

The Friday Book: Essays and Other Nonfiction
How to Make a Universe (p. 17)

Blount, Sir Thomas Pope

Whoever surveys the curious fabric of the universe, can never imagine that so noble a structure should be framed for no other use, than barely for mankind to live and breathe in. It was certainly the design of the great Architect that his creatures should afford not only necessaries and accommodations to our animal part, but also instructions to our intellectual.

A Natural History
Preface

Bruno, Giordano

The universe is then one, infinite, immobile...It is not capable of comprehension and therefore is endless and limitless, and to that extent infinite and indeterminable, and consequently immobile.

Cause, Principle, and Unity
Fifth Dialogue (p. 135)

We can securely declare that the universe is all centre, or that the universe's centre is everywhere and the circumference is nowhere insofar as it differs from the centre...

Cause, Principle, and Unity
Fifth Dialogue (p. 137)

Camus, Albert

...I laid my heart open to the benign indifference of the universe.

The Outsider
Part II
Chapter V (p. 127)

Clark, Roger N.

To stand beneath a dark, crystal-clear, moonless country sky is an awe-inspiring experience. Those thousands of stars, many larger than our own Sun, can make us feel small indeed. It seems possible to see to infinity, though we cannot reach beyond arm's length. The beauty of the universe defies description.

Visual Astronomy of the Deep Sky
Preface (p. xi)

Yet the heavens are subtle. Imagine that the fuzzy patch at the threshold of visibility is really a trillion suns—a galaxy larger than our own, in which *our* Sun is but a tiny speck. Incomprehensible; yet somehow we try. Seeing that galaxy first-hand, even through a small telescope, is much more inspiring than the large, beautiful photograph in the astronomy book back indoors. Nothing can compare to viewing the universe directly.

Visual Astronomy of the Deep Sky
Preface (p. xi)

Clarke, Arthur C.

Many and strange are the universes that drift like bubbles in the foam of the river of time.

The Collected Stories of Arthur C. Clarke
The Wall of Darkness (p. 104)

. . . the universe has no purpose and no plan. . .

The Collected Stories of Arthur C. Clarke
The Star (p. 521)

There is no reason to assume that the universe has the slightest interest in intelligence—or even in life. Both may be random accidental by-products of its operations like the beautiful patterns on a butterfly's wings. The insect would fly just as well without them. . .

The Lost Worlds of 2001
Chapter 16 (p. 109)

Cook, Peter

I am very interested in the Universe—I am specializing in the Universe and all that surrounds it.

Beyond the Fringe
Disc 2
Sitting on the Bench

Copernicus, Nicolaus

But they say that beyond the heavens there isn't any body or place or void or anything at all; and accordingly it is not possible for the heavens to move outward: in that case it is rather surprising that something can be held together by nothing. But if the heavens were infinite and were finite only with respect to a hollow space inside, then it will be said with more truth that there is nothing of heaven, since anything which occupied any space would be in them, but the heavens will remain immobile. For movement is the most powerful reason wherewith they try to conclude that the universe is finite.

On Revolutions of the Heavenly Spheres
Book 1, Chapter 8 (p. 519)

Darling, David
In giving birth to us, the universe has performed its most astonishing creative act. Out of a hot, dense mêlee of subatomic particles...it has fashioned intelligence and consciousness...Somehow the anarchy of genesis has given way to exquisite, intricate order, so that now there are portions of the universe that can reflect upon themselves...

Equations of Eternity
Introduction (p. xiii)

Davy, Humphry
The more the phenomena of the universe are studied, the more distinct their connection appears, the more simple their causes, the more magnificent their design, and the more wonderful the wisdom and power of their author.

Elements of Chemical Philosophy (p. 60)

Dawkins, Richard
The universe we observe has precisely the properties we should expect if there is, at bottom, no design, no purpose, no evil, no good, nothing but blind, pitiless indifference.

River Out of Eden
Chapter 4 (p. 133)

Day, Clarence
Is it possible that our race may be an accident, in a meaningless universe, living its brief life uncared for, on this dark, cooling star: but so—and all the more—what marvelous creatures we are! What fairy story, what tale from the Arabian Nights of the Jinns, is a hundredth part as wonderful as this story of simians! It is so much more heartening too, than the tales we invent. A universe capable of giving birth to so many accidents is—blind or not—a good world to live in, a promising universe.

This Simian World
XIX (p. 91)

de Fontenelle, Bernard Le Bovier

...when the heavens appeared to me as a little blue vault, stuck with stars, methought the universe was too straight and close, I was almost stifled for want of air; but now, it is enlarged in height and breadth, and a thousand and a thousand vortexes taken in, I begin to breathe with more freedom, and think the universe to be incomparably more magnificent than it was before.

Conversations on the Plurality of Worlds
The Fifth Evening (pp. 151–2)

de Sitter, W.

Our conception of the structure of the Universe bears all the marks of a transitory structure. Our theories are decidedly in a state of continuous and just now very rapid evolution.

In J.H.F. Umbgrove
The Pulse of the Earth
Chapter I (p. 1)

de Vries, Peter

Anyone informed that the universe is expanding and contracting in pulsations of eighty billion years has a right to ask, "What's in it for me?"

The Glory of the Hummingbird
Chapter 1 (p. 6)

The universe is like a safe to which there is a combination but the combination is locked up in the safe.

Let Me Count The Ways
Chapter 22 (p. 307)

Deutsch, Karl W.

Any universe uneven enough to sustain the life of a flatworm should perhaps be uneven enough to be eventually known by man.

Philosophy of Science
Mechanism, Organism, and Society:
Some Models in Natural and Social Science (p. 231)
Volume 18, Number 3, July 1951

Dyson, Freeman J.

I have found a universe growing without limit in richness and complexity, a universe of life surviving forever and making itself known to its neighbors across unimaginable gulfs of space and time. Whether the details of my calculations turn out to be correct or not, there are good scientific reasons for taking seriously the possibility that life and intelligence can succeed in molding this universe of ours to their own purposes.

Infinite in All Directions
Part I
Chapter 6 (p. 117)

The hypothesis is that the universe is constructed according to a principle of maximum diversity. The principle of maximum diversity operates both at the physical and at the mental level. It says that the laws of nature and the initial conditions are such as to make the universe as interesting as possible. As a result, life is possible but not too easy. Always when things are dull, something new turns up to challenge us and to stop us from settling into a rut. Examples of things which make life difficult are all around us: comet impacts, ice ages, weapons, plagues, nuclear fission, computers, sex, sin and death. Not all challenges can be overcome, and so we have tragedy. Maximum diversity often leads to maximum stress. In the end we survive, but only by the skin of our teeth.

Infinite in All Directions
Part II
Chapter 17 (p. 298)

Eddington, Sir Arthur Stanley
I would feel more content that the universe should accomplish some great scheme of evolution and, having achieved whatever may be achieved, lapse back into chaotic changelessness, than its purpose should be banalised by continual repetition. I am an Evolutionist, not a Multiplicationist. It seems rather stupid to keep doing the same thing over and over again.

The Nature of the Physical World
Chapter IV (p. 86)

Ehrmann, Max
You are a child of the universe, no less than the trees and the stars; you have a right to be here. And whether or not it is clear to you, no doubt the universe is unfolding as it should.

Desiderata

Emerson, Ralph Waldo
We are taught by great actions that the universe is the property of every individual in it.

Essays and Lectures
Nature, Addresses, and Lectures
Beauty (p. 16)

Engard, Charles J.

We accept the universe as far as we know it, but we do not attempt to explain why it exists. It is difficult enough to understand *how*!

In Bertha Mueller
Goethe's Botanical Writings
Introduction (p. 14)

Estling, Ralph

I do not know what, if anything, the Universe has in its mind, but I am quite, quite sure that, whatever it has in its mind, it is not at all like what we have in ours. And, considering what most of us have in ours, it is just as well.

The Skeptical Inquirer
Spring Issue 1993

There is no question about there being design in the Universe. The question is whether this design is imposed from the Outside or whether it is inherent in the physical laws governing the Universe. The next question is, of course, who or what made these physical laws.

The Skeptical Inquirer
Spring 1993

Farmer, Phillip José

The universe is a big place, perhaps the biggest.

Venus on the Half-Shell

Flammarion, Camille

May we conclude, then, that in these successive endings the universe will one day become an immense and dark tomb. No: otherwise it would already have become so during a past eternity. There is in nature something else besides blind matter; an intellectual law of progress governs the whole creation; the forces which rule the universe cannot remain inactive. The stars will rise from their ashes. The collision of ancient wrecks causes new flames to burst forth, and the transformation of motion into heat creates nebulae and worlds. Universal death shall never reign.

Popular Astronomy
Book I, Chapter VII (p. 80)

When the last human eyelid closes here below, and our globe—after having been for so long the abode of life with its passions, its labour, its pleasures and its pains, its loves and its hatred, its religious and political expectations and all its vain finalities—is enshrouded in the winding-sheet of a profound night, when the extinct sun wakes no more; well, then—then, as to-day, the universe will be as complete, the stars will continue to shine in the sky, other suns will illuminate other worlds, other springs will bring round the bloom of flowers and the illusions of youth, other mornings and other evenings will follow in succession, and the universe will move on as at present; for creation is developed in infinity and eternity.

Popular Astronomy
Book II, Chapter VI (p. 164)

France, Anatole

The universe which science reveals to us is a dispiriting monotony. All the suns are drops of fire and all the planets drops of mud.

In Stanley L. Jaki
Chance or Reality and Other Essays
Chapter 4 (p. 46)

If desire lends a grace to whatsoever be the object of it, then the desire of the unknown makes beautiful the Universe.

My Friend's Book
Later Exploits
The Grove of Myrtle (p. 167)

Frayn, Michael

The complexity of the universe is beyond expression in any possible notation.

Lift up your eyes. Not even what you see before you can ever be fully expressed.

Close your eyes. Not even what you see now.

Constructions
Number 1

Galilei, Galileo

Philosophy is written in this grand book, the universe, which stands continually open to our gaze. But the book cannot be understood unless one first learns to comprehend the language and read the letters in which it is composed. It is written in the language of mathematics, and its characters are triangles, circles, and other geometric figures without which it is humanly impossible to understand a single word of it.

Discoveries and Opinions of Galileo (pp. 237–8)
[Dialogues Concerning Two Chief World Systems]

Giraudoux, Jean

COUNTESS: I know perfectly well that at this moment the whole universe is listening to us—and that every word we say echoes to the remotest star.

The Madwoman of Chaillot
Act II (p. 94)

Guth, Alan
Steinhardt, Paul

The inflationary model of the universe provides a possible mechanism by which the observed universe could have evolved from an infinitesimal region. It is then tempting to go one step further and speculate that the entire universe evolved from literally nothing.

Scientific American
The Inflationary Universe (p. 128)
Volume 250, Number 5, May 1984

Halacy, D.S. Jr

Our universe operates not at the whims of those who live in it, but by inexorable natural laws.

They Gave Their Names to Science
Prologue (p. 9)

Harrison, Edward

We do not know what sets limits to the Great Chain of hierarchical structures, nor do we know what unifies it. We are clueless as to why atoms exist and why the Universe is structured the way it is. Of course, if the Universe were structured in any other way, we would not be here asking these pertinent questions; or so we are told. But I am a heretic and inclined to think the other way: without us this Universe would not be here.

Quarterly Journal of the Royal Astronomical Society
A Twinkle in the Eye of the Universe (p. 428)
Volume 25, Number 4, December 1984

The universes are our models of the Universe. They are great schemes of intricate thought—grand cosmic pictures—that rationalize human experience; these universes harmonize and invest with meaning the rising and setting Sun, the waxing and waning Moon, the jeweled lights of the night sky, the landscape of rocks and trees and clouds.

Masks of the Universe
Chapter 1 (p. 1)

We cannot doubt the existence of an ultimate reality. It is the Universe forever masked. We are part of an aspect of it, and the masks figured by us are the Universe observing and understanding itself from a human point of view. When we doubt the Universe we doubt ourselves. The Universe thinks, therefore it is.

Masks of the Universe
Chapter I (p. 14)

From the outset we must decide whether to use *Universe* or *universe*. This in not so trivial a matter as it might seem. We know of only one planet called Earth; similarly, we know of only one Universe. Surely then the proper word is *Universe*?

Cosmology: The Science of the Universe
Chapter 1 (p. 10)

Haught, James A.
The universe is a vast, amazing, seething dynamo which has no discernable purpose except to keep on churning. From quarks to quasars, it's alive with incredible power. But it seems utterly indifferent to any moral laws. It destroys as blindly as it nurtures.

2000 Years of Disbelief
Afterthought (p. 324)

Hinshelwood, C.N.
To some men knowledge of the universe has been an end possessing in itself a value that is absolute: to others it has seemed a means of useful application.

The Structure of Physical Chemistry
Chapter I (p. 2)

Hogan, John
...cosmologists—and the rest of us—may have to forego attempts at understanding the universe and simply marvel at its infinite complexity and strangeness.

Scientific American
Universal Truths (p. 117)
Volume 263, Number 4, October 1990

Hoyle, Fred
...if there is one important result that comes out of our inquiry into the nature of the Universe it is this: when by patient inquiry we learn the answer to any problem, we always find, both as a whole and in detail, that the answer thus revealed is finer in concept and design than anything we could ever have arrived at by a random guess.

The Nature of the Universe
Chapter 7 (p. 140)

Perhaps the most majestic feature of our whole existence is that while our intelligences are powerful enough to penetrate deeply into the evaluation of this quite incredible Universe, we still have not the smallest clue to our own fate.

The Nature of the Universe
Chapter 7 (p. 142)

There is a coherent plan in the universe, though I don't know what it's a plan for.

Wired
2/98 (p.174)

Hubble, Edwin Powell

Equipped with his five senses, man explores the universe around him and calls the adventure science.

The Nature of Science and Other Lectures
Part I
The Nature of Science (p. 6)

Huygens, Christianus

What a wonderful and amazing Scheme have we here of the magnificent Vastness of the Universe! So many Suns, so many Earths...!

New Conjectures Concerning the Planetary Worlds, Their Inhabitants and Productions
Kosmotheoros (p. 222)

Ionesco, Eugene

...the universe seems to me infinitely strange and foreign. At such a moment I gaze upon it with a mixture of anguish and euphoria; separate from the universe, as though placed at a certain distance outside it; I look and I see pictures, creatures that move in a kind of timeless time and spaceless space, emitting sounds that are a kind of language I no longer understand or ever register.

Notes and Counter Notes: Writing on the Theatre
Part II
Interviews
Brief Notes for Radio (p. 136)

James, William

Whatever universe a professor believes in must at any rate be a universe that lends itself to lengthy discourse. A universe definable in two sentences is something for which the professorial intellect has no use. No faith in anything of that cheap kind!

Writings, 1902–1910
Pragmatism
Lecture 1
The Present Dilemma in Philosophy (p. 487)

Joad, C.E.M.

When the scientist leaves his laboratory and speculates about the universe as a whole, the resultant conclusions are apt to tell us more about the scientist than about the universe.

Philosophical Aspects of Modern Science
Chapter XI (p. 339)

Kant, Immanual

...God has deposited in the forces of Nature a certain secret art so that it may develop by itself from the chaos into a perfect world system...

Universal Natural History and Theory of the Heavens
Opening Discourse (p. 87)

The world-edifice puts one into a quiet astonishment by its immeasurable greatness and by the infinite manifoldness and beauty which shine forth from it on all sides.

Universal Natural History and Theory of the Heavens
Seventh Section (p. 148)

Keillor, Garrison

We wondered if there is a God or is the universe only one seed in one apple on a tree in another world where a million years of ours is only one of their moments and what we imagine as our civilization is only a tiny charge of static electricity and the great truth that our science is slowly grasping is the fact the apple in which we are part of one seed is *falling*, has been falling for a million years and in one one-millionth of a second it will hit hard-frozen ground in that other world and split open and lie on the ground and a bear will come along and gobble it up, everything, the Judeo-Christian heritage, science, democracy, the Renaissance, art, music, sex, sweet corn—all disappear into that black hole of a bear.

The Atlantic Monthly
Leaving Home (p. 48)
Volume 260, Number 3, September 1987

Kepler, Johannes

The diversity of the phenomena of Nature is so great, and the treasures hidden in the heavens so rich, precisely in order that the human mind shall never be lacking in fresh nourishment.

Mysterium Cosmographicum
Original Dedication (p. 55)

Kirshner, Robert P.

Although the Universe is under no obligation to make sense, students in pursuit of the Ph.D. are.

Quarterly Journal of the Royal Astronomical Society
Exploding Stars and the Expanding Universe (p. 240)
Volume 32, Number 3, September 1991

Koestler, Arthur

In my youth I regarded the universe as an open book, printed in the language of physical equations, whereas now it appears to me as a text written in invisible ink, of which in our rare moments of grace we are able to decipher a small fragment.

Bricks to Babel
Epilogue (pp. 682–3)

There are no longer any absolute directions in space. The universe has lost its core. It no longer has a heart, but a thousand hearts.

The Sleepwalkers

Mach, Ernst

The universe is like a machine in which the motion of certain parts is determined by that of others, only nothing is determined about the motion of the whole machine.

History and Root of the Principle of the Conservation of Energy
Chapter IV (p. 62)

Melville, Herman

It's too late to make any improvements now. The universe is finished; the copestone is set on, and the chips were carted off a million years ago.

Moby Dick
Chapter 2

Muir, John

When we try to pick out anything by itself, we find it hitched to everything else in the universe.

My First Summer in the Sierra
July 27 (p. 211)

How hard to realize that every camp of men or beast has this glorious starry firmament for a roof! In such places standing alone on the mountaintop it is easy to realize that whatever special nests we make of leaves and moss like marmots and birds, or tents or piled stone we all dwell in a house of one room the world with the firmament for its roof and are sailing the celestial spaces without leaving any track.

The Wilderness World of John Muir
The Philosophy of John Muir (p. 312)

The clearest way into the Universe is through a forest wilderness.

The Wilderness World of John Muir
The Philosophy of John Muir (p. 312)

Ovenden, M.W.

Pythagoras, if he could but be with us, would (I hope) smile indulgently upon our endeavors. But I think that he would be inclined to say that he knew that the Universe would turn out to be harmonious, for harmony was for him an axiom, a definition of the way in which he chose to organize his experience of the world.

Nature
Bode's Law—Truth or Consequences?
Volume 239, 1972

Pascal, Blaise

By space the universe encompasses and swallows me up like an atom; by thought I comprehend the world.

Pensées
Number 348

Plato

...had we never seen the stars, and the sun, and the heaven, none of the words which we have spoken about the universe would ever have been uttered.

Timaeus
Section 47

Poe, Edgar Alan

I design to speak of the *Physical, Metaphysical and Mathematical—of the Material and Spiritual Universe:—of its Essence, its Origin, its Creation, its Present Condition and its Destiny.*

Eureka (p. 1)

...the perceptible universe exists as a cluster of clusters, irregularly disposed.

Eureka (p. 4)

Telescopic observations, guided by the laws of perspective, enables us to understand that the perceptible Universe exists as *a roughly spherical cluster of clusters irregularly disposed.*

Eureka (p. 96)

Polanyi, Michael
The universe is still dead, but it already has the capacity of coming to life.

Personal Knowledge (p. 404)
In Freeman J. Dyson
Infinite in All Directions
Part I
Chapter 3 (p. 53)

Ramón y Cajal, Santiago
As long as the brain is a mystery, the universe will also be a mystery.

In Victor Cohn
The Washington Post
Charting 'the Soul's Frail Dwelling-House'
September 5, 1982
Final Edition (p. A1)

Reade, Winwood
The universe is anonymous; it is published under secondary laws; these at least we are able to investigate, and in these perhaps we may find a partial solution of the great problem.

The Martyrdom of Man
Chapter IV (p. 521)

Reichenbach, Hans
Instead of asking for a cause of the universe, the scientist can ask only for the cause of the present state of the universe; and his task will consist in pushing farther and farther back the date from which he is able to account for the universe in terms of laws of nature.

The Rise of Scientific Philosophy
Chapter 12 (p. 208)

Renard, Maurice
Man, peering at the Universe through only a few tiny windows—his senses—catches mere glimpses of the world around him. He would do well to brace himself against unexpected surprises from the vast unknown; from that immeasurable sector of reality that has remained a closed book.

In Charles Noël Martin
The Role of Perception in Science (p. 8)

Richards, Rheodore William
No one can predict how far we shall be enabled by means of our limited intelligence to penetrate into the mysteries of a universe immeasurably

vast and wonderful; nevertheless, each step in advance is certain to bring new blessing to humanity and new inspiration to greater endeavor.

In Faraday Lectures
Lectures Delivered Before the Chemical Society
The Fundamental Properties of the Elements (p. 236)

Rothman, Tony
When confronted with the order and beauty of the universe and the strange coincidences of nature, it's very tempting to take the leap of faith from science into religion. I am sure many physicists want to. I only wish they would admit it.

In J.L. Casti
Paradigms Lost (pp. 482–3)

Russell, Bertrand
The Universe may have a purpose, but nothing that we know suggests that, if so, this purpose has any similarity to ours.

Why I am Not a Christian
Do We Survive Death? (p. 92)

So far as scientific evidence goes, the universe has crawled by slow stages to a somewhat pitiful result on this earth, and is going to crawl by still more pitiful stages to a condition of universal death.

Why I am Not a Christian
Has Religion Made Useful Contributions to Civilization (p. 32)

Santayana, George
The universe, as far as we can observe it, is a wonderful and immense engine; its extent, its order, its beauty, its cruelty, make it alike impressive. If we dramatize its life and conceive its spirit, we are filled with wonder, terror, and amusement, so magnificent is that spirit, so prolific, inexorable, grammatical and dull.

In Logan Pearsall Smith
Little Essays
Piety (p. 85)

Shelley, Percy Bysshe
Its easier to suppose that
the universe has existed from
all eternity than to conceive a
Being beyond its limits capable of creating it.

The Complete Poetical Works of Shelley
Queen Mab

Siegel, Eli

The universe is Why, How, and What, in any order, and all at once.

Damned Welcome
Aesthetic Realism Maxims
Part 1, No. 69 (p. 28)

The weight of the universe is at one with all its space.

Damned Welcome
Aesthetic Realism Maxims
Part 1, No. 70 (p. 28)

The universe, being clever, has given scientists trouble. Damned Welcome

Aesthetic Realism Maxims
Part 1, No. 71 (p. 28)

Spenser, Edmund

Why then should witless man so much misween,
That nothing is, but that which he hath seene?
What if in the Moones faire shining speheare?
What if in every other starre unseene,
Of other worldes he happily should heare?
That nothing is, but that which he hath seene?

The Complete Works in Verse and Prose of Edmund Spenser
Volume 8
Faerie Queene
Book the Second
Introduction

Stern, S. Alan

The place we call our Universe is, for the most part, cold and dark and all but endless. It is the emptiest of empties. It is old, and yet very young. It contains much that is dead, and yet much that is alive, forever reinventing itself, and sometimes inventing something wholly new.

Our Universe
The Frontier Universe: At the Edge of the Night (p. 1)

Swann, W.F.G.

There is one great work of art; it is the universe. Ye men of letters find the imprints of its majesty in your sense of the beauty of words. Ye men of song find it in the harmony of sweet sounds. Ye painters feel it in the design of beauteous forms, and in the blending of rich soft colors do your souls mount on high to bask in the brilliance of nature's sunshine. Ye lovers are conscious of its beauties in forms ye can but ill define. Ye men of science find it in the rich harmonies of nature's mathematical design.

The Architecture of the Universe
Chapter 12 (p. 424)

Swenson, May

What
is it about
the universe,
the universe about us stretching out?
We, within our brains,
within I,
think
we must unspin it.

In John Osborne and David Paskow
Looking Back on Tomorrow
The Universe

Tennyson, Lord Alfred

This truth within thy mind rehearse,
That in a boundless universe
Is boundless better, boundless worse.

The Complete Poetical Works of Tennyson
The Two Voices

Thom, Rene

Next we must concede that the universe we see is a ceaseless creation, evolution, and destruction of forms and that the purpose of science is to foresee this change of form and, if possible, explain it.

Structural Stability and Morphogenesis
Chapter 1 (p. 1)

Thompson, Francis

The universe is his box of toys. He dabbles his fingers in the day-fall. He is gold-dusty with tumbling amidst the stars. He makes bright mischief with the moon. The meteors nuzzle their noses in his hand.

Shelley (pp. 45–6)

Thoreau, Henry David

The universe is wider than our views of it.

Walden
Chapter XVIII (p. 317)

Toynbee, Arnold

Huddled together in our little earth we gaze with frightened eyes into the dark universe.

Lectures on the Industrial Revolution of the 18th Century in England
Notes and Jottings (p. 256)

Tyron, E.P.

If it is true that our Universe has a zero net value for all conserved quantities, then it may simply be a fluctuation of the vacuum of some larger space in which our Universe is imbedded. In answer to the question of why it happened, I offer the modest proposal that our Universe is simply one of those things which happen from time to time.

Nature
Is the Universe a Vacuum Fluctuation? (p. 397)
Volume 246, Number 5433, December 14, 1973

Tzu, Lao

In the universe the difficult things are done as if they are easy.

In Gia-Fu Geng and Jane English
Tao Te Ching
Sixty-Three

Unknown

To be rational is to look the universe in the face and not flinch.

Unknown source

Weinberg, Steven

The more the universe seems comprehensible, the more it also seems pointless.

The First Three Minutes
Epilogue (p. 154)

...the urge to trace the history of the universe back to its beginning is irresistible.

The First Three Minutes
Chapter I (p. 4)

Wheeler, John A.

The Universe is a self-excited circuit.

Frontiers of Time (p. 13)
In Freeman J. Dyson
Infinite in All Directions
Chapter 3 (p. 53)

...this is *our* Universe, our museum of wonder and beauty, our cathedral.

A Journey into Gravity and Spacetime
Opening

We will first understand how simple the universe is when we recognize how strange it is.

Cosmic Search Magazine
From the Big Bang to the Big Crunch
Volume 1, Number 4, Fall 1979

Whitman, Walt
Let your soul stand cool and composed before a million universes.

In James E. Miller, Jr (ed.)
Complete Poetry and Selected Prose
Song of Myself

Wiechert, Emil
The universe is infinite in all directions.

Schriften der Physikalisch-Ökonomischen Gesellschaft zu Königsberg in Preussen
Die Theorie der Elektrodynamik und die Röntgensche Entdeckung
Volume 37, 1896
In Freeman J. Dyson
Infinite in All Directions
Part I
Chapter 3 (p. 53)

Young, Louise B.
The universe is unfinished, not just in the limited sense of an incompletely realized plan but in the much deeper sense of a creation that is a living reality of the present. A masterpiece of artistic unity and integrated Form, infused with meaning, is taking shape as time goes by. But its ultimate nature cannot be visualized, its total significance grasped, until the final lines are written.

The Unfinished Universe
Conclusion (p. 207-8)

COSMOGENESIS

Barrow, John D.
One day we may be able to say something about the origins of our own cosmic neighborhood. But we can never know the origins of the universe. The deepest secrets are the ones that keep themselves.

The Origin of the Universe
Chapter 8 (p. 137)

Bowyer, Stuart
Ultimately, the origin of the universe is, and always will be, a mystery.

In Henry Margenau and Roy Abraham Varghese (eds)
Cosmos, Bios, Theos
Chapter 2 (p. 32)

Gamow, George
Before we can discuss the basic problem of the origin of our universe, we must ask ourselves whether such a discussion is necessary.

The Creation of the Universe
Chapter I (p. 6)

Genesis 1:1
In the beginning God created the heaven and the earth.

The Bible

Kipling, Rudyard
Before the High and Far-Off Times, O my Best Beloved, came the Time of the Very Beginnings; and that was in the days when the Eldest Magician was getting Things ready. First he got the Earth ready; then he got the Sun ready; and then he told all the Animals that they could come out and play.

Just So Stories
The Crab that Played with the Sun (p. 123)

Reeves, Hubert
In the beginning was the absolute rule of the flame: The universe was in limbo. Then after countless eras, the fires slowly abated like the sea at the outgoing tide. Matter awoke and organized itself; the flame gave way to music.

Atoms of Silence
Introduction (p. 5)

Singer, Isaac Bashevis
"Who created the world?"
"There was matter somewhere in the cosmos and for a long time it lay there and stank. That stench was the origin of life."
"Where did the matter come from?"
"What is the difference? The main thing is that we have no responsibility—neither to ourselves nor to others. The secret of the universe is apathy. The earth, the sun, the rocks, they're all indifferent, and this is a kind of passive force. Perhaps indifference and gravitation are the same."
He spoke and yawned. He ate and smoked.
"Why do you smoke so much?" I asked.
"It keeps me indifferent."

A Crown of Feathers
The Captive (p. 47)

Spenser, Edmund
Through knowledge we behold the world's creation,

How in his cradle first he fostered was;
And judge of Natures cunning operation,
How things she formed of a formless mass...

The Complete Poetical Works of Edmund Spenser
The Tears of the Muses
L. 499–502

Sturluson, Snorri
Erst was the age when nothing was:
Nor sand nor sea, nor chilling stream-waves;
Earth was not found, nor Ether-Heaven,—
A Yawning Gap, but grass was none.

The Prose Edda
Here Begins the Beguiling of Gylfi (p. 16)

Tagore, Rabindranath
It seems to me that, perhaps, creation is not fettered by rules,
That all the hubbub, meeting and mingling are
blind happenings of fate...

Our Universe
Chapter 4 (p. 75)

Townes, Charles H.
I do not understand how the scientific approach alone, as separated from a religious approach, can explain an origin of all things. It is true that physicists hope to look behind the 'big bang,' and possibly to explain the origin of our universe as, for example, a type of fluctuation. But then, of what is it a fluctuation and how did this in turn begin to exist? In my view the question of origin seems always left unanswered if we explore from a scientific view alone.

In Henry Margenau and Roy Abraham Varghese (eds)
Cosmos, Bios, Theos
Chapter 25 (p. 123)

Unknown
In the beginning there was nothing...which exploded.

Source unknown

DYING

Balfour, A.J.
...the energies of our system will decay, the glory of the sun will be dimmed, and the earth, tideless and inert, will no longer tolerate the race

which has for a moment disturbed its solitude. Man will go down into the pit, and all his thoughts will perish.

The Foundations of Belief
Part I, Chapter I, Section III (p. 33)

Byron, George Gordon
I had a dream, which was not all a dream
The bright sun was extinguish'd, and the stars
Did wander darkling in the eternal space,
Rayless, and pathless, and the icy earth
Swung blind and blackening in the moonless air...

The Complete Poetical Works of Byron
Darkness

Davies, Paul
Many billions of years will elapse before the smallest, youngest stars complete their nuclear burning and shrink into white dwarfs. But with slow, agonizing finality perpetual night will surely fall.

The Last Three Minutes
Chapter 5 (p. 50)

A universe that came from nothing in the big bang will disappear into nothing at the big crunch. Its glorious few zillion years of existence not even a memory.

The Last Three Minutes
Chapter 9 (p. 123)

de Goncourt, Jules
At some particular state in scientific development will the good Lord, with a flowing white beard, arrive on earth with his chain of keys and say to humanity, just like they do at the Art Gallery at 5:00... "Gentlemen it's closing time."

In *Bartlett's Familiar Quotations* (p. 146)

Dyson, Freeman J.
Since the universe is on a one-way slide toward a state of final death in which energy is maximally degraded, how does it manage, like King Charles, to take such an unconsciously long time a-dying.

Quoted by John D. Barrow and Frank J. Tipler in
The Anthropic Cosmological Principle
Chapter 6 (p. 385)

Eddington, Sir Arthur Stanley
...the universe will finally become a ball of radiation, becoming more and more rarified and passing into longer and longer wave-lengths.

The longest waves of radiation are Hertzian waves of the kind used in broadcasting. About every 1500 million years this ball of radio waves will double in diameter; and it will go on expanding in geometrical progression for ever. Perhaps then I may describe the end of the physical world as—one stupendous broadcast.

New Pathways in Science
Chapter III (p. 71)

Eliot, T.S.
This is the way the world ends
Not with a bang but a whimper.

Complete Poems and Plays of T.S. Eliot
The Hollow Men
v

Frost, Robert
Some say the world will end in fire,
Some say in ice.
From what I've tasted of desire
I hold with those who favor fire.
But if it had to perish twice,
I think I know enough of hate
To say that for destruction ice
Is also great
And would suffice.

Complete Poems of Robert Frost
Fire and Ice

Harrison, Edward
The stars begin to fade like guttering candles and are snuffed out one by one. Out of the depths of space the great celestial cities, the galaxies, cluttered with the memorabilia of ages, are gradually dying. Tens of billions of years pass in the growing darkness. Occasional flickers of light pierce the fall of cosmic night, and spurts of activity delay the sentence of a universe condemned to become a galactic graveyard.

Cosmology: The Science of the Universe
Chapter 18 (p. 360)

Huxley, Julian
And all about the cosmic sky,
The black that lies beyond our blue,
Dead stars innumerable lie,
And stars of red and angry hue

Not dead but doomed to die.

Cosmic Death
Source unknown

James, William

Though the *ultimate* state of the universe may be its vital and psychical extinction, there is nothing in physics to interfere with the hypothesis that the *penultimate* state might be the millennium—in other words a state in which a minimum of difference of energy-level might have its exchanges so skillfully *canalises* that a maximum of happy and virtuous consciousness would be the only result. In short, the last expiring pulsation of the universe's life might be, 'I am so happy and perfect that I can stand it no longer.'

The Atlantic Monthly
Letter to Henry Adams dated June 17, 1910 (p. 316)
September 1920

Jeffers, Robinson

Time will come no doubt,
When the sun shall die, the planets will freeze, and
the air on them; frozen gases, white flakes of air
Will be the dust: which no wind ever will stir: this very dust in dim starlight glistening
Is dead wind, the white corpse of wind.
Also, the galaxy will die; the glitter of the Milky Way,
our universe, all the stars that have names are dead.
Vast is the night. How you have grown, dear night,
walking your empty halls, how tall!

The Double Axe
Part II of The Double Axe
The Inhumanist
Stanza 11 (p. 58)

Nicholson, Norman

And if the universe
Reversed and showed
The colour of its money;
If now observable light
Flowed inward, and the skies snowed
A blizzard of galaxies,
The lens of night would burn
Brighter than the focused sun,
And man turn blinded

With white-hot darkness in his eyes.

In Neil Curry (ed.)
Norman Nicholson Collected Works
The Expanding Universe

Russell, Bertrand
...all the labours of the ages, all the devotion, all the inspiration, all the noonday brightness of human genius, are destined to extinction in the vast death of the solar system, and that the whole temple of man's achievements must inevitably be buried beneath the debris of a universe in ruins—all these things, if not quite beyond dispute, are yet so nearly certain, that no philosophy which rejects them can hope to stand. Only within the scaffolding of these truths, only on the firm foundation of unyielding despair, can the soul's habitation henceforth be safely built.

Why I am not a Christian
A Free Man's Worship (p. 107)

Wells, H.G.
...a steady twilight brooded over the Earth...All traces of the moon had vanished. The circling of the stars, growing slower and slower, had given place to creeping points of light...the sun, red and very large, [had] halted motionless upon the horizon, a vast dome glowing with a dull heat...The rocks about me were of a harsh reddish colour, and all the traces of life that I could see at first was the intensely green vegatation...the same rich green that one sees on forest moss or on the lichen in caves: plants which like these grow in a perpetual twilight...I cannot convey the sense of abominable desolation that hung over the world.

The Time Machine

Yeats, William Butler
When shall the stars be blown about the sky,
Like the sparks blown out of a smithy, and die?

Collected Poems
The Secret Rose

SAINT AUGUSTINE ERA

Cardenal, Ernesto
Before the big explosion
there wasn't even empty space,
for space and time, and matter and energy, emerged from the explosion,
neither was there any "outside" into which the universe could explode

for the universe embraced it all, even the whole of empty space.

Cosmic Canticle
Cantiga 1
Big Bang (p. 12)

Pagels, H.R.

The nothingness "before" the creation of the universe is the most complete void that we can imagine—no space, time, or matter existed. It is a world without place, without duration or eternity, without number—it is what mathematicians call "the empty set." Yet this unthinkable void converts itself into a plenum of existence—a necessary consequence of physical laws. Where are these laws written into that void? What "tells" the void that it is pregnant with a possible universe? It would seem that even the void is subject to law, a logic that exists prior to time and space.

Perfect Symmetry
Part 3
Chapter 5 (p. 347)

Saint Augustine

See, I answer him that asketh, "What did God before He made heaven and earth?" I answer not as one is said to have done merrily (eluding the pressure of the question): "He was preparing hell (saith he) for pryers into mysteries."

The Confessions
Book XI, XII, 14

Tzu, Lao

Something mysteriously formed,
Born before heaven and earth.
In the silence and the void,
Standing alone and unchanging, Ever present and in motion.

In Gia-Fu Feng and Jane English
Tao Te Ching
Twenty-five

UNKNOWN

Asimov, Isaac
If it is exciting to probe the unknown and shed light on what was dark before, then more and more excitement surely lies ahead of us.

The Universe
The Edge of the Universe (p. 302)

Huxley, Thomas H.
The known is finite, the unknown infinite; intellectually we stand on an islet in the midst of an illimitable ocean of inexplicability. Our business in every generation is to reclaim a little more land.

In Francis Darwin (ed.)
The Life and Letters of Charles Darwin
Volume 1
On the Reception of the Origin of Species

Lindberg, Charles H.
Whether outwardly or inwardly, whether in space or in time, the farther we penetrate the unknown, the vaster and more marvelous it becomes.

Autobiography of Values
Chapter 15 (p. 402)

Nicholson, Norman
No man has seen it; nor the lensed eye
That pin-points week by week the same patch of sky
Records even a blur across its pupil; only
The errontry of Saturn, the wry
Retarding of Uranus, speak
Of the pull beyond the pattern:—
The unknown is shown
Only by a bend in the known.

In Neil Curry (ed.)
Norman Nicholson Collected Works
The Undiscovered Planet

Nietzsche, Friedrich

To trace something unknown back to something known is alleviating, soothing, gratifying and gives moreover a feeling of power, Danger, disquiet, anxiety attend the unknown—the first instinct is to eliminate these distressing states. First principle: any explanation is better than none...

In Alexander Tille (ed.)
The Works of Friedrich Nietzsche
Volume XI
Twilight of the Idols
The Four Great Errors
Section 5 (p. 138)

Rabi, I.I.

To science the unknown is a problem full of interest and promise; in fact science derives its sustenance from the unknown; all the good things have come from that inexhaustible realm.

The Atlantic Monthly
Faither in Science (p. 28)
Volume 187, Number 1, January 1951

VACUUM

Bacon, Roger

For vacuum rightly conceived of is merely a mathematical quantity extended in the three dimensions, existing per se without heat and cold, soft and hard, rare and dense, and without any natural quality, merely occupying space, as the philosophers maintained before Aristotle, not only within the heavens, but beyond.

Opus Majus
Volume II
Part 5, Ninth Distinction, chapter II (p. 485)

Huygens, Christiaan

...but what God has bin pleas'd to place beyond the Region of the Stars, is as much above our Knowledge, as it is our Habitation.

Or what if beyond such a determinate space he has left an infinite Vacuum; to show, how inconsiderable is all that he has made is, to what his Power could, had he so pleas'd, have produc'd?

The Celestial Worlds Discover'd
Book the Second (p. 156)

VERNAL EQUINOX

Cuppy, Will

Among things you might be thinking about today is the vernal equinox—it's March 21, you know. The vernal equinox is the point at which the sun apparently crosses the celestial equator toward the north, or you can say it is the moment at which this occurs, or you can simply say: "Hooray! Spring is here!" Exactly why the sun does this on March 21 is a long story.

How to Get From January to December
March 21 (p. 61)

VERNIER

Langley, Samuel Pierpoint
That little Vernier, on whose slender lines
The midnight taper trembles as it shines,
Tells through the mist where dazzled Mercury burns,
And marks the point where Uranus returns.

The New Astronomy
Chapter I (p. 3)

WAVE

Gamow, George
In wave mechanics there are no impenetrable barriers, and as the British physicist R.H. Fowler put it after my lecture on that subject at the Royal Society of London... "Anyone at present in this room has a finite chance of leaving it without opening the door, or, of course, without being thrown out through the window."

My World Line
Chapter 3 (p. 60)

Thomson, G.P.
The wind catches the filaments and the spider is carried where the filaments take it. In much the same way the point which represents the energy of the electron is guided by the waves which surround it, and extend possibly to an indefinite distance in all directions. If the waves pass over an obstacle like an atom their direction is modified and the modification is transmitted back to the electron and enable it to guide its path in accordance with the distribution of matter which it finds around it.

The Atom
Chapter VII (p. 110)

WISDOM

Descartes, Rene

...human wisdom...always remains one and the same, however applied to different subjects, and suffers no more differentiation proceeding from them than the light of the sun experiences from the variety of things which it illumines...

Rules for the Direction of the Mind
Rule 1 (p. 1)

Milton, John

To know
That which before us lies in daily life
Is the prime Wisdom...

Paradise Lost
Book VIII, L. 192–3

WORK

Littlewood, John E.

Most of the best work starts in hopeless muddle and floundering, sustained on the 'smell' that something is there.

In Béla Bollabás (ed.)
Littlewood's Miscellany
Academic Life (p. 144)

Lowell, Percival

Gauge your work by its truth to nature, not by the plaudits it receives from man. In the end the truth will prevail and though you may never live to see it, your work will be recognized after you are gone.

In William Graves Hoyt
Lowell and Mars
Chapter 15 (p. 300)

WORLD

Bruno, Giordana
God is infinite, so His universe must be too. Thus is the excellence of God magnified and the greatness of His kingdom made manifest; He is glorified not in one, but in countless suns; not in a single earth, a single world, but in a thousand thousand, I say in an infinity of worlds.

On the Infinite Universe and Worlds

Pascal, Blaise
The whole visible world is only an imperceptible atom in the ample bosom of nature. No idea approaches it. We may enlarge our conceptions beyond all imaginable space; we only produce atoms in comparison with the reality of things.

Pensées
Aphorism 72

Regnault, Pére
We have not...Eyes piercing enough to penetrate so far as the Surface of the World; we don't see the external Figure of it: But if we judge it by the common Persuasion, and by what is offered to our Senses, when the Weather is serene, and the Heavens sparkles with Stars, the World is round: It is a Sphere.

Philosophical Conversations
Volume I
Conversation XIII (p. 158)

Till you have discovered to me the Mysteries of the Loadstone, I shall be no more at Quiet than a Loadstone itself, which is not in its natural Situation, and which is seeking out the Poles of the Earth.

Philosophical Conversations
Volume I
Conversation XV (p. 196)

Shelley, Percy Bysshe

Worlds on worlds are rolling ever
From creation to decay
Like the bubbles on a river
Sparkling, bursting, borne away.

The Complete Poetical Works of Shelley
Worlds on Worlds

Tennyson, Alfred

Come, my friends,
'Tis not too late to seek a newer world.
To sail beyond the sunset, and the baths
Of all the western stars.

The Complete Poetical Works of Tennyson
Ulysses

Wordsworth, William

... worlds unthought of till the searching mind
Of Science laid them open to mankind.

The Complete Poetical Works of Wordsworth
To the Moon
Rydal, L. 40

WRITING

Einstein, Albert

Your exposition is of matchless clarity and perspicuity. You did not dodge any problems but took the bull by the horns, said all that is essential, and omitted all that is inessential.

The Collected Papers of Albert Einstein
Volume 8
Letter 297
Letter to Moritz Schlick
6 February 1917 (p. 284)

Emerson, William

I am very sensible how difficult a thing it is to write well upon the science of Astronomy; by reason the subject is so comprehensive, and consists of so many parts, and is connected with so many other sciences, which it requires the perfect knowledge of; and besides, is a work of so much time, that a man had need have the life of Mathusalem, to go thro' the whole of it.

A System of Astronomy
The Preface (p. iii)

Lowell, Percival

I believe that all writing should be a collection of precious stones of truth which is beauty. Only the arrangement differs with the character of the book. You string them into a necklace for the world at large, pigeon hole them in drawers for the scientist. In the necklace you have the cutting of your thought, i.e., the expressing of it and the arrangement of the thoughts among themselves.

In Ferris Greenslet
The Lowells and Their Seven Worlds
Book VII
Chapter I (p. 355)

Rossi, Hugo

It is extremely hard for mathematicians to do expository writing. It is not in our nature. In fact, the very nature of mathematical meaning and grammar militates against it. However, this puts us at a distinct disadvantage relative to other sciences...Good exposition should be valued, not only for the success in communication but also as evidence of real mathematical insight. It is no accident that among our greatest mathematicians are our greatest teachers and expositors.

Notices of the American Mathematical Society
From the Editor (p. 4)
Volume 42, Number 1, January 1995

von Braun, Wernher

When a good scientific paper earns a student as much glory as we shower upon the halfback who scored the winning touchdown, we shall have restored the balance that is largely missing from our schools.

The New York Times
Text of the Address by von Braun Before the Publishers' Group Meeting Here
29 April 1960
L20, column 5

BIBLIOGRAPHY

Now let us come to those references to authors which other books have, and you want for yours. The remedy for this is very simple: You have only to look out for some book that quotes them all, from A to Z as you say yourself, and then insert the very same alphabet in your book, and though the imposition may be plain to see, because you have so little need to borrow from them, that is no matter; there will probably be some simple enough to believe that you have made use of them all in this plain, artless story of yours. At any rate, if it answers no other purpose, this long catalogue of authors will serve to give a surprising look of authority to your book.

de Cervantes, Miguel
The History of Don Quixote de la Mancha
Preface (p. xiii)

Abbey, Edward. *The Monkey Wrench Gang.* J.B. Lippincott Company, Philadelphia. 1975.

Ackerman, Diane. *A Natural History of the Senses.* Random House, New York. 1990.

Ackerman, Diane. *The Planets.* William Morrow and Company, Inc., New York. 1976.

Adams, Ansel. *An Autobiography.* Little, Brown and Company, Boston. 1985.

Adams, Douglas. *So Long, and Thanks for All the Fish.* Harmony Books, New York. 1985.

Adams, Douglas. *The Long Dark Tea-Time of the Soul.* Heinemann, London. 1988.

Adams, Douglas. *The Restaurant at the End of the Universe.* Pocket Books, New York. 1980.

Adams, Henry. *The Degradation of the Democratic Dogma.* The Macmillan Company, New York. 1919.

Addison, Joseph. *Cato.* J. Dicks, London. 1883.

Addison, Joseph. *Interesting Anecdotes, Memoirs, Allegories, Essays, and Poetical Fragments.* Printed by the Author, London. 1794.

Aeschylus. *The Agamemnon*. Translated by Gilbert Murray. George Allen & Unwin Ltd, London. 1920.

Albers, Donald J.; Alexanderson, Gerald L. and Constance Reid. *More Mathematical People*. Harcourt Brace Jovanovich, Boston. 1990.

Albran, Kehlog. *The Profit*. Price/Stern/Sloan, Los Angeles. 1974.

Alighieri, Dante. 'The Divine Comedy of Dante Alighieri' in *Great Books of the Western World*. Volume 21. Encyclopaedia Britannica, Inc., Chicago. 1952.

Allamandola, L.J. and Tielens, A.G.G.M. *Interstellar Dust*. Proceedings of the 135th Symposium of the International Astronomical Union. Santa Clara, California. 26–30 July, 1988. Kluwer Academic Press, Dordrecht. 1989.

Allen, Elizabeth Akers. *Rock Me to Sleep, Mother*. H.M. Caldwell Co., New York. 1882.

Allen, P.S. *The Praise of Folly*. At the Clarendon Press, Oxford. 1913.

Amaldi, Ginestra Giovene. *Our World and the Universe Around Us*. Volume I. Abradale Press, New York. 1966.

Aristotle. *On the Heavens in Great Books of the Western World*. Volume 8. Encyclopaedia Britannica, Inc., Chicago. 1952.

Aristotle. 'Physics' in *Great Books of the Western World*. Volume 8. Encyclopaedia Britannica, Inc., Chicago. 1952.

Aristotle. 'Posterior Analytics' in *Great Books of the Western World*. Volume 8. Encyclopaedia Britannica, Inc., Chicago. 1952.

Asimov, Isaac. *The Collapsing Universe*. Walker and Company, New York. 1977.

Asimov, Isaac. *The Universe*. Avon Books, New York. 1966.

Atkins, P.W. *The Creation*. W.H. Freeman & Company, Oxford. 1981.

Atwood, Margaret. *The Handmaid's Tale*. McClelland and Stewart, Toronto. 1985.

Aurelius, Marcus. 'Meditations' in *Great Books of the Western World*. Volume 12. Encyclopaedia Britannica, Inc., Chicago. 1952.

Ayres, C.E. *Science: The False Messiah*. The Bobbs–Merrill Company, Indianapolis. 1927.

Bachrach, Arthur J. *Psychological Research*. Random House, New York. 1972.

Bacon, Francis. 'Novum Organum' in *Great Books of the Western World*. Volume 30. Encyclopaedia Britannica, Inc., Chicago. 1952.

Bacon, Francis. *Opus Majus*. Translated by Robert Belle Burke. Volume II. Russell & Russell, New York. 1962.

Bailey, Janet. *The Good Servant*. Simon & Schuster, New York. 1995.

Bailey, M.E. and Williams, D.A. *Dust in the Universe*. Cambridge University Press, Cambridge. 1988.

Balfour, Arthur James. *The Foundations of Belief.* Longmans, Green, and Co., London. 1912.

Ball, Robert S. *The Story of the Heavens.* Cosell and Company Limited, London. 1910.

Barnett, Lincoln. *The Universe and Dr Einstein.* Bantam Books, New York. 1968.

Barrow, John D. *The Origin of the Universe.* Basic Books, New York. 1994.

Barrow, John D. and Tipler, Frank J. *The Anthropic Cosmological Principle.* Clarendon Press, Oxford. 1986.

Barth, John. *The Friday Book: Essays and Other Nonfiction.* G.P. Putnam's Sons, New York. 1984.

Bayliss, William Maddock. *Principles of General Physiology.* Longmans, Green, and Co., London. 1920.

Bean, William B. *Aphorisms from Latham.* The Prairie Press, Iowa City. 1962.

Beaver, Harold. *The Science Fiction of Edgar Allan Poe.* Penguin Books Ltd, Harmondsworth. 1976.

Beer, Arthur. *Vistas in Astronomy.* Volume 1. Pergamon Press, Oxford. 1960.

Beer, Arthur. *Vistas in Astronomy.* Volume 2. Pergamon Press, Oxford. 1960.

Bell, J.S. *Speakable and Unspeakable in Quantum Mechanics.* Cambridge University Press, Cambridge. 1987.

Bennett, Arnold. *Clayhanger.* E.P. Dutton & Company, New York. 1910.

Berendzen, Richard. *Life Beyond Earth & the Mind of Man.* NASA, Washington, D.C. 1973.

Bergaust, Erik. *Wernher von Braun.* National Space Institute, Washington, D.C. 1976.

Berlinski, David. *A Tour of the Calculus.* Pantheon Books, New York. 1995.

Bernstein, Jeremy. *Elementary Particles and Their Currents.* W.H. Freeman and Company, San Francisco. 1968.

Bernstein, Jeremy. *Experiencing Science.* Basic Books, Inc., New York. 1978.

Bichat, Xavier. *Physiological Researches on Life and Death.* Arno Press, New York. 1977.

Bierce, Ambrose. *The Devil's Dictionary.* Doubleday, Garden City. 1967.

Birks, J.B. *Rutherford at Manchester.* W.A. Benjamin Inc., New York. 1963.

Bisland, Elizabeth. *The Life and Letters of Lafcadio Hearn.* Volume I. Houghton, Mifflin and Company, Boston. 1900.

Black, Max. *Critical Thinking.* Prentice-Hall, New York. 1952.

Blackall, Eric A. *Wilhelm Meister's Apprenticeship.* Translated by Eric A. Blackall. Suhrkamp, New York. 1989.

Blaise, Clark. *Time Lord.* Weidenfeld & Nicolson, London. 2000.

Bloom, Allan. *The Closing of the American Mind.* Simon and Schuster, New York. 1987.

Blount, Sir Thomas Pope. *The Natural History*. Printed for R. Bentley, London. 1693.

Boethius. *On The Consolation of Philosophy*. Translated by W.W. Cooper. J.M. Dent & Sons, Ltd, London. 1902.

Bohm, David. *Causality and Chance in Modern Physics*. University of Pennsylvania Press, Philadelphia. 1957.

Bohr, Niels. *Atomic Theory and the Description of Nature*. At the University Press, Cambridge. 1961.

Bolles, Edmund Blair. *A Second Way of Knowing*. Prentice-Hall, New York. 1991.

Bollobás, Béla. *Littlewood's Miscellany*. Cambridge University Press, Cambridge. 1986.

Bonetti, A.; Greenberg, J.M. and S. Aiello. *Evolution of Interstellar Dust and Related Topics*. North- Holland, Amsterdam. 1989.

Born, Max. *Experiment and Theory in Physics*. At the University Press, Cambridge. 1944.

Born, Max. *My Life & My Views*. Charles Scribner's Sons, New York. 1968.

Born, Max. *The Born–Einstein Letters*. Translated by Irene Born. Macmillan Press Ltd, London. 1971.

Bourdillon, Francis William. *Among the Flowers, and Other Poems*. Marcus Ward & Co., London. 1878.

Boyle, Robert. A Free Enquiry into the Vulgarly Received Notion of Nature. Cambridge University Press, Cambridge. 1996.

Bradbury, Ray; Clarke, Arthur C.; Murray, Bruce; Sagan, Carl and Sullivan, Walter. *Mars and the Mind of Man*. Harper & Row, New York. 1973.

Bradley, John Hodgdon. *Autobiography of Earth*. Coward–McCann, New York. 1953.

Bradley, John Hodgson. *Parade of the Living*. Coward–McCann, Inc., New York. 1930.

Brandt, John C. and Chapman, Robert D. *Introduction to Comets*. Cambridge University Press, Cambridge. 1981.

Brewton, Sara and Blackburn, John Brewton. *Of Quarks, Quasars, and Other Quirks*. Thomas Y. Crowell Company, New York. 1977.

Bridgman, P.W. *Reflections of a Physicist*. Philosophical Library, New York. 1950.

Bridgman, P.W. *The Logic of Modern Physics*. The Macmillan Company, New York. 1927.

Bridgman, P.W. *The Nature of Physical Theory*. Princeton University Press, Princeton. 1936.

Bridgman, P.W. *The Way Things Are*. Harvard University Press, Cambridge. 1959.

Bronowski, J. *The Ascent of Man*. Little, Brown and Company, Boston. 1973.

Brontë, Charlotte. *Life and Works of The Sisters Brontë.* Volume III. AMS Press, New York. 1973.

Brown, Fredric. *The Lights in the Sky are Stars.* E.P. Dutton & Company, Inc., New York. 1953.

Browne, Thomas. *Religio Medici.* Macmillan and Co., London. 1881.

Browning, Robert. *The Complete Poetical Works of Browning.* Houghton Mifflin Company, Boston. 1895.

Bruner, Jerome S. *On Knowing: Essays for the Left Hand.* Harvard University Press, Cambridge. 1964.

Bruno, Giordano. *Cause, Principle, and Unity.* Translated by Jack Lindsay. International Publishers, New York. 1962.

Büchner, Ludwig. *Force and Matter.* Trübner I Co., London. 1864.

Bunge, Mario. *Studies in the Foundations Methodology and Philosophy of Science.* Volume 1. Springer-Verlag, Berlin. 1967.

Bunting, Basil. *Collected Poems.* Oxford University Press, Oxford. 1978.

Burke, Edmund. *A Philosophical Enquiry into the Origin of Our Ideas of the Sublime and Beautiful.* London. 1812.

Burnet, Thomas. *The Sacred Theory of the Earth.* Southern Illinois University Press, Carbondale. 1965.

Burnham, Robert Jr. *Burnham's Celestial Handbook.* Celestial Handbook Publications, Flagstaff. Volume 1.

Burns, Marilyn. *About Teaching Mathematics.* Math Solutions Publications, Sausalito. 1992.

Burns, Robert. *The Poems and Songs of Robert Burns.* Methuen and Company, London. 1896.

Burroughs, William. *The Adding Machine.* Seaver Books, New York. 1985.

Burton, Sir Richard. *The Kasîdah.* H.J. Cook. London. 1900.

Bury, J.B. *The Idea of Progress.* Dover Publications, Inc., New York. 1955.

Bush, Vannevar. *Endless Horizons.* Public Affairs Press, Washington, D.C. 1946.

Bush, Vannevar. *Science is Not Enough.* William Morrow & Company, Inc., New York. 1967.

Butler, Samuel. *Hudibras.* At the Clarendon Press, Oxford. 1967.

Butterfield, H. *The Origins of Modern Science 1300–1800.* The Macmillian Company, New York. 1961.

Buzzati-Traverso, Adriano. *The Scientific Enterprise, Today and Tomorrow.* United Nations Educational, Scientific and Cultural Organization, Paris. 1977.

Byron, George Gordon. *The Complete Poetical Works of Byron.* Houghton, Mifflin and Company, Boston. 1905.

Cabell, James Branch. *Jurgen.* Crown Publishers, Inc., New York. 1919.

Caithness, James Balharrie. *Pastime Poems.* Erskine MacDonald, Ltd, London. 1924.

Calder, Nigel. *Violent Universe*. The Viking Press, New York. 1969.

Calvin, Melvin. *Following the Trail of Light*. American Chemical Society, Washington, D.C. 1992.

Calvin, Melvin and Gazenko, Oleg G. *Foundations of Space Biology and Medicine*. Volume I. National Aeronautics and Space Administration, Washington, D.C. 1975.

Campbell, Norman. *Physics, The Elements*. At the University Press, Cambridge. 1920.

Campbell, Norman. *What is Science?* Dover Publications, Inc., New York. 1952.

Campbell, Thomas. *The Poetical Works of Thomas Campbell*. Crosby, Nichols, Lee & Co., Boston. 1860.

Camus, Albert. *The Myth of Sisyphus and Other Essays*. Translated by Justin O'Brien. Alfred A. Knopf, New York. 1969.

Camus, Albert. *The Outsider*. Translated by Stuart Gilbert. Hamish Hamilton, London. 1957.

Card, Orson Scott. *Seventh Son*. Tom Doherty Associates, Inc., New York. 1987.

Cardenal, Ernesto. *Cosmic Canticle*. Translated by John Lyons. Curbstone Press, Willimantic. 1993.

Carlyle, Thomas. *Sartor Resartus, On Heroes, Hero-Worship and the Heroic in History*. J.M. Dent, London. 1929.

Carrel, Alexis. *Man the Unknown*. Harper & Brothers, New York. 1939.

Carroll, Lewis. *Alice's Adventure in Wonderland*. Random House, Inc., New York. 1965.

Carroll, Lewis. *Through the Looking Glass*. Random House, Inc., New York. 1965.

Carson, Rachel. *The Sense of Wonder*. Harper & Row, New York. 1956.

Carter, John. *Urne Buriall and The Garden of Cyrus*. At the University Press, Cambridge. 1958.

Casti, John. *Paradigms Lost*. William Morrow and Company, Inc., New York. 1989.

Cayley, Arthur. *The Collected Mathematical Papers of Arthur Cayley*. Volume XI. At the University Press, Cambridge. 1896.

Chaisson, Eric. *Cosmic Dawn*. Little, Brown and Company, Boston. 1981.

Chandrasekhar, S. *Truth and Beauty: Aesthetics and Motivation in Science*. The University of Chicago Press, Chicago. 1987.

Chapman, Clark R. *The Inner Planets*. Charles Scribner's Sons, New York. 1877.

Chargaff, Edwin. *Serious Questions*. Birkhäuser, Boston. 1986.

Chargaff, Erwin. *Heraclitean Fire*. The Rockefeller University Press, New York. 1978.

Chaucer, Geoffrey. 'The Canterbury Tales' in *Great Books of the Western World*. Volume 22. Encyclopaedia Britannica, Inc., Chicago. 1952.

Chesterton, G.K. *Come to Think of It*. Methuen & Co., Ltd, London. 1932.
Chesterton, G.K. *The Coloured Lands*. Sheed & Ward, New York. 1938.
Church, Peggy Pond. *Ultimatum for Man*. Stanford University Press, Stanford. 1946.
Clark, Roger N. *Visual Astronomy of the Deep Sky*. Cambridge University Press, Cambridge. 1990.
Clarke, Arthur C. *2001: A Space Odyssey*. William Heinemann Ltd, London. 1985.
Clarke, Arthur C. *The Collected Stories of Arthur C. Clarke*. A Tom Doherty Associates Book, New York. 2000.
Clarke, Arthur C. *The Lost Worlds of 2001*. Gregg Press, Boston. 1979.
Clerke, Agnes M. *A Popular History of Astronomy during the Nineteenth Century*. Adam and Charles Black, London. 1908.
Cloos, Hans. *Conversation with the Earth*. Alfred A. Knopf, New York. 1953.
Coffin, Charles M. *The Complete Poetry and Selected Prose of John Donne*. The Modern Library, New York. 1952.
Cohen, Morris R. *Reason and Nature*. The Free Press, Glencoe. 1953.
Cole, William. *Philosophical Remarks on the Theory of Comets*. B.J. Holdsworth, London. 1823.
Coleridge, Samuel Taylor. *The Poems of Samuel Taylor Coleridge*. Oxford University Press, London. 1912.
Collins, Wilkie. *The Moonstone*. International Collectors Library, Garden City. 1900.
Collins, Wilkie. *The Woman in White*. The Heritage Press, New York. 1964.
Conrad, Joseph. *The Secret Agent: A Simple Tale*. Doubleday, Page & Company, Garden City. 1916.
Cooper, Necia Grant. *From Cardinals to Chaos*. Cambridge University Press, Cambridge. 1989.
Copernicus, Nicolaus. 'On the Revolutions of the Heavenly Spheres' in *Great Books of the Western World*. Volume 16. Encyclopaedia Britannica, Inc., Chicago. 1952.
Cuppy, Will. *How to Get From January to December*. Holt, New York. 1951.
Curie, Marie. *Pierre Curie*. Translated by Charlotte and Vernon Kellogg. The Macmillan Company, New York. 1923.
Curry, Neil. *Norman Nicholson Collected Poems*. Faber and Faber, London. 1994.

d'Abro, A. *The Evolution of Scientific Thought*. Dover Publication, Inc., New York. 1950.
Darling, David. *Equations of Eternity*. Hyperion, New York. 1993.
Darwin, Erasmus. *The Botanic Garden*. Jones & Company, London. 1825.
Dash, Joan. *A Life of One's Own*. Harper & Row, New York, 1973.
Davies, Paul. *Superforce*. Simon and Schuster, New York. 1984.

Davies, Paul. *The Last Three Minutes*. Basic Books, New York. 1994.

Davies, Paul. *The Mind of God*. Simon & Schuster, New York. 1992.

Davies, P.C.W. and Brown, Julian. *Superstrings: A Theory of Everything?* Cambridge University Press, Cambridge. 1988.

Davis, Philip J. and Hersh, Reuben. *The Mathematical Experience*. Birkhäuser, Boston. 1981.

Dawkins, Richard. *River Out of Eden*. Basic Books, Inc., New York. 1995.

Day, Clarence, Jr. *This Simian World*. Alfred A. Knopf, New York. 1922.

De Cervantes, Miguel. 'Don Quixote de la Mancha' in *Great Books of the Western World*. Volume 29. Encyclopaedia Britannica, Inc., Chicago. 1952.

de Fontenelle, Bernard. *Conversations on the Plurality of Worlds*. Printed for Peter Wilson, Dublin. 1761.

de Morgan, Augustus. *A Budget of Paradoxes*. Volume I. The Open Court Publishing Co., Chicago. 1915.

de Saint-Exupéry, Antoine. *Southern Mail*. Quinn and Boden, Rahway. 1933.

de Saint-Exupéry, Antoine. *The Little Prince*. Translated by Katherine Woods. Harcourt, Brace & World, New York. 1943.

de Sola Pinto, Vivian and Roberts, Warren. *The Complete Poems of D.H. Lawrence*. Volume 1. The Viking Press, New York. 1964.

de Tabley, Lord. *The Collected Poems of Lord de Tabley*. Chapman and Hall Limited, London. 1903.

de Vries, Peter. *Let Me Count the Ways*. Little, Brown and Company, Boston. 1965.

de Vries, Peter. *The Glory of the Hummingbird*. Little, Brown and Company, Boston. 1974.

Debus, Allen G. *The French Paracelsians*. Cambridge University Press, Cambridge. 1991.

Dennett, Daniel C. *Consciousness Explained*. Little, Brown and Company, Boston. 1991.

Dennett, Daniel C. *Darwin's Dangerous Idea*. Simon & Schuster, New York. 1995.

Descartes, Rene. 'Rules for the Direction of the Mind' in *Great Books of the Western World*. Volume 31. Encyclopaedia Britannica, Inc., Chicago. 1952.

Deudney, Daniel. *Space: The High Frontier in Perspective*. Worldwatch Paper 50. August 1982.

Devaney, James. *Where The Wind Goes*. Angus and Robertson, Sydney. 1939.

Diamond, Jared. *The Third Chimpanzee*. HarperCollins, New York. 1992.

Dick, Steven J. *Extraterrestrial Life and Our World View at the Turn of the Millennium*. Smithsonian Institution Libraries. May 2, 2000.

Dick, Thomas. *The Sidereal Heavens and Other Subjects Connected With Astronomy*. R. Worthington, New York. 1884.
Dickerson, Richard E. *Molecular Thermodynamics*. W.A. Benjamin, Inc., New York. 1969.
Dickinson, Emily. *Collected Poems of Emily Dickinson*. Avenel Books, New York. 1982.
Dietrich, Marlene. *Marlene Dietrich's ABC*. Doubleday & Company, Inc., Garden City. 1962.
Dixon, Bernard. *From Creation to Chaos*. Basil Blackwell Ltd, Oxford. 1989.
Donne, John. *Donne's Sermons*. At the Clarendon Press, Oxford. 1932.
Dorozynski, Alexandre. *The Man They Wouldn't Let Die*. The Macmillan Company, New York. 1965.
Doyle, Sir Arthur Conan. *The Adventures of Sherlock Holmes*. The Reader's Digest Association, Pleasantville. 1987.
Drake, Frank and Sobel, Dava. *Is Anyone Out There?* Delacorte Press, New York. 1992.
Dryden, John. *The Poetical Works of Dryden*. The Riverside Press, Cambridge. 1937.
du Noüy, Pierre Lecomte. *The Road to Reason*. Longmans, Green and Co., New York. 1949.
Duhem, Pierre. *The Aim and Structure of Physical Theory*. Princeton University Press, Princeton. 1954.
Dukas, Helen and Hoffman, Banesh. *Albert Einstein: The Human Side*. Princeton University Press, Princeton. 1979.
Dunnington, G. Waldo. *Inaugural Lecture on Astronomy and Papers on the Foundations of Mathematics*. Louisiana State University Press, Baton Rouge. 1937.
Durant, Will. *The Story of Civilization*. Part I. Our Oriental Heritage. Simon and Schuster, New York. 1954.
Durant, Will. *The Story of Philosophy*. Ernest Benn Limited, London. 1920.
Durham, Frank and Purrington, Robert D. *Some Truer Method*. Columbia University Press, New York. 1990.
Dürrenmatt, Friedrich. *The Physicists*. Translated by James Kirkup. Grove Press, Inc., New York. 1964.
Dyson, Freeman J. *Disturbing the Universe*. Harper & Row, New York. 1979.
Dyson, Freeman J. *Infinite in all Directions*. Harper & Row, New York. 1988.

Eco, Umberto. *The Name of the Rose*. Harcourt Brace Jovanovich, San Diego. Translated by William Weaver. 1983.
Eddington, Sir Arthur. *New Pathways in Science*. At the University Press, Cambridge. 1935.
Eddington, Sir Arthur. *The Internal Constitution of the Stars*. At the University Press, Cambridge. 1926.

Eddington, Sir Arthur. *The Mathematical Theory of Relativity.* At the University Press, Cambridge. 1954.
Eddington, Sir Arthur. *The Nature of the Physical World.* The Macmillan Company, New York. 1948.
Eddington, Sir Arthur. *The Philosophy of Physical Science.* The Macmillan Company, New York. 1939.
Egler, Frank E. *The Way of Science.* Hafner Publishing Company, New York. 1970.
Ehrmann, Max. *Desiderata.* Brook House, Los Angeles. 1972.
Einstein, Albert. *Cosmic Religion.* Covici Friede, New York. 1931.
Einstein, Albert. *Ideas and Opinions.* Bonanza Books, New York. 1954.
Einstein, Albert. *The Collected Papers of Albert Einstein.* Volume 8. Translated by Ann M. Hentschel. Princeton University Press, Princeton. 1998.
Einstein, Albert and Infeld, Leopold. *The Evolution of Physics.* Simon and Schuster, New York. 1966.
Eiseley, Loren. *The Immense Journey.* Random House, New York. 1957.
Eliot, George. *Middlemarch.* The Zodiac Press, London. 1982.
Emerson, Edward Waldo. *Journals of Ralph Waldo Emerson.* Houghton Mifflin Company, Boston. 1911.
Emerson, Ralph Waldo. *Essays and Lectures.* Literary Classics of the United States, New York. 1983.
Emerson, Ralph Waldo. *Society and Solitude.* Fields, Osgood & Co., Boston. 1870.
Emerson, William. *A System of Astronomy.* Printed for J. Nourse, London. 1769.
Empson, William. *Collected Poems.* Harcourt, Brace and Company, New York. 1949.
Epictetus. *The Discourses and Enchiridions.* Translated by Thomas Higginson. Walter J. Black, New York. 1944.
Esar, Evan. *Esar's Comic Dictionary.* Fourth Edition. Doubleday & Company, Garden City. 1983.

Falconer, William. *The Poetical Works of Beattie, Blair, and Falconer.* James Nichol, Edinburgh. 1854.
Feng, Gia-Fu and English, Jane. *Tao Te Ching.* Alfred A. Knopf, New York. 1974.
Ferguson, Kitty. *Measuring the Universe.* Walker and Company, New York. 1999.
Ferris, Timothy. *Coming of Age in the Milky Way.* William Morrow and Company, Inc., New York. 1988.
Ferris, Timothy. *Galaxies.* Sierra Club Books, San Francisco. 1980.
Feyerabend, Paul. *Against Method: Outline of an Anarchistic Theory of Knowledge.* Humanities Press, London. 1975.

Feynman, Richard P. *QED: The Strange Theory of Light and Matter.* Princeton University Press, Princeton. 1985.

Feynman, Richard. *The Character of Physical Law.* The MIT Press, Cambridge. 1989.

Feynman, Richard P. *What Do You Care What Other People Think?* W.W. Norton & Company, New York. 1988.

Feynman, Richard P.; Leighton, Robert B. and Matthew Sands. *The Feynman Lectures on Physics.* Volume I. Addison-Wesley Publishing Company, Reading. 1963.

Feynman, Richard P.; Leighton, Robert B. and Matthew Sands. *The Feynman Lectures on Physics.* Volume III. Addison-Wesley Publishing Company, Reading. 1963.

Firsoff, V.A. *Exploring the Planets.* A.S. Barnes and Company, South Brunswick. 1964.

Fischer, Ernst Peter. *Beauty and the Beast.* Translated by Elizabeth Oehlkers. Plenum Trade, New York. 1999.

Fisher, H.A.L. and others. *Essays in Honour of Gilbert Murray.* George Allen & Unwin Ltd, London. 1936.

Fitzgerald, Edward. *The Rubaiyat of Omar Khayyam.* Thomas Y. Crowell Company, New York. 1964.

Flammarion, Camille. *Popular Astronomy.* Chatto & Windus, London. 1894.

Flecker, James Elroy. *The Collected Poems of James Elroy Flecker.* Alfred A. Knopf, New York. 1921.

France, Anatole. *My Friend's Book.* Translated by J. Lewis May. Parke, Austin & Lipscomb, Inc., New York. 1923.

Frängsmyr, Tore. *Nobel Lectures: Chemistry 1971–1980.* World Scientific, Singapore. 1993.

Frankel, Felice and Whitesides, George M. *On the Surface of Things.* Chronicle Books, San Francisco. 1997.

Frauenfelder, H. and Henley, E.M. *Subatomic Physics.* Prentice-Hall, Inc., Englewood Cliffs. 1959.

Frayn, Michael. *Constructions.* Wildwood House, London. 1974.

French, A.P. and Taylor, Edwin F. *An Introduction to Quantum Physics.* W.W. Norton & Company, Inc., New York. 1978.

Friedman, Alan J. and Donley, Carol C. *Einstein As Myth and Muse.* Cambridge University Press, Cambridge. 1985.

Friedman, Herbert. *The Amazing Universe.* National Geographic Society, Washington, D.C. 1980.

Frost, Robert. *Complete Poems of Robert Frost.* Holt, Rinehart and Winston, New York. 1949.

Frothingham, N.L. *Metrical Pieces.* Crosby, Nichols, and Company, Boston. 1855.

Fry, Christopher. *The Lady's Not For Burning, A Phoenix Too Frequent and an Essay An Experience of Critics*. Oxford University Press, New York. 1977.
Fulbright, J. William. *Old Myths and New Realities*. Random House, New York. 1964.

Galilei, Galileo. *Discoveries and Opinions of Galileo*. Translated by Stillman Drake. Doubleday Anchor Books, Garden City. 1957.
Galilei, Galileo. 'The Two New Sciences' in *Great Books of the Western World*. Volume 28. Encyclopaedia Britannica, Inc., Chicago. 1952.
Gamow, George. *My World Line*. The Viking Press, New York. 1970.
Gamow, George. *The Creation of the Universe*. The Viking Press, New York. 1961.
Gardner, Martin. *Fads and Fallacies*. Dover Publications, Inc., New York. 1957.
Gardner, Martin. *Order and Surprise*. Prometheus Books, Buffalo. 1983.
Gardner, Martin. *Science: Good, Bad and Bogus*. Prometheus Books, Buffalo. 1981.
Gauss, Carl Friedrich. *Briefwechsel zwischen Gauss und Bessel*. Wilhelm Engelmann, Leipzig. 1880.
Gibran, Kahlil. *Sand and Foam*. Alfred A. Knopf, New York. 1952.
Gibran, Kahlil. *The Prophet*. Alfred A. Knopf, New York. 1969.
Gilbert, William and Sullivan, Arthur. *The Complete Plays of Gilbert and Sullivan*. The Modern Library, New York. 1936.
Giraudoux, Jean. *The Madwoman of Chaillot*. English adaptation by Maurice Valency. Random House, New York. 1947.
Gleick, James. *Chaos: Making a New Science*. Viking Penguin Inc., New York. 1987.
Gleick, James. *Genius: The Life and Science of Richard Feynman*. Pantheon Books, New York. 1992.
Godwin, Parke. *The Poetical Works of William Cullen Bryant*. Russell & Russell, New York. 1967.
Goodenough, Ursula. *The Sacred Depths of Nature*. Oxford University Press, New York. 1998.
Gordon, Isabel S. and Sorkin, Sophie. *The Armchair Science Reader*. Simon and Schuster, New York. 1959.
Gore, George. *The Art of Scientific Discovery*. Longmans, Green, and Co., London. 1878.
Gould, Stephen Jay. *Hen's Teeth and Horse's Toes*. W.W. Norton & Company, Inc., New York. 1983.
Green, Celia. *The Decline and Fall of Science*. Hamish Hamilton, London. 1976.
Greenslet, Ferris. *The Lowells and Their Seven Worlds*. Houghton Mifflin Company, Boston. 1946.
Greenstein, George. *Frozen Star*. Freundlich Books, New York. 1983.

Greenstein, George. *The Symbiotic Universe.* William Morrow and Company, Inc., New York. 1988.

Gregory, R.A. *Discovery.* Macmillan and Co., Limited, London. 1918.

Grondal, Florence Armstrong. *The Music of the Spheres.* The Macmillan Company, New York. 1926.

Grossland, Maurice. *Gay-Lussac: Scientist and Bourgeois.* Cambridge University Press, Cambridge. 1978.

Guiterman, Arthur. *Gaily the Troubadour.* E.P. Dutton & Co., Inc., New York. 1936.

Haber, Heinz. *Stars, Men and Atoms.* Golden Press, New York. 1962.

Habington, William. *The Poems of William Habington.* The University Press of Liverpool, London. 1948.

Halacy, D.S. Jr. *They Gave Their Names to Science.* G.P. Putnam's Sons, New York. 1967.

Haldane, J.B.S. *On Being the Right Size and Other Essays.* Oxford University Press, Oxford. 1985.

Haldane, J.B.S. *The Inequality of Man and Other Essays.* Penguin Books Limited, Harmondsworth. 1937.

Haliburton, Thomas Chandler. *The Old Judge.* Clarke, Irwin & Company Limited, Toronto. 1968.

Halmos, Paul R. *I Want to be a Mathematician.* Springer, New York. 1985.

Hannabuss, Keith. *An Introduction to Quantum Theory.* Oxford University Press, Oxford. 1997.

Hanson, Norwood Russell. *Patterns of Discovery.* At the University Press, Cambridge. 1965.

Hardy, G.H. *A Mathematician's Apology.* Cambridge University Press, Cambridge. 1969.

Hardy, Thomas. *Collected Papers of Thomas Hardy.* The Macmillan Company, New York. 1925.

Hardy, Thomas. *Far From the Madding Crowd.* Nelson Doubleday, Inc., Garden City. 1900.

Harrison, Edward. *Cosmology: The Science of the Universe.* Cambridge University Press, Cambridge. 2000.

Harrison, Edward. *Masks of the Universe.* Macmillan Publishing Company, New York. 1985.

Hastie, W. *Kant's Cosmogony.* James Maclehouse and Sons, Glasgow. 1900.

Haught, James A. *2000 Years of Disbelief.* Prometheus Books, Amherst. 1996.

Hawking, Stephen. *A Brief History of Time.* Bantam Books, New York. 1988.

Hawking, Stephen and Penrose, Roger. *The Nature of Space and Time.* Princeton University Press, Princeton. 1996.

Hawkins, Michael. *Hunting Down the Universe*. Addison-Wesley, Reading. 1977.

Hearn, Lafcadio. *The Writings of Lafcadio Hearn*. Volume VIII. Houghton Mifflin Company, Boston. 1923.

Hecht, Eugene. *Optics*. Addison-Wesley Publishing Company, Reading. 1987.

Heidegger, Martin. *The Question Concerning Technology and Other Essays*. Translated by William Lovitt. Harper & Row, New York. 1977.

Heidmann, Jean. *Extragalactic Adventure*. Translated by Maureen Schaeffer and Ann Boesgaard. Cambridge University Press, Cambridge. 1982.

Heine, Heinrich. *The Romantic School and Other Essays*. Continuum, New York. 1985.

Heinlein, Robert A. *Stranger in a Strange Land*. G.P. Putnam's Sons, New York. 1991.

Heisenberg, Werner. *Physics and Beyond*. Harper & Row, New York. 1971.

Heisenberg, Werner. *Physics and Philosophy*. Harper & Brothers, New York. 1958.

Heisenberg, Werner. *The Physicist's Conception of Nature*. Translated by Arnold J. Pomerans. Greenwood Press, Westport. 1970.

Hellman. C. Doris. *The Comet of 1577: Its Place in the History of Astronomy*. Columbia University Press, New York. 1944.

Hemans, Felicia. *The Complete Works of Mrs Hemans*. Volume I. D. Appleton & Co., New York. 1863.

Herbert, George. *The Temple*. Medieval & Renaissance Texts & Studies, Binghamton. 1995.

Herschel, J.F.W. *The Cabinet of Natural Philosophy*. Longman, Rees, Orme, Brown, and Green, London. 1831.

Hinshelwood, C.N. *The Structure of Physical Chemistry*. At the Clarendon Press, Oxford. 1951.

Hippocrates. 'The Law' in *Great Books of the Western World*. Volume 10. Encyclopaedia Britannica, Inc., Chicago. 1952.

Hodgson, Leonard. *Theology in an Age of Science*. At the Clarendon Press, Oxford. 1944.

Hodgson, Ralph. *Collected Poems*. Macmillan & Co., London. 1961.

Hoffmann, Roald. *The Metamict State*. University of Central Florida Press, Orlando. 1987.

Hollenbach, D.J. and Thronson, H.A. *Interstellar Processes*. D. Reidel Publishing Company, Dordrecht. 1987.

Holmes, Oliver Wendell. *The Complete Poetical Works of Oliver Wendell Holmes*. Houghton, Mifflin and Company, Boston. 1899.

Holton, Gerald. *Thematic Origins of Scientific Thought: Kepler to Einstein*. Harvard University Press, Cambridge. 1973.

Holton, Gerald and Roller, Duane H.D. *Foundations of Modern Physical Science*. Addison-Wesley Publishing Company, Inc., Reading. 1958.

Homer. 'The Iliad' in *Great Books of the Western World.* Volume 4. Encyclopaedia Britannica, Inc., Chicago. 1952.

Hooke, Robert. *Micrographia.* J. Martyn and J Allestry, London. 1665.

Horowitz, Norman H. *To Utopia and Back: The Search for Life in the Solar System.* W.H. Freeman and Company, New York. 1986.

Hoyle, Fred. *Frontiers of Astronomy.* Harper & Brothers, New York. 1955.

Hoyle, Fred. *Home is Where the Wind Blows.* University Science Books, Mill Valley. 1994.

Hoyle, Fred. *The Nature of the Universe.* Harper & Brothers, New York. 1950.

Hoyt, William Graves. *Lowell and Mars.* The University of Arizona Press, Tucson. 1976.

Hoyt, William Graves. *Planets X and Pluto.* The University of Arizona Press, Tucson. 1980.

Hubble, Edwin. *The Nature of Science and Other Lectures.* The Huntington Library, San Marino. 1964.

Hubble, Edwin. *The Realm of the Nebulae.* Yale University Press, New Haven. 1936.

Huggins, William. *The Scientific Papers of Sir William Huggins.* W. Wesley, London. 1909.

Hugo, Victor. *Les Misérables.* Translated by Isabel F. Hapgood. Thomas Y. Crowell Company, New York. 1915.

Hutchins, Robert M. and Adler, Mortimer J. *The Great Ideas Today 1974.* Encyclopaedia Britannica, Inc., Chicago. 1974.

Huxley, Aldous. *Brave New World.* Harper & Brothers, New York. 1946.

Huxley, Julian. *The Captive Shrew.* Harper & Brothers, New York. 1933.

Huxley, Thomas. *Collected Essays.* Volume II. D. Appleton and Company, New York. 1912.

Huygens, Christian. *The Celestial Worlds Discover'd.* Printed for Timothy Childs, London. 1698.

Inge, William Ralph. *God and the Astronomers.* Longmans, Green and Co., London. 1934.

Ionesco, Eugene. *Notes and Counter Notes: Writings on the Theatre.* Translated by Donald Watson. Grove Press, Inc., New York. 1964.

Jaki, Stanley L. *Chance or Reality and Other Essays.* University Press of America, Lantham. 1986.

James, William. *The Will to Believe and Other Essays in Popular Philosophy.* Longmans Green and Co., 1899.

James, William. *Writings: 1902–1910.* Library of America, New York. 1987.

Jastrow, Robert. *God and the Astronomers.* W.W. Norton & Company, Inc., New York. 1978.

Jeans, Sir James. *Astronomy and Cosmogony*. Dover Publications, Inc., New York. 1961.
Jeans, Sir James. *Physics and Philosophy*. Dover Publications, Inc., New York. 1981.
Jeans. Sir James. *The Mysterious Universe*. At the University Press, Cambridge. 1930.
Jeans, Sir James. *The New Background of Science*. At the University Press, Cambridge. 1953.
Jeans, Sir James. *The Universe Around Us*. The Macmillan Company, New York. 1929.
Jeffers, Robinson. *The Double Axe*. Liveright, New York. 1977.
Jeffers, Robinson. *The Selected Poetry of Robinson Jeffers*. Random House, New York. 1959.
Jevons, W. Stanley. *The Principles of Science*. Macmillan and Co., London. 1887.
Joad, C.E.M. *Philosophical Aspects of Modern Science*. George Allen & Unwin Ltd, London. 1932.
Jones, H. Bence. *The Life and Letters of Faraday*. Longmans, Green, and Co., London. 1870.
Jones, Sir Harold Spencer. *Life on Other Worlds*. The Macmillan Company, New York. 1954.
Jonson, Ben. *The Poetaster*. D.C. Heath and Company, Boston. 1913.
Joubert, Joseph. *Pensées and Letters of Joseph Joubert*. Translated by H.P. Collins. Book for Libraries Press, Freeport. 1972.
Jourdain, Philip E.B. *The Nature of Mathematics*. T.C. & E.C. Jack, London. nd.
Joyce, James. *Ulysses*. The Bodley Head, London. 1937.

Kahn, Fritz. *Design of the Universe*. Crown Publishers, Inc., New York. 1954.
Kant, Immanuel. *Universal Natural History and Theory of the Heavens*. Translated by Stanley L. Jaki. Scottish Academic Press, Edinburgh. 1981.
Kasner, Edward and Newman, James. *Mathematics and the Imagination*. Simon and Schuster, New York. 1940.
Keats, John. *The Complete Poetical Works of Keats*. Houghton Mifflin Company, Boston. 1899.
Keill, John. *An Introduction to the True Astronomy*. B. Lintot, London. 1721.
Kelley, Kevin W. *The Home Planet*. Addison-Wesley Publishing Company, Reading. 1988.
Kelly, Fred C. *Miracle at Kitty Hawk*. Farrar, Straus and Young, New York. 1951.
Kendall, Phebe Mitchell. *Maria Mitchell: Life, Letters, and Journals*. Lee and Shepard, Boston. 1896.

Kepler, Johann. *Dioptrice*. Translated by E.S. Carlos. London. 1880.

Kepler, Johann. *Kepler's Conversation with Galileo's Sidereal Messenger*. Johnson Reprint Corporation, New York. 1965.

Kepler, Johann. *Mysterium Cosmographicum*. Translated by A.M. Duncan. Abaris Books, New York. 1981.

Kepler, Johann. *New Astronomy*. Translated by William H. Donahue. Cambridge University Press, Cambridge. 1992.

Keyser, Cassius J. *Mathematics*. The Columbia University Press, New York. 1907.

Kiepenheuer, Karl. *The Sun*. The University of Michigan Press, Ann Arbor. 1959.

Kilmer, Joyce. *Main Street and Other Poems*. George H. Doran Company, New York. 1917.

King, Stephen. *The Dark Tower III*. Donald M. Grant, Hampton Falls. 1991.

Kingsley, Charles. *Poems*. J.M. Dent & Sons Ltd, London. 1927.

Kipling, Rudyard. *Just So Stories*. International Collectors Library, Garden City. 1912.

Kirk, G.S., and Raven, J.E. *The Presocratic Philosophers*. At the University Press, Cambridge. 1962.

Kline, Morris. *Mathematics in Western Culture*. Oxford University Press, New York. 1953.

Kline, Morris. *Mathematics: The Loss of Certainty*. Oxford University Press, New York. 1980.

Koestler, Arthur. *Bricks to Babel*. Hutchison, London. 1980.

Koestler, Arthur. *The Sleepwalkers*. Hutchinson of London, London. 1959.

Kone, Eugene H. and Jordan, Helene J. *The Greatest Adventure*. The Rockefeller University Press, New York. 1974.

Körner, T.W. *Fourier Analysis*. Cambridge University Press, Cambridge. 1988.

Kragh, Helge. *Dirac: A Scientific Biography*. Cambridge University Press, Cambridge. 1990.

Kraus, John. *Big Ear*. Cygnus-Quasar Books, Powell. 1976.

Krutch, Joseph Wood. *The Twelve Seasons*. William Sloane Associates, New York. 1949.

Kuhn, Thomas S. *The Structure of Scientific Revolutions*. Second Edition. The University of Chicago Press, Chicago. 1970.

Kundera, Milan. *The Unbearable Lightness of Being*. Translated by Michael Henry Heim. Harper & Row, New York. 1987.

Lahti, Pekka and Mittelstaedt, Peter. *Symposium on the Foundations of Modern Physics*. World Scientific Publishing, Singapore. 1985.

Lamar, René. *Satires and Miscellaneous Poetry and Prose*. Cambridge University Press, Cambridge. 1928.

Lambert, Johann Heinrich. *The System of the World.* Vernor and Hood, London. 1800.

Lanczos, Cornelius. *Linear Differential Operators.* D. Van Nostrand Company Limited, London. 1961.

Langley, Samuel Pierpoint. *The New Astronomy.* Houghton Mifflin and Company, Boston. 1889.

Laplace, Pierre Simon. *A Philosophical Essay on Probabilities.* John Wiley & Sons, New York. 1902.

Laplace, Pierre Simon. *The System of the World.* Volume II. Translated by Henry H. Harte. Longman, Rees, Orme, Brown, and Green, Dublin. 1830.

Laszlo, Ervin. *The Systems View of the World.* George Braziller, New York. 1972.

Lavoisier, Antoine. 'Elements of Chemistry' in *Great Books of the Western World.* Volume 45. Encyclopaedia Britannica, Inc., Chicago. 1952.

Lawrence, D.H. *Apocalypse.* The Viking Press, New York. 1932.

Lawrence, D.H. *Pansies.* Martin Secker, London. 1930.

Lawrence, Louise de Kiriline. *The Lovely and the Wild.* McGraw-Hill Book Company, New York. 1968.

Lee, Oliver Justin. *Measuring Our Universe: From the Inner Atom to Outer Space.* The Ronald Press Company, New York. 1950.

Legge, James. *The Chinese Classics.* Volume I. Agency Publications, New York. 1967.

Levy, David H. *Clyde Tombaugh: Discoverer of Planet Pluto.* The University of Arizona Press, Tucson. 1991.

Levy, David H. *Comets.* Simon & Schuster, New York. 1998.

Lewin, Roger. *Complexity.* Macmillan Publishing Company, New York. 1992.

Lewis, John S. *Physics and Chemistry of the Solar System.* Academic Press, San Diego. 1995.

Lightman, Alan. *Einstein's Dreams.* Pantheon Books, New York. 1993.

Lindberg, Charles. *Autobiography of Values.* Harcourt Brace Jovanovich, New York. 1978.

Litz, Francis A. *The Poetry of Father Tabb.* Dodd, Mead & Company, New York. 1928.

Lockyer, J. Norman. *Elements of Astronomy.* D. Appleton and Company, New York. 1885.

London. Jack. *The Cruise of the Snark.* The Macmillan Company, New York. 1913.

Long, Roger. *Astronomy, in Five Books.* Volume I. Printed for the Author, Cambridge. 1742.

Longfellow, Henry Wadsworth. *The Poetical Works of Henry Wadsworth Longfellow.* Houghton Mifflin and Company, Boston. 1894.

Lowell, Amy. *The Complete Poetical Works of Amy Lowell*. Houghton Mifflin Co., Boston. 1955.

Lowell, Percival. *Mars*. Houghton, Mifflin and Company, Boston. 1895.

Lowell, Percival. *Mars and Its Canals*. The Macmillan Company, New York. 1906.

Lubbock, Sir John. *The Beauties of Nature*. Macmillan and Co., New York. 1893.

Lucretius. 'On the Nature of Things' in *Great Books of the Western World*. Volume 12. Encyclopaedia Britannica, Inc., Chicago. 1952.

Luminet, Jean-Pierre. *Black Holes*. University Press, Cambridge. 1991.

Mach, Ernst. *Analysis of Sensations and the Relation of the Physical to the Psychical*. Dover Publications, Inc., New York. 1959.

Mach, Ernst. *History and Root of the Principle of the Conservation of Energy*. The Open Court Publishing Company, Chicago. 1911.

Mach, Ernst. The *Science of Mechanics*. The Open Court Publishing Co., La Salle. 1942.

Mackay, Charles. *The Collected Songs of Charles Mackay*. G. Routledge & Co., London. 1859.

MacLaurin, Colin. *An Account of Sir Isaac Newton's Philosophical Discoveries*. Printed for the Author's Children, London. 1748.

MacLeish, Archibald. *Riders on the Earth*. Houghton Mifflin Company, Boston. 1978.

MacLeish, Archibald. *Songs for Eve*. Houghton Mifflin Company, Boston. 1954.

MacLennan, Hugh. *Scotchman's Return and other Essays*. Charles Scribner's Sons, New York. 1960.

Macpherson, James. *The Poems of Ossian*. Edward Kearney, New York. nd.

Maffei, Paolo. *Beyond the Moon*. Translated by D.J.K. O'Connell. The MIT Press, Cambridge. 1978.

Mandino, Og. *The Greatest Salesman in the World*. Frederick Fell, Inc., New York. 1968.

Mann, Thomas. *The Magic Mountain*. Translated by H.T. Lowe-Porter. Alfred A. Knopf, New York. 1958.

Manning, Henry Parker. *Geometry of Four Dimensions*. Dover Publications, Inc., 1914.

Marcet, Jane. *Conversations on Chemistry*. Sidney's Press, New Haven. 1814.

Margenau, Henry and Bergamini, David. *The Scientist*. Time Incorporated, New York. 1964.

Margenau, Henry and Varghese, Roy Abraham. *Cosmos, Bios, Theos*. Open Court, La Salle. 1992.

Margulis, Lynn and Sagan, Dorion. *Microcosmos*. Simon & Schuster, New York. 1986.

Maritain, Jacques. *Distinguish to Unite or The Degrees of Knowledge*. Charles Scribner's Sons, New York. 1959.
Marlow, Christopher. *Tamberlaine the Great*. Blandford Press, London. 1948.
Marquardt, Martha. *Paul Ehrlich*. Henry Schuman, New York. 1951.
Marschall, Laurence A. *The Supernova Story*. Plenum Press, New York. 1988.
Martin, Charles Noël. *The Role of Perception in Science*. Translated by A.J. Pomerans. Hutchinson of London, London. 1963.
Matthaei, Rupprecht. *Goethe's Color Theory*. Van Nostrand Reinhold Company, New York. 1971.
Maunder, E. Walter. *Are the Planets Inhabited?* Harper & Brothers, London. 1913.
Maxwell, James Clerk. *The Scientific Papers of James Clerk Maxwell*. Volume II. At the University Press, Cambridge. 1890.
McCormmach, Russell. *Historical Studies in the Physical Sciences*. Fourth Annual Volume. Princeton University Press, Princeton. 1974.
McLennan, Evan. *Cosmical Evolution*. Donohue, Henneberry & Co., Chicago. 1890.
McLuhan, Marshall and Fiore, Quentin. *The Medium is the Message*. Random House, New York. 1967.
McNally, Derek. *The Vanishing Universe*. Cambridge University Press, Cambridge. 1994.
Mead, George Herbert. *The Philosophy of the Act*. The University of Chicago Press, Chicago. 1938.
Medawar, Peter. *Pluto's Republic*. Oxford University Press, Oxford. 1982.
Melville, Herman. 'Moby Dick' in *Great Books of the Western World*. Volume 48. Encyclopaedia Britannica, Inc., Chicago. 1952.
Mencken, H.L. *Prejudices: Third Series*. Alfred A. Knopf, New York. 1922.
Mendeléef, D. *Principles of Chemistry*. Volume I. Translated by George Kamensky. Longmans, Green and Co., London. 1891.
Meredith, George. *A Reading of Earth*. Macmillan, London. 1888.
Meredith, George. *Diana of the Crossways*. Charles Scribner's Sons, New York. 1924.
Meredith, George. *Poems of George Meredith*. Charles Scribner's Sons, New York. 1918.
Meredith, Owen. *The Complete Poetical Works of Owen Meredith*. The American News Company, New York. 1905.
Meyerson, Emile. *Identity & Reality*. Translated by Kate Loewenberg. George Allen & Unwin Ltd, London. 1930.
Michelson, A.A. *Light Waves and Their Uses*. The University of Chicago Press, Chicago. 1903.
Miller, Douglas. *Scientific Studies*. Translated by Douglas Miller. Volume 12. Suhrkamp, New York, Inc. 1988.

Miller, Henry. *Tropic of Cancer*. Grove Press, Inc., New York. 1961.
Miller, Hugh. *Geology Versus Astronomy*. James R. Macnair, Glasgow. 1855.
Miller, James E. *Complete Poetry and Selected Prose*. Houghton, Mifflin Company, Boston. 1959.
Miller, Perry. *The New England Mind*. The Macmillan Company, New York. 1939.
Milne, A.A. *Winnie-the-Pooh*. E.P. Dutton and Co., Inc., New York. 1954.
Milton, John. 'Miscellaneous Poems' in *Great Books of the Western World*. Volume 32. Encyclopaedia Britannica, Inc., Chicago. 1952.
Milton, John. 'Paradise Lost' in *Great Books of the Western World*. Volume 32. Encyclopaedia Britannica, Inc., Chicago. 1952.
Misner, Charles W.; Thorne, Kip S. and John Archibald Wheeler. *Gravitation*. W.H. Freeman and Company, San Francisco. 1973.
Montague, Basil. *The Works*. Volume 3. Parry & MacMillan, Philadelphia. 1854.
Morey, Janet Namoura and Dunn, Wendy. *Famous Asian Americans*. Cobblehill Books, Dutton. 1992.
Moore, James R. *The Post-Darwinian Controversies*. Cambridge University Press, Cambridge. 1979.
Morrell, J.M. *Four English Tragedies*. Penguin Books Ltd, Harmondsworth. 1953.
Moulton, Forest Ray. *Astronomy*. The Macmillan Company, New York. 1931.
Mueller, Bertha. *Goethe's Botanical Writings*. University of Hawaii Press, Honolulu. 1952.
Muir, John. *My First Summer in the Sierra*. Houghton Mifflin Company, Boston. 1911.
Muir, John. *Steep Trails*. Houghton Mifflin Company, Boston. 1918.
Muir, John. *The Wilderness World of John Muir*. Houghton Mifflin Company, Boston. 1954.

Nabokov, Vladimir. *Bend Sinister*. Henry Holt and Company, New York. 1947.
Needham, Joseph. *Science and Civilisation in China*. Volume 3. At the University Press, Cambridge. 1959.
Needham, Joseph and Pagel, Walter. *Background to Modern Science*. The Macmillan Company, New York. 1940.
Newton, Isaac. 'Optics' in *Great Books of the Western World*. Volume 34. Encyclopaedia Britannica, Inc., Chicago. 1952.
Nietzsche, Friedrich. *Thus Spoke Zarathustra*. Translated by R.J. Hollingdale. Penguin Books. Middlesex. 1969.
Nobel Foundation. *Nobel Lectures*. Physics. 1901–1921. Elsevier Publishing Company, Amsterdam. 1965.

Nobel Foundation. *Nobel Lectures*. Physics. 1922–1941. Elsevier Publishing Company, Amsterdam. 1965.
Nobel Foundation. *Nobel Lectures*. Physics. 1942–1962. Elsevier Publishing Company, Amsterdam. 1962.
Nobel Foundation. *Nobel Lectures*. Physics. 1963–1970. Elsevier Publishing Company, Amsterdam. 1972.
Noyes, Alfred. *The Torch-Bearers*. Frederick A. Stokes Company, New York. 1922.

Oliver, David. *The Shaggy Steed of Physics*. Springer, New York. 1994.
Oliver, Mary. *Blue Pastures*. Harcourt Brace & Company, New York. 1991.
Olson, Richard. *Science Deified and Science Defied: The Historical Significance of Western Culture*. University of California Press, Berkeley. 1982.
Orgel, Irene. *The Odd Tales of Irene Orgel*. The Eakins Press, New York. 1966.
Orwell, Sonia and Angus, Ian. *The Collected Essays, Journalism and Letters of George Orwell*. Volume IV. Harcourt, Brace & World, Inc., New York. 1968.
Osborne, John Paskow, David. *Looking Back on Tomorrow*. Addison-Wesley, New York. 1974.
Ovid. *Metamorphoses*. Translated by Rolfe Humphries. Indiana University Press, Bloomington. 1955.

Pagels, Heinz, R. *Perfect Symmetry*. Simon and Schuster, New York. 1985.
Paine, Albert Bigelow. *Mark Twain's Notebook*. Harper & Brothers, New York. 1935.
Paine, Thomas. *The Age of Reason*. Watts & Co., London. 1938.
Palazzo, Tony. *Edward Lear's Nonsense Book*. Garden City Books, Garden City. 1956.
Pallister, William. *Poems of Science*. Playford Press, New York. 1931.
Panek, Richard. *Seeing and Believing*. Viking, New York. 1998.
Parker, E.N. *Cosmical Magnetic Fields*. Clarendon Press, Oxford. 1979.
Pascal, Blaise. *Pensées*. J.M. Dent & Sons Ltd, London. 1901.
Patch, Blanche. *Thirty Years with G.B.S.* Dodd, Mead & Company, New York. 1951.
Patrick, J. Max. *The Complete Poetry of Robert Herrick*. New York University Press, New York. 1963.
Peacock, Thomas Love. *Gryll Grange*. Penguin Books, Harmondsworth. 1949.
Peltier, Leslie C. *Starlight Nights*. Harper & Row, New York. 1965.
Peterson, Ivars. *Islands of Truth: A Mathematical Mystery Cruise*. W.H. Freeman and Company, New York. 1990.
Pippard, A.B. *Elements of Classical Thermodynamics*. At the University Press, Cambridge. 1957.

Planck, Max. *Where Is Science Going?* W.W. Norton & Company, Inc., New York. 1932.

Plato. 'Gorgias' in *Great Books of the Western World.* Volume 7. Encyclopaedia Britannica, Inc., Chicago. 1952.

Plato. 'Phaedo' in *Great Books of the Western World.* Volume 7. Encyclopaedia Britannica, Inc., Chicago. 1952.

Plato. 'The Republic' in *Great Books of the Western World.* Volume 7. Encyclopaedia Britannica, Inc., Chicago. 1952.

Plato. 'Timaeus' in *Great Books of the Western World.* Volume 7. Encyclopaedia Britannica, Inc., Chicago. 1952.

Poe, Edgar. *Eureka.* G.P. Putnam, New York. 1848.

Poincaré, Henri. *The Foundations of Science.* The Science Press, New York. 1913.

Poincaré, Lucien. *The New Physics and Its Evolution.* Paul, Trench, Trübner, London. 1907.

Polkinghorne, J.C. *The Quantum World.* Princeton University Press, Princeton. 1984.

Polya, George. *Mathematical Discovery.* Volume II. John Wiley & Sons, Inc., New York. 1965.

Pope, Alexander. *The Complete Poetical Works of Pope.* Houghton, Mifflin and Co., Boston. 1903.

Popper, Karl. *Conjectures and Refutations.* Harper & Row, New York. 1965.

Popper, Karl. *The Logic of Scientific Discovery.* Basic Books, Inc., New York. 1959.

Popper, Karl. *The Open Universe.* Towman and Littlefield, Totowa. 1982.

Popper, Karl. *The Poverty of Historicism.* The Beacon Press, Boston. 1957.

Pratchett, Terry. *Reaper Man.* Victor Gollancz, Ltd, London. 1991.

Pratchett, Terry. *Witches Abroad.* Victor Gollancz, Ltd, London. 1991.

Price, Lucien. *Dialogues of Alfred North Whitehead.* Little, Brown and Company, Boston. 1954.

Rankine, William John Macquorn. *Songs and Fables.* James Maclehose, Glasgow. 1874.

Raymo, Chet. *The Virgin and the Mousetrap.* Viking, New York. 1991.

Reade, Winwood. *The Martyrdom of Man.* Kegan Paul, Trench, Trubner & Co., Ltd, London. Nd.

Raether, H. *Electron Avalanches and Breakdown in Gases.* Butterworths, London. 1964.

Rees, Martin. *Before the Beginning.* Addison-Wesley, Reading. 1997.

Reeves, Hubert. *Atoms of Silence.* Translated by Ruth A. Lewis and John S. Lewis. The MIT Press, Cambridge. 1984.

Regnault, Pére. *Philosophical Conversations.* Volume I. Printed for W. Innys, London. 1731.

Reichenbach, Hans. *The Rise of Scientific Philosophy.* University of California Press, Berkeley. 1951.

Reid, Thomas. *Essays on the Intellectual Powers of Man.* The MIT Press, Cambridge. 1969.

Reiss, Howard. *Methods of Thermodynamics.* Blaisdell Publishing Company, New York. 1965.

Renan, Ernest. *The Future of Science.* Roberts Brothers, Boston. 1893.

Rexroth, Kenneth. *The Collected Shorter Poems.* New Direction Publishing Corporation, New York. 1966.

Rich, Adrienne. *Leaflets: Poems 1965–1968.* W.W. Norton & Company, New York. 1969.

Rosseland, S. *Theoretical Astrophysics.* At the Clarendon Press, Oxford. 1936.

Rothman, Tony. *A Physicist on Madison Avenue.* Princeton University Press, Princeton. 1991.

Rowan-Robinson, Michael. *Our Universe: An Armchair Guide.* W.H. Freeman and Company, New York. 1990.

Rowland, Henry Augustus. *The Physical Papers of Henry Augustus Rowland.* The Johns Hopkins Press, Baltimore. 1902.

Rukeyser, Muriel. *The Speed of Darkness.* Random House, New York. 1968.

Russell, Bertrand. *An Essay on the Foundations of Geometry.* Dover Publications, Inc., New York. 1956.

Russell, Bertrand. *Mysticism and Logic.* George Allen & Unwin Ltd, London. 1917.

Russell, Bertrand. *Religion and Science.* Henry Holtz and Company, New York. 1935.

Russell, Bertrand. *The Autobiography of Bertrand Russell.* Little, Brown and Company, Boston. 1967.

Russell, Bertrand. *Why I Am Not a Christian.* Simon and Schuster, New York. 1957.

Saaty, Thomas L. *Modern Nonlinear Equations.* McGraw-Hill Book Company, New York. 1967.

Sagan, Carl. *Pale Blue Dot.* Random House, New York. 1994.

Sagan, Carl. *The Cosmic Connection.* Anchor Books, Garden City. 1973.

Sagan, Carl. *The Demon Haunted World.* Random House, New York. 1995.

Saint Augustine. 'Confessions' in *Great Books of the Western World.* Volume 18. Encyclopaedia Britannica, Inc., Chicago. 1952.

Sarlemijn, A. and Sparnaay, M.J. *Physics in the Making.* North-Holland, Amsterdam. 1989.

Sayers, Dorothy L. and Eustace, Robert. *The Documents in the Case.* Victor Gollancz Ltd, London. 1978.

Schaaf, Fred. *The Starry Room.* John Wiley & Sons, Inc., New York. 1988.

Schlipp, Paul Arthur. *Albert Einstein: Philosopher–Scientist*. The Library of Living Philosophers, Inc., Evanston. 1949.

Schmidt, Franz and Stäckel, Paul. *Briefwechsel zwischen Carl Friedrich Gauss und Wolfgang Bolyai*. B.G. Teubner, Leipzig. 1899.

Schwarzschild, Martin. *Structure and Evolution of the Stars*. Dover Publications, Inc., New York. 1958.

Scott, R. *Scott's Last Expedition*. Beacon Press, Boston. 1957.

Scott, Sir Walter. *The Complete Poetical Work of Sir Walter Scott*. Houghton Mifflin Company, Boston. 1900.

Seneca. *Physical Science in the Time of Nero*. Macmillan, London. 1910.

Serrano, Miguel. *C.G. Jung and Hermann Hesse*. Schocken Books Inc., New York. 1966.

Service, Robert. *Collected Poems of Robert Service*. Dodd, Mead & Company, New York. 1961.

Shakespeare, William. 'A Midsummer-Night's Dream' in *Great Books of the Western World*. Volume 26. Encyclopaedia Britannica, Inc., Chicago. 1952.

Shakespeare, William. 'Hamlet, Prince of Denmark' in *Great Books of the Western World*. Volume 27. Encyclopaedia Britannica, Inc., Chicago. 1952.

Shakespeare, William. 'Julius Caesar' in *Great Books of the Western World*. Volume 26. Encyclopaedia Britannica, Inc., Chicago. 1952.

Shakespeare, William. 'King Lear' in *Great Books of the Western World*. Volume 27. Encyclopaedia Britannica, Inc., Chicago. 1952.

Shakespeare, William. 'Love's Labour Lost' in *Great Books of the Western World*. Volume 26. Encyclopaedia Britannica, Inc., Chicago. 1952.

Shakespeare, William. 'Othello, The Moor of Venice' in *Great Books of the Western World*. Volume 27. Encyclopaedia Britannica, Inc., Chicago. 1952.

Shakespeare, William. 'The First Part of King Henry the Fourth' in *Great Books of the Western World*. Volume 26. Encyclopaedia Britannica, Inc., Chicago. 1952.

Shakespeare, William. 'The First Part of King Henry the Sixth' in *Great Books of the Western World*. Volume 26. Encyclopaedia Britannica, Inc., Chicago. 1952.

Shakespeare, William. 'The Merchant of Venice' in *Great Books of the Western World*. Volume 26. Encyclopaedia Britannica, Inc., Chicago. 1952.

Shakespeare, William. 'The Second Part of King Henry the Sixth' in *Great Books of the Western World*. Volume 26. Encyclopaedia Britannica, Inc., Chicago. 1952.

Shapley, Harlow. *Of Stars and Men*. Beacon Press, Boston. 1964.

Shapley, Harlow. *The View from a Distant Star*. Dell Publishing Co., New York. 1963.

Shapley, Harlow, Wright, Helen and Samuel Rapport. *Readings in the Physical Sciences*. Appleton-Century-Crofts, Inc., New York. 1948.

Shelley, Percy Bysshe. *The Complete Poetical Works of Shelley*. Houghton Mifflin Company, Boston. 1901.

Sherrod, P. Clay. *A Complete Manual of Amateur Astronomy*. Prentice-Hall, Inc., Englewood Cliffs. 1981.

Shore, Jane. *Eye Level*. The University of Massachusetts Press, Amherst. 1977.

Siegel, Eli. *Damned Welcome*. Definition Press, New York. 1972.

Silesius, Angelus. *The Book of Angelus Silesius*. Translated by Frederick Franck. Alfred A. Knopf, New York. 1976.

Sillman, Benjamin. *Elements of Chemistry*. Volume I. Hezekiah Howe, New Haven. 1830.

Sime, James. *William Herschel and His Work*. C. Scribner's Sons, New York. 1900.

Singer, Isaac Basheivs. *A Crown of Feathers*. Farrar, Straus and Giroux, New York. 1973.

Skinner, B.F. *Cumulative Record*. Appleton–Century–Crofts, Inc., New York. 1959.

Smith, Logan Pearsall. *Little Essays*. Books for Libraries Press, Freeport. 1967.

Smith, Logan Pearsall. *Trivia*. Doubleday, Page & Company, Garden City. 1917.

Smolin, Lee. *The Life of the Cosmos*. Oxford University Press, New York. 1997.

Smoot, George and Davidson, Keay. *Wrinkles in Time*. William Morrow and Company, Inc., New York. 1993.

Snow, C.P. *The Two Cultures: And a Second Look*. At the University Press, Cambridge. 1964.

Soddy, Frederick. *Science and Life*. John Murray, London. 1920.

Spenser, Edmund. *The Complete Works in Verse and Prose of Edmund Spenser*. Printed for the Spenser Society, Manchester. 1882–1884.

Sprat, Thomas. *The History of the Royal-Society of London*. Printed by T.R., Printers to the Royal Society. 1667.

Standage, Tom. *The Neptune File*. The Penguin Press, London. 2000.

Standen, Anthony. *Science is a Sacred Cow*. E.P. Dutton and Company, Inc., New York. 1950.

Stapledon, Olaf. *Last and First Men and Star Maker*. Dover Publications, Inc., New York. 1968.

Starchild, Adam. *The Science Fiction of Konstantin Tsiolkovsky*. University Press of the Pacific, Inc., Seattle. 1979.

Steele, J. Dorman. *Popular Physics*. American Book Company, New York. 1896.

Steinbeck, John. *The Log from the Sea of Cortez.* Bantam Books, New York. 1971.

Stenger, Victor J. *Physics and Psychics.* Prometheus Books, Buffalo. 1990.

Stephenson, F. Richards. *Historical Eclipses and Earth's Rotation.* Cambridge University Press, Cambridge. 1997.

Stern, Laurence. *The Life & Opinions of Tristram Shandy.* The Heritage Press, New York. 1935.

Stern, S. Alan. *Our Universe.* Cambridge University Press, Cambridge. 2001.

Stevenson, Robert Louis. *Travels with a Donkey in the Cevennes.* Wm. H. Wise & Co., New York. 1928.

Stoll, Clifford. *The Cuckoo's Egg.* Doubleday, New York. 1989.

Stoppard, Tom. *Arcadia.* Faber and Faber, London. 1993.

Struve, Otto. *The Universe.* The MIT Press, Cambridge. 1962.

Sturluson, Snorri. *The Prose Edda.* Translated by Arthur Gilchrist Brodeur. The American–Scandinavian Foundation, New York. 1929.

Sullivan, J.W.N. *The Bases of Modern Science.* Doubleday, Doran & Company, Inc., Garden City. 1929.

Sullivan, W.T. *The Early Years of Radio Astronomy.* Cambridge University Press, Cambridge. 1984.

Sullivan, W.T. *Classics in Radio Astronomy.* D. Reidel Publishing Company, Dordrecht. 1982.

Swann, W.F.G. *The Architecture of the Universe.* The Macmillan Company, New York. 1934.

Swift, Jonathan. 'Gulliver's Travels' in *Great Books of the Western World.* Volume 36. Encyclopaedia Britannica, Inc., Chicago. 1952.

Swinburne, Richard. *The Existence of God.* At the Clarendon Press, Oxford. 1979.

Tagore, Rabindranath. *Our Universe.* Translated by Indu Dutt. Jaico Publishing House, Bombay. 1969.

Taube, M. *Evolution of Matter and Energy.* Springer, New York. 1985.

Taylor, Bayard. *The Poetical Works of Bayard Taylor.* Houghton, Mifflin and Company, Boston. 1882.

Taylor, Edwin F. and Wheeler, John Archibald. *Spacetime Physics.* W.H. Freeman and Company, San Francisco. 1966.

Teasdale. *The Collected Poems of Sara Teasdale.* The Macmillan Company, New York. 1937.

Teilhard de Chardin, Pierre. *The Future of Man.* Translated by Norman Denny. Harper & Row, New York. 1964.

Teilhard de Chardin, Pierre. *The Phenomenon of Man.* Harper & Brothers, New York. 1959.

Tennyson, Alfred. *The Complete Poetical Works of Tennyson.* Houghton, Mifflin and Company, Boston. 1898.

Thom, René. *Structural Stability and Morphogenesis.* Translated by D.H. Fowler. W.A. Benjamin, Inc., Reading. 1975.

Thomas, Dylan. *The Poems of Dylan Thomas.* New Directions Publishing Company, New York. 1971.

Thomas, Lewis. *Late Night Thoughts on Listening to Mahler's Ninth Symphony.* The Viking Press, New York. 1983.

Thompson, Francis. *Complete Poetical Works of Francis Thompson.* Boni and Liveright, Inc., New York. nd.

Thompson, Francis. *Shelley.* Burns and Oates, London. 1909.

Thompson, W.R. *Science and Common Sense.* Longmans Green and Co., London. 1937.

Thomson, George. *The Atom.* Oxford University Press, London. 1956.

Thomson, George. *The Inspiration of Science.* Oxford University Press, London. 1961.

Thomson, J. Arthur. *Concerning Evolution.* Yale University Press, New Haven. 1925.

Thomson, James. *Seasons.* Phillips, Sampson, and Company, Boston. 1850.

Thoreau, Henry David. *The Writings of Henry David Thoreau.* Volume 4. Houghton, Mifflin and Company, Boston. 1894–95.

Thoreau, Henry David. *The Writings of Henry David Thoreau.* Volume 5. Houghton, Mifflin and Company, Boston. 1894–95.

Thoreau, Henry David. *The Writings of Henry David Thoreau.* Volume 6. Houghton, Mifflin and Company, Boston. 1894–95.

Thoreau, Henry David. *Walden.* Time Incorporated, New York.1962.

Thorne, Kip. *Black Holes and Time Warps.* W.W. Norton & Company, New York. 1994.

Tikhonravov, M.K. *Works on Rocket Technology by K.E. Tsiolkovsky.* NASA translation TT F-243. National Aeronautics and Space Administration, Washington, D.C. 1965.

Tille, Alexander. *The Works of Friedrich Nietzsche.* Translated by Thomas Common. Volume XI. The Macmillan Company, New York. 1908.

Tolman, Richard C. *Relativity, Thermodynamics and Cosmology.* At the Clarendon Press, Oxford. 1934.

Tolstoy, Leo. 'War and Peace' in *Great Books of the Western World.* Volume 51. Encyclopaedia Britannica, Inc., Chicago. 1952.

Toogood, Hector B. *The Outline of Everything.* Little, Brown, and Company, Boston. 1923.

Toulmin, Stephen. *The Philosophy of Science.* Hutchinson's University Library, London. 1953.

Toulmin, Stephen and Goodfield, June. *The Architecture of Matter.* Harper & Row, New York. 1962.

Toynbee, Arnold. *Lectures on the Industrial Revolution of the 18th Century in England.* Rivingtons, London. 1887.

Travers, P.L. *Mary Poppins.* Harcourt, Brace & World, Inc., New York. 1963.

Trevelyan, George Macaulay. *Clio, A Muse and Other Essays Literary and Pedestrian*. Longmans, Green and Co., London. 1913.

Truesdell, C. *The Tragicomical History of Thermodynamics: 1822–1854*. Springer, New York. 1980.

Tucker, Abraham. *The Light of Nature Pursued*. Volume I. Hilliard and Brown, Cambridge. 1831.

Twain, Mark. *Following the Equator*. Volume I. The Ecco Press, Hopewell. 1992.

Twain. Mark. *Letters From the Earth*. Harper & Row, New York. 1962.

Twain, Mark. *The Complete Works of Mark Twain*. Volume 14. Harper & Brothers, New York. 1911.

Twain, Mark. *The Complete Works of Mark Twain*. Volume 24. Harper & Brothers, New York. 1911.

Twain, Mark. *The Dairies of Adam and Eve*. Oxford University Press, New York. 1996.

Twain, Mark. *The Mark Twain Papers*. University of California Press, Berkeley. 1995.

Tyndall, John. *New Fragments*. Longmans, Green, London. 1892.

Umbgrove, J.H.F. *The Pulse of the Earth*. Martinus Nijhoff, The Hague. 1947.

Updike, John. *Collected Poems 1953–1993*. Alfred A. Knopf, New York. 1993.

Updike, John. *The Poorhouse Fair*. The Modern Library, New York. 1965.

Valéry, Paul. *The Collected Works of Paul Valéry*. Edited by Jackson Mathews. Volume 14. Princeton University Press, Princeton. nd.

van Laarhoven, Jan. *Entheticus Maior and Minor*. Volume 1. E.J. Brill, Leiden. 1987.

Vaughan, Henry. *Poetry and Selected Prose*. Oxford University Press, London. 1963.

vas Dias, Robert. *Inside Outer Space: New Poems of the Space Age*. Doubleday & Company, Inc., Garden City. 1970.

Veblen, Horstein. *The Place of Science in Modern Civilization and Other Essays*. Transaction Publishers, New Brunswick. 1990.

Vehrenberg, Hans. *Atlas of Deep Sky Splendors*. Sky Publishing Company, Cambridge. 1978.

Velikovsky, Immanuel. *Earth in Upheaval*. Dell Publishing Co., Inc., New York. 1968.

Verne, Jules. *A Journey to the Cent4er of the Earth*. The Limited Editions Club, New York. 1966.

Verne, Jules. *From Earth to the Moon*. A.L. Burt Company, New York. 1890.

Virgil. *The Works of Virgil*. Translated by John Dryden. Oxford University Press, London. 1906.

Vizinczey, Stephen. *Truth and Lies in Literature.* The Atlantic Monthly Press, Boston. 1986.

von Goethe, Johann Wolfgang. 'Faust' in *Great Books of the Western World.* Volume 47. Encyclopaedia Britannica, Inc., Chicago. 1952.

von Humboldt, Alexander. *Cosmos.* Volume 1. Translated by E.C. Otté. Henry G. Bohn, London. 1848.

von Schelling, F.W.J. *Ideas for a Philosophy of Nature.* Translated by Errol E. Harris and Peter Heath. Cambridge University Press, Cambridge. 1988.

Walker, Kenneth. *Meaning and Purpose.* Jonathan Cape, London. 1944.

Warner, Aaron W.; Morse, Dean and Cooney, Thomas E. *The Environment of Change.* Columbia University Press, New York. 1969.

Weaver, Warren. *Science and Imagination.* Basic Books, Inc., New York. 1967.

Weber, Brom. *The Complete Poems and Selected Letters and Prose of Hart Crane.* Liveright Publishing Corporation, New York. 1966.

Weil, Simone. *Gravity and Grace.* Translated by Arthur Willis. G.P. Putnam's Sons, New York. 1952.

Weinberg, Steven. *Dreams of a Final Theory.* Vintage Books, New York. 1993.

Weinberg, Steven. *Gravitation and Cosmology: Principles and Applications of the General Theory of Relativity.* John Wiley & Sons, New York. 1972.

Weinberg, Steven. *The First Three Minutes.* Basic Books, Inc., New York. 1977.

Weisskopf, Victor F. *Physics in the Twentieth Century: Selected Essays.* The MIT Press, Cambridge. 1972.

Wells, H.G. *Seven Famous Novels by H.G. Wells.* Alfred A. Knopf, New York. 1934.

Wells, H.G. *Tono-Bungay.* The Limited Editions Club, New York. 1960.

Weyl, Hermann. *Philosophy of Mathematics and Natural Science.* Princeton University Press, Princeton. 1949.

Weyl, Hermann. *Symmetry.* Princeton University Press, Princeton. 1952.

Weyl, Hermann. *Space–Time–Matter.* E.P. Dutton and Company, New York. 1921.

Wheeler, John Archibald. *A Journey into Gravity and Spacetime.* Scientific American Library, New York. 1990.

Wheeler, John Archibald and Zurek, Wojciech Hubert. *Quantum Theory and Measurement.* Princeton University Press, Princeton. 1983.

Whewell, William. *Novum Organon Renovatium.* John W. Parker and Son, London. 1858.

Whewell, William. *The Philosophy of the Inductive Sciences.* Volume II. Johnson Reprint Corporation, New York. 1967.

Whipple, Fred L. *Earth, Moon, and Planets.* Harvard University Press, Cambridge. 1968.

Whitehead, Alfred North. *Adventures of Ideas.* At the University Press, Cambridge. 1961.

Whitehead, Alfred North. *An Introduction to Mathematics.* Oxford University Press, New York. 1948.

Whitehead, Alfred North. *Modes of Thought.* At the University Press, Cambridge. 1938.

Whitehead, Alfred North. *Nature and Life.* Greenwood Press, New York. 1934.

Whitehead, Alfred North. *Science and the Modern World.* The Macmillian Company, New York. 1929.

Whitehead, Alfred North. *The Aims of Education.* The Macmillan Company, New York. 1929.

Whitrow, G.J. *The Structure and Evolution of the Universe.* Hutchinson of London, London. 1959.

Wickham, Anna. *The Contemplative Quarry.* London. 1915.

Wiener, Philip P. and Noland, Aaron. *Roots of Scientific Thought.* Basic Books, New York. 1957.

Wilde, Oscar. *Oscar Wilde Selected Writing.* Oxford University Press, London. 1961.

Wilde, Oscar. *The Picture of Dorian Gray.* The Modern Library, New York. 1992.

Wilfred, John Noble. *We Reach the Moon.* W.W. Norton & Company, New York. 1971.

Wilson, Edward O. *Consilience: The Unity of Knowledge.* Alfred A. Knopf, New York. 1998.

Wisdom, John Oulton. *Foundations of Inference in Natural Science.* Methuen & Co., Ltd, London. 1952.

Wolf, Fred Alan. *Parallel Universes.* Simon and Schuster, New York. 1988.

Wolfe, Linnie Marsh. *John of the Mountains.* Houghton Mifflin Company, Boston. 1938.

Woodruff, L.L. *The Development of the Sciences.* Yale University Press, New Haven. 1923.

Wordsworth, William. *The Complete Poetical Works of Wordsworth.* Houghton Mifflin Company, Boston. 1904.

Wright, Helen. *Palomar: The World's Largest Telescope.* The Macmillan Company, New York. 1952.

Wright, Helen. *Sweeper in the Sky.* The Macmillan Company, New York. 1949.

Yeats, William Butler. *Collected Poems.* Macmillan, New York. 1942.

Young, J.Z. *Doubt and Certainty in Science.* Oxford University Press, New York. 1960.

Young, Louise B. *The Unfinished Universe.* Simon and Schuster, New York. 1986.

Zee, A. *Fearful Symmetry.* Macmillian Publishing Company, New York. 1986.

Ziman, J.M. *Public Knowledge: An Essay Concerning the Social Dimension of Science.* At the University Press, Cambridge. 1968.

Ziman, John. *Reliable Knowledge.* Cambridge University Press, Cambridge. 1978.

Zirin, Harold. *Astrophysics of the Sun.* Cambridge University Press, Cambridge. 1988.

Zukav, Gary. *The Dancing Wu Li Masters.* William Morrow, New York. 1979.

PERMISSIONS

Grateful acknowledgment is made to the following for their kind permission to reprint copyright material. Every effort has been made to trace copyright ownership but if, inadvertently, any mistake or omission has occurred, full apologies are herewith tendered.

Full reference to authors and the titles of their works are given under the appropriate quotations.

A FREE ENQUIRY INTO THE VULGARLY RECEIVED NOTION OF NATURE by Robert Boyle. Copyright 1996. Reprinted by permission of the publishers, Cambridge University Press, Cambridge, UK.

A MATHEMATICIAN'S APOLOGY by G.H. Hardy. Copyright 1969. Reprinted by permission of the publishers, Cambridge University Press, Cambridge, UK.

ADVENTURES OF IDEAS by Alfred North Whitehead. Copyright 1961. Reprinted by permission of the publishers, at the University Press, Cambridge, UK.

ASTROPHYSICS OF THE SUN by Harold Zirin. Copyright 1988. Reprinted by permission of the publishers, Cambridge University Press, Cambridge, UK.

ATOMIC THEORY AND THE DESCRIPTION OF NATURE by Neils Bohr. Copyright 1961. Reprinted by permission of the publishers, at the University Press, Cambridge, UK.

ATOMS OF SILENCE by Hurbert Reeves. Translated by Ruth A. Lewis and John S. Lewis. Copyright 1984. Reprinted by permission of the publishers, The MIT Press, Cambridge, MA.

BEYOND THE MOON by Paolo Maffei. Translated by D.J.K. O'Connel. Copyright 1978. Reprinted by permission of the publishers, The MIT Press, Cambridge, MA.

BLACK HOLES by Jean-Pierre Luminet. Copyright 1991. Reprinted by permission of the publishers, at the University Press, Cambridge, UK.

COSMOLOGY: THE SCIENCE OF THE UNIVERSE by Edward Harrison. Copyright 2000. Reprinted by permission of the publishers, Cambridge University Press, Cambridge, UK.

CRITICAL THINKING by Max Black. Copyright 1952. Reprinted by permission of the publishers, Pearson Education, Inc., Upper Saddle River.

DIRAC: A SCIENTIFIC BIBLIOGRAPHY by Helge Kragh. Copyright 1990. Reprinted by permission of the publishers, Cambridge University Press, Cambridge, UK.

DUST IN THE UNIVERSE by M.E. Bailey and D.A. Williams. Copyright 1988. Reprinted by permission of the publishers, Cambridge University Press, Cambridge, UK.

EARTH, MOON, AND PLANETS by Fred L. Whipple. Copyright 1941, 1963, 1968 by the President and Fellows of Harvard College. Copyright renewed 1986, 1991 by Fred L. Whipple.

EINSTEIN AS MYTH AND MUSE by Alan J. Friedman and Carol C. Donley. Copyright 1985. Reprinted by permission of the publishers, Cambridge University Press, Cambridge, UK.

ELEMENTS OF CLASSICAL THERMODYNAMICS by A.B. Pippard. Copyright 1957. Reprinted by permission of the publishers, at the University Press, Cambridge, UK.

EXPERIMENT AND THEORY IN PHYSICS by Max Born. Copyright 1944. Reprinted by permission of the publishers, at the University Press, Cambridge, UK.

EXTRAGALACTIC ADVENTURE by Jean Heidmann. Copyright 1982. Reprinted by permission of the publishers, Cambridge University Press, Cambridge, UK.

FOURIER ANALYSIS by T.W. Körner. Copyright 1988. Reprinted by permission of the publishers, Cambridge University Press, Cambridge, UK.

FROM CARDINALS TO CHAOS by Necia Grant Cooper. Copyright 1989. Reprinted by permission of the publishers, Cambridge University Press, Cambridge, UK.

GAY-LUSSAC: SCIENTIST AND BOURGEOIS by Maurice Grossland. Copyright 1978. Reprinted by permission of the publishers, Cambridge University Press, Cambridge, UK.

HISTORICAL ECLIPSES AND THE EARTH'S ROTATION by F. Richards Stephenson. Copyright 1997. Reprinted by permission of the publishers, Cambridge University Press, Cambridge, UK.

I WANT TO BE A MATHEMATICIAN by Paul R. Halmos. Copyright 1985. Reprinted by permission of the publishers, Springer, New York.

IDEAS FOR A PHILOSOPHY OF NATURE by F.W.J. von Schelling. Copyright 1988. Reprinted by permission of the publishers, Cambridge University Press, Cambridge, UK.

INTRODUCTION TO COMETS by John C. Brandt and Robert D. Chapman. Copyright 1981. Reprinted by permission of the publishers, Cambridge University Press, Cambridge, UK.

LITTLEWOOD'S MISCELLANY by Béla Bollobás. Copyright 1986. Reprinted by permission of the publishers, Cambridge University Press, Cambridge, UK.

MIND AND MATTER by Erwin Schrödinger. Copyright 1958. Reprinted by permission of the publishers, at the University Press, Cambridge, UK.

MODES OF THOUGHT by Alfred North Whitehead. Copyright 1938. Reprinted by permission of the publishers, at the University Press, Cambridge, UK.

MORE HEAT THAN LIGHT by Philip Mirowski. Copyright 1989. Reprinted by permission of the publishers, Cambridge University Press, Cambridge, UK.

NEW ASTRONOMY by Johann Kepler. Copyright 1992. Reprinted by permission of the publishers, Cambridge University Press, Cambridge, UK.

NEW PATHWAYS IN SCIENCE by Sir Arthur Eddington. Copyright 1935. Reprinted by permission of the publishers, at the University Press, Cambridge, UK.

OLD MYTHS AND NEW REALITIES by J. William Fulbright. Copyright 1964. Reprinted by permission of the publishers, Random House, New York.

ON KNOWING: ESSAYS FOR THE LEFT HAND by Jerome S. Bruner. Copyright 1962, 1979. Reprinted by permission of the publishers, Harvard University Press, Cambridge, MA.

OUR UNIVERSE by S. Alan Stern. Copyright 2001. Reprinted by permission of the publishers, Cambridge University Press, Cambridge, UK.

PATTERNS OF DISCOVERY by Norwood Russell Hanson. Copyright 1965. Reprinted by permission of the publishers, at the University Press, Cambridge, UK.

PHILOSOPHY IN A NEW KEY: A STUDY IN THE SYMBOLISM OF REASON, RITE, AND ART by Susan K. Langer. Copyright 1942, 1951, 1957 by the President and Fellows of Harvard College, Renewed 1970, 1979 by Susan K. Langer, 1985 by Leonard C.R. Langer.

PHYSICS IN THE TWENTIETH CENTURY: SELECTED ESSAYS by Victor F. Weisskopf. Copyright 1972. Reprinted by permission of the publishers, The MIT Press, Cambridge, MA.

PHYSICS, THE ELEMENTS by Norman Campbell. Copyright 1920. Reprinted by permission of the publishers, at the University Press, Cambridge, UK.

PROOFS AND REFUTATIONS by Imre Lakatos. Copyright 1976. Reprinted by permission of the publishers, Cambridge University Press, Cambridge, UK.

PSYCHOLOGICAL RESEARCH by Arthur Bachrach. Copyright 1972. Reprinted by permission of the publishers, Random House, New York.

PUBLIC KNOWLEDGE: AN ESSAY CONCERNING THE SOCIAL DIMENSION OF SCIENC by J.M. Ziman. Copyright 1968. Reprinted by permission of the publishers, at the University Press, Cambridge, UK.

RELIABLE KNOWLEDGE by John Ziman. Copyright 1978. Reprinted by permission of the publishers, Cambridge University Press, Cambridge, UK.

SATIRES AND MISCELLANEOUS POETRY AND PROSE by René Lamar. Copyright 1928. Reprinted by permission of the publishers, Cambridge University Press, Cambridge, UK.

SCIENCE AND CIVILISATION IN CHINA by Joseph Needham. Volume 3. Copyright 1959. Reprinted by permission of the publishers, at the University Press, Cambridge, UK.

SPEAKABLE AND UNSPEAKABLE IN QUANTUM MECHANICS by J.S. Bell. Copyright 1987. Reprinted by permission of the publishers, Cambridge University Press, Cambridge, UK.

SUPERSTRINGS: A THEORY OF EVERYTHING? by P.C.W. Davies and Julian Brown. Copyright 1988. Reprinted by permission of the publishers, Cambridge University Press, Cambridge, UK.

THE COLLECTED MATHEMATICAL PAPERS OF ARTHUR CAYLEY by Arthur Cayley. Copyright 1896. Reprinted by permission of the publishers, at the University Press, Cambridge, UK.

THE EARLY YEARS OF RADIO ASTRONOMY by W.T. Sullivan. Copyright 1984. Reprinted by permission of the publishers, Cambridge University Press, Cambridge, UK.

THE FRENCH PARACELSIANS by Allen G. Debus. Copyright 1991. Reprinted by permission of the publishers, Cambridge University Press, Cambridge, UK.

THE IMMENSE JOURNEY by Loren Eiseley. Copyright 1957. Reprinted by permission of the publishers, Random House, New York.

THE INTERNAL CONSTITUTION OF THE STARS by Sir Arthur Stanley. Copyright 1926. Reprinted by permission of the publishers, at the University Press, Cambridge, UK.

THE MATHEMATICAL THEORY OF RELATIVITY by Sir Arthur Stanley. Copyright 1954. Reprinted by permission of the publishers, at the University Press, Cambridge, UK.

THE MEDIUM IS THE MESSAGE by Marshall McLuhan and Quentin Fiore. Copyright 1967. Reprinted by permission of the publishers, Random House, New York.

THE MYSTERIOUS UNIVERSE by Sir James Jeans. Copyright 1930. Reprinted by permission of the publishers, at the University Press, Cambridge, UK.

THE NATURAL HISTORY OF THE SENSES by Diane Ackerman. Copyright 1990. Reprinted by permission of the publishers, Random House, New York.

THE NEW BACKGROUND OF SCIENCE by Sir James Jeans. Copyright 1953. Reprinted by permission of the publishers, at the University Press, Cambridge, UK.

THE POST-DARWINIAN CONTROVERSIES by James R. Moore. Copyright 1979. Reprinted by permission of the publishers, at the University Press, Cambridge, UK.

THE SHAGGY STEED OF PHYSICS by David Oliver. Copyright 1994. Reprinted by permission of the publishers, Springer, New York.

THE TRAGICOMICAL HISTORY OF THERMODYNAMICS: 1822–1854 by C. Truesdell. Copyright 1980. Reprinted by permission of the publishers, Springer, New York.

THE TWO CULTURES: AND A SECOND LOOK by C.P. Snow. Copyright 1964. Reprinted by permission of the publishers, Cambridge University Press, Cambridge, UK.

THE UNIVERSE by Otto Struve. Copyright 1962. Reprinted by permission of the publishers, The MIT Press, Cambridge, MA.

THE VANISHING UNIVERSE by Derek McNally. Copyright 1994. Reprinted by permission of the publishers, Cambridge University Press, Cambridge, UK.

THE WAY THINGS ARE by P.W. Bridgman. Copyright 1959. Reprinted by permission of the publishers, Harvard University Press, Cambridge, MA.

THE PRESOCRATIC PHILOSOPHERS by G.S. Kirk and J.E. Raven. Copyright 1962. Reprinted by permission of the publishers, at the University Press, Cambridge, UK.

URNE BURIALL AND THE GARDEN OF CYRUS by John Carter. Copyright 1958. Reprinted by permission of the publishers, at the University Press, Cambridge, UK.

VISUAL ASTRONOMY OF THE DEEP SKY by Roger N. Clark. Copyright 1990. Reprinted by permission of the publishers, Cambridge University Press, Cambridge, UK.

SUBJECT BY AUTHOR INDEX

accurate
Mitchell, Maria
...might become wonderfully accurate in results..., 1
age
Dirac, P.A.M.
Age is of course a fever chill..., 2
Eliot, George
The young ones always have a claim to the old..., 2
Hardy, G.H.
...mathematics...is a young man's game, 2
Yudowitch, K.L.
...work done by a man at very nearly their own age, 2
alien
de Saint-Exupéry, Antoine
Which is your planet?, 115
Diamond, Jared
If there really are any radio civilizations..., 115
Dietrich, Marlene
Until they come to see us from their planet..., 117
Koch, Howard
...something's wriggling out of the shadow like a grey snake, 119
Sagan, Carl
Questions to Ask an Alien,
The civilizations vastly more advanced than we..., 121
Wells, H.G.
Those who have never seen a living Martian..., 123
analogy
Bernstein, Jeremy
...all theoretical physics proceeds by analogy, 3
Campbell, Norman Robert
...analogies are not "aids" to the establishment of theories..., 3
Heinlein, Robert
Analogy is even slipperier than logic, 3
Melville, Herman
...beyond all autterance are your linked analogies!, 3
Andromeda
Keats, John
Andromeda! Sweet woman!, 70
Kingsley, Charles
High for a star in the heavens..., 70
anti-chance
Eddington, Sir Arthur Stanley
We have swept away the anti-chance..., 55
arbitrary
Poincaré, Henri
Conventions yes, 4
Arcturus
Dickinson, Emily

"Arcturus" is his other name. . . , 70
Teasdale, Sara
Arcturus, bringer of spring. . . , 70
Job 28:32
Canst thou guide Arcturus. . . , 346
Aries
Longfellow, Henry Wadsworth
And the Ram that bore unsafely the burden of Helle. . . , 71
asteroid
Asphaung, Erik
Neither rocks nor planets. . . , 5
astrology
Byron, George Gordon
. . . we would read the fate/Of men and empires. . . , 6
Durant, Will
. . . astrology antedated-and perhaps will survive-astronomy. . . , 6
Emerson, Ralph Waldo
Astronomy to the selfish becomes astrology, 6
Johnson, Severance
. . . astrology/ For simpletons, 6
Shakespeare, William
The stars above us, govern our condition, 7
The fault. . . is not in the stars. . . , 7
astronaut
Apollo 11
Here men from the planet Earth. . . , 8
Armstrong, Neil
That's one small step for man. . . , 8
Cernan, Gene
I'm on the footpad, Godspeed the crew of Apollo 17, 9
Conrad, Pete
. . . that may have been a small one for Neil. . . , 8
Scott, Dave
Man must explore, 8
Shepherd, Alan
It's been a long way. . . , 8
Swigert, Jack
. . . we've had a problem, 8
Young, John
Apollo 16 is gonna change your image, 8
astronomer
Calder, Nigel
When astronomers express dissatisfaction with. . . , 10
Cunningham, Clifford J.
Today's astronomers live and die by journals and conferences, 10
Donne, John
If then th' Astronomers. . . , 10
Friedman, Herbert
To the astronomer of today. . . , 10
Gibran, Kahlil
The astronomer may speak to you of. . . , 11
Grondal, Florence Armstrong
. . . the fledgling astronomer prods about in the depths of the gloom. . . , 11
Halley, Edmond
. . . to the curious investigators of the stars. . . , 11
Herbert, George
The fleet Astronomer Can bore. . . , 11
Hoyle, Fred
The astronomer seems at first sight to be. . . , 12
Jeans, Sir James

The task of the observational astronomer is..., 12
Jeffers, Robinson
The learned astronomer/Analyzing the light..., 12
Jones, Sir Harold Spencer
The task of the astronomer is..., 12
Keats, John
...some watcher of the skies..., 13
Kühnert, Franz
...on account of the support they give their Astronomers..., 13
Mackay, Charles
O lonely Sage..., 13
Milton, John
...when they come to model heaven..., 13
Mitchell, Maria
I cannot expect to make astronomers..., 13
The Astronomer breaks up starlight..., 13
Osiander, Andrew
It is the job of the astronomer to..., 14
Rees, Martin
Everything astronomers observe turns out to be..., 14
Sayers, Dorothy L.
The astronomer goes back untold millions of years..., 14
Shakespeare, William
These earthly godfathers of heaven's lights..., 15
Stoll, Clifford
The astronomer's rule of thumb..., 15
Thompson, Francis
Starry amorist, starward gone..., 15
Twain, Mark
I do not see how astronomers can help feeling..., 15
...the Christian astronomer has known..., 16
Unknown
Astronomers seem to be able to predict more and more precisely—except..., 16
Walcott, Derek
...I study the stars, 16
astronomer's drinking song
Unknown
...would search the starry sky..., 20
astronomical
Paracelsus
Everything which astronomical theory has searched..., 17
astronomical computation
Boethius
You know from astronomical computation that..., 17
astronomical rhyme
Cook, Joseph
Father's gone star-hunting..., 17
astronomical science
Hoyle, Fred
...the sober facts that have been unearthed by astronomical science, 17
astronomical song
Jedicke, Peter
The stars go nova..., 19
Krisciunas, Kevin
Give me a supernova..., 19
astronomy
Bennett, Arnold
...he was as exquisitely ignorant as of astronomy, 22

Bichot, Xavier
...astronomy is the physiology of the stars, 22
Bronowski, Jacob
Astronomy is not the apex of science or of invention, 22
Burnham, Robert Jr
No one can date that remote epoch when astronomy "began"..., 22
Chargaff, Erwin
...astronomy was probably the first exact science..., 23
Clerke, Agnes M.
[Astronomy] is a science of hairbreadths..., 23
Conrad, Joseph
What do you think of having a go at Astronomy?, 23
Emerson, Ralph Waldo
...astronomy induces a dignity of mind..., 23
Emerson, William
Astronomy is the science which treats..., 23
Huxley, Julian
Reel after reel is all astronomy..., 24
Keill, John
...been always given to Astronomy, 24
Laplace, Pierre Simon
Astronomy...is the most beautiful monument of..., 24
Long, Roger
Astronomy is a science which..., 24
Mitchell, Maria
Astronomy is not star gazing, 25
Murdin, Paul
The aims of astronomy are nothing less than..., 25
Newcomb, Simon
[astronomy] seems to have the strongest hold on minds which are..., 25
Penrose, Roger
Nature does not always prefer conventional explanations...in astronomy, 25
Penzias, Arno
Astronomy leads us to a unique event..., 25
Plato
The words of astronomy are about..., 26
We shall treat astronomy as..., 26
Sagan, Carl
...astronomy is a humbling and character building experience, 26
Shapley, Harlow
...the most interesting feature of this science astronomy is..., 26
Sherrod, P. Clay
Astronomy is a unique science..., 26
Sillman, Benjamin
Astronomy is not without reason, 27
Struve, Otto
Astronomy has had three great revolutions..., 27
Twain, Mark
I study astronomy more than any other foolishness..., 27
Unknown
...the great desiderata of astronomy..., 27
Virgil
Give me the ways of wandering stars..., 28

Whitehead, Hal
Studying the behavior of large whales has been likened to astronomy, 28
astrophysicist
Douglas, Vibert
...the astrophysicist roams the universe from atom to atom..., 32
Luminet, Jean-Pierre
Astrophysicists have the formidable privilege of..., 32
Spenser, Edmund
For who so list into the heavens looke..., 32
astrophysics
Greenstein, J.L.
In astrophysics...theories have only seldon had predictive usefulness..., 32
atom
Davies, Paul
...we are to an atom..., 34
Eddington, Sir Arthur
The atom is as porous as the solar system, 34
The physical atom is..., 34
Esar, Evan
[Atom] The smallest thing in the world..., 34
Feynman, Richard P.
I, a universe of atoms..., 34
...the atoms that are in the brain..., 34
Nabokov, Vladimir
But the individual atom is free..., 35
Rowland, Henry
The round hard atom of Newton..., 35
Rukeyser, Muriel
...Universe is made of stories, not atoms, 35
Unknown
When an atom is so small..., 35
atomic power
Church, Peggy Pond
We had thought the magicians were all dead..., 37
Laurence, William L.
...at that instant there rose as if from the bowels of the earth a light not of this world..., 37
atomic theory
Mach, Ernst
The atomic theory has in physical science..., 35
atomism
Democritus
...we apprehend nothing exactly..., 38
Lucretius
...but what bodies depart at any given time..., 38
aurora borealis
Ayouton, William
All night the northern streamers shot..., 39
Burns, Robert
Her lights, wi' hissing eerie din, 39
Haliburton, T.C.
...ere the aurora borealis mimics its setting beams..., 39
Kingsley, Charles
Night's son was driving..., 39
Scott, Robert F.
The Eastern sky was massed with swaying auroral light..., 40
Scott, Sir Walter

...spirits were riding the northern lights, 40
Service, R.W.
And the Northern Lights in the crystal nights..., 40
...the Northern Lights are the glare of the Arctic ice..., 40
Taylor, Bayard
The amber midnight smiles in dreams of dawn, 41
Wilde, Oscar
They are like the Aurora Borealis..., 41
axial tilt
Milton, John
Some say he bid his angels turn askance..., 42
axiom
Doyle, Sir Arthur Conan
It has long been an axiom of mine..., 43
Frayn, Michael
...in interlocking axioms and theorems, 43
Planck, Max
Axioms are instruments which are..., 43
beautiful
Steensen, Niels
Beautiful are the things we see..., 47
beauty
Cayley, Arthur
...beauty can be perceived but not explained, 44
Chandrasekhar, Subrahmanyan
...a search after the beautiful in mathematics..., 44
Collins, Wilkie
Admiration of those beauties of the inanimate world..., 44
Copernicus, Nicolaus
...to do with things that are very beautiful, 45
Dirac, P.A.M.
...beauty does depend on one's culture..., 45
...taking mathematical beauty as our guiding beacon..., 45
Duhem, Pierre
...without being charmed by the beauty..., 45
Emerson, Ralph Waldo
...the necessity of beauty under which the universe lies, 46
...Beauty is the creator of the Universe, 46
Hilbert, David
...an unexpected view which was pleasing to our eyes..., 46
Leibniz, Gottfried Wilhelm
The beauty of nature is so great..., 46
Misner, Charles W.
...the glittering central mechanism of the world in all its beauty..., 46
Poincaré, Henri
...the mathematical entities to which we attribute this character of beauty..., 46
big bang
Guth, Alan
The classic big bang theory describes..., 48
Hoyle, Fred
Big-bang cosmology is a form of religious fundamentalism..., 48
Maddox, John
...compelling evidence for the Big Bang..., 48
Poe, Edgar Alan

...on account of Matter's haging being irradiated...into a limited sphere of Space..., 49
black hole
Asimov, Isaac
...nothing more than a black hole, 50
Gardner, Martin
The healthy side of the black hole craze is..., 50
...disappear into a black hole..., 50
Koestler, Arthur
Went down to explore a Black Hole, 50
Lasota, Jean-Pierre
Black holes may still be black..., 51
Thorne, Kip
...the most fantastic, perhaps, is the black hole, 51
book
Bernstein, Jeremy
...a physics book, unlike a novel..., 52
Eco, Umberto
Books are not made to be believed..., 53
Mitchell, Maria
A book is a very good institution!, 53
Newton rolled up the cover of a book..., 53
Canis Major
Frost, Robert
The Great Overdog..., 71
Canopus
Carlyle, Thomas
Canopus shining down over the desert..., 340
Capricornus
Aratus
...when the Goat/With the sun rises..., 71
celestial motion
Milton, John
And their motions harmony divine..., 54
celestial spaces
Muir, John
...sailing the celestial spaces..., 401
chance
Jevons, W. Stanley
...the Universe in which we dwell is not the result of chance..., 55
John of Salisbury
Chance blows together the atoms..., 55
change
Burns, Robert
Nature's law is change, 56
Emerson, Ralph Waldo
There are no fixtures in nature, 56
Mitchell, Maria
...and the changes are infinitely more startling..., 56
Ovid
All things are fluent..., 56
chaos
Bradley, John Hodgdon Jr
Chaos and caprice do not exist, 57
Miller, Henry
Chaos is the score upon which reality is written, 57
Kant, Immanuel
...to fashion itself out of chaos..., 400
comet
Babylonian Inscription
A comet arose whose body was bright..., 58

Byron, George Gordon
...Or wild cold of a comet..., 58
de Fontenelle, Bernard Le Bovier
These foreign planets, with their tails..., 58
...the misfortune is the comet's, 58
Dick, Thomas
...the physical constitution of comets..., 58
...we ought to contemplate the approach of a comet..., 59
Donne, John
Who vagrant transitory comets sees..., 59
Halley, Edmond
...comets were nothing else than sublunary vapors..., 59
Holmes, Oliver Wendell
The Comet! He is on his way..., 59
Lee, Oliver Justin
...it is quite possible that the great number of comets were..., 60
Levy, David H.
Comets are like cats..., 60
Maunder, E. Walter
Comets cannot be homes of life..., 60
Peltier, Leslie C.
I had watched a dozen comets..., 60
...the age old allure of the comets, 60
Seneca
If a rare [comet] and one of unusual shape appears..., 60
How many other bodies besides these comets..., 61
Shakespeare, William
Comets, importing change of time and states..., 61
Thomson, James
The rushing comet to the sun descends..., 61
Tolstoy, Leo
...the radiant star which, after traveling in its orbit..., 61
von Humboldt, Alexander
...possible evils threatened by comets..., 61
communication
Archytas
...he had not a Friend to communicate it to, 63
Mach, Ernst
Science is communicated by instruction..., 63
Neal, Patricia
Gort, Klaatu berada nikto!, 63
compulsory
Bilaniuk, O.
...anything which is not prohibited is compulsory, 64
concept
Einstein, Albert
Physical concepts are free creations of..., 65
Milne, A.A.
A huge great enormous thing..., 51
Weaver, Warren
It is by no means clear that our present concepts..., 65
constellation
Aratus
...devised their titles,/Forming the constellations, 67
Burns, Robert
I'd heeze thee up a constellation!, 67

de Cervantes, Miguel
...the muzzle of the Bear is at the top of his head..., 67
Donne, John
And in these Constellations then arise..., 68
Homer
...the signs that glorify the face of the heaven..., 68
Noyes, Alfred
And wondered at the mystery of it all, 68
Sagan, Carl
...organize into patterns these separate and distinct points of light, 68
Twain, Mark
Constellations have always been troublesome things to name, 69
Whitman, Walt
I do not want the constellations any nearer..., 70
cosmochemistry
Frost, Robert
Tell us what elements you blend, 75
Fuller, Buckminster
All the chemistries of the Universe are essential..., 75
Marcet, Jane
...she is incessantly employed in chemical operations, 75
cosmogenesis
Bowyer, Stuart
Ultimately, the origin of the universe is..., 408
Gamow, George
...discuss the basic problem of the origin of the universe..., 409
Kipling, Rudyard
Before the High and Far-Off Times..., 409
Reeves, Hubert
In the beginning was the absolute rule of the flame, 409
Sturluson, Snorri
Erst was the age when nothing was..., 410
Townes, Charles H.
...to explain the origin of our universe..., 410
Unknown
In the beginning there was nothing..., 410
cosmogony
Bridgman, Percy
...the most striking thing about cosmogony is..., 76
cosmological
Chaisson, Eric
There are no larger thoughts than cosmological ones, 77
cosmologists
Turok, Neil G.
Cosmologists...may have to forego attempts at..., 78
cosmology
Görtniz, Thomas
Modern cosmology is a myth..., 77
Hawking, Stephen
Cosmology used to be considered a pseudoscience..., 77
Popper, Karl R.
All science is cosmology..., 77
Tolman, R.C.
It is appropriate to approach the problems of cosmology with feelings of..., 78
Turok, Neil G.

Maybe the problems of cosmology has set for itself..., 78
cosmos
Ferris, Timothy
The history of the cosmos is..., 79
Lawrence, D.H.
We and the cosmos are one, 79
Mencken, H.L.
The cosmos is a gigantic fly-wheel..., 79
Plato
The universe is called Cosmos..., 79
Reeves, Hubert
Knowledge of the cosmos is..., 80
Santayana, George
...the cosmos has its own way of doing things..., 80
Shapley, Harlow
Cosmography is to the Cosmos what..., 80
Tomlinson, C.
Where Cosmos is his temple..., 80
creation
Bush, Vannevar
...we may be looking at only a small part of a grand creation, 81
Epictetus
No great thing is created suddenly..., 81
Fitzgerald, Edward
There was a door to which I found no key..., 81
Hoyle, Fred
Without continuous creation..., 81
curiosity
Amaldi, Ginestra Giovene
...man's curiosity is excited by the wonderful sights..., 82
Haber, Heinz
The curiosity of man must forever find its greatest challenge in..., 83
Pittendreigh, W. Maynard Jr
I burn with curiosity about what lies beyond the sky, 83
Weisskopf, Victor
Curiosity without compassion is inhuman..., 83
Wright, Helen
The curiosity of Alice..., 83
dark matter
de Saint-Exupéry, Antoine
Anything essential is invisible to the eye, 84
data
Greenstein, George
Data in isolation are meaningless..., 85
Lowell, Percival
...deduction rests ultimately upon the data derived from experience, 85
Russell, Bertrand
...when our data are confined to a finite part of the universe, 85
depletion
Jenkins, E.B.
...when we study depletions..., 86
design
Davies, Paul
The physical universe is put together with an ingenuity..., 87
Russell, Bertrand
...the argument from design, 87

determined
Einstein, Albert
Everything is determined. . . , 88
determinism
Popper, Karl
The intuitive idea of determinism. . . , 88
differential equation
Eddington, Sir Arthur Stanley
. . . to reduce God to a system of differential equations, 89
Haldane, J.B.S.
. . . a man falling in love with a differential equation. . . , 89
Lanczos, Cornelius
Our symbolic mechanism is eminently useful. . . , 89
Whitehead, Alfred North
. . . relapses into the study of differential equations, 89
dimension
Einstein, Albert
His universe will be two-dimensional, 91
discover
Milne, A.A.
It's just a thing you discover. . . , 94
discovery
Bolyai, John
Mathematical discoveries. . . have their season. . . , 92
Bruner, Jerome
. . . clear about what the act of discovery entails, 92
Clerke, Agnes M.
. . . the course of astronomical discovery. . . , 92
Curie, Marie
A great discovery does not leap completely achieved from the brain. . . , 93
Eddington, Sir Arthur Stanley
. . . we have not to discover the properties of a thing. . . , 93
Glass, Bentley
. . . the endless horizons no longer exist, 93
Holton, G.
. . . the drive toward discovery. . . , 93
Kepler, Johannes
. . . chance occurrences by which I the author first came upon that understanding, 93
Körner, T.W.
. . . the great discovery of the nineteenth century was. . . , 94
Lowell, Percival
The road to discovery is not an easy one to travel, 94
Thomson, J.J.
In the distance tower still higher peaks. . . , 94
Thoreau, Henry David
Do not engage to find anything as you think they are, 94
Unknown
Pioneers occupy new land, 95
von Lenard, Philipp E.A.
. . . those who pluck the fruit. . . , 95
distance
Coblentz, Stanton
Whose measuring rods are light-years. . . , 96
Heidmann, Jean
The distances we are going to embrace. . . , 96
distribution
Kapteyn, A.J.
. . . the real distribution of the stars. . . , 99

doctrines
Abbey, Edward
Let us practice from our doctrine..., 97
dogmatize
Huxley, Julian
The undevout astronomer is mad..., 97
Draco
Darwin, Erasmus
With vast convolutions Draco holds..., 71
dust
Greenberg, J. Mayo
Astronomers no longer consider interstellar dust a nuisance, 98
Earth
Ackerman, Diane
...Earth bunched its granite to form the continents..., 261
Albran, Kehlog
The Earth is like a grain of sand..., 261
Cloos, Hans
Earth beautiful, round, colorful planet, 261
Coleridge, Samuel Taylor
...O Earth, whom the comets forget not..., 262
Eddington, Sir Arthur Stanley
...but it unites the earth, 262
Guiterman, Arthur
/Our Earth, a paltry little mommet,/..., 262
Irwin, James
The earth reminded us of a Christmas tree ornament..., 262
MacLeish, Archibald
To see the earth as we now see it..., 263
Sagan, Carl
If we are to understand the Earth..., 263
The surface of the Earth is the shore of the cosmic ocean, 263
Teilhard de Chardin, Pierre
...concentrate our attention on the planet we call Earth..., 263
Thomas, Lewis
...the queerest structure we know about...is the earth, 264
Vizinczey, Stephen
Is it possible that...the centre of man's universe is the earth?, 264
Whipple, Fred L.
Our Earth seems so large, so substantial..., 264
eclipse
Amos 8:9
I will make the sun go down at noon..., 100
Archilochus
...now that Zeus...has made night out of noonday..., 100
Caithness, James Balharrie
I watched the shadow of our globe..., 100
Flammarion, Camille
Eclipses...have always been interpreted as..., 100
Hardy, Thomas
At a Lunar eclipse..., 101
Joel 2:30, 31
...the sun shall be turned into darkness..., 101
Plato
...gazing on the sun during an eclipse..., 101
Poincaré, Henri

...ridiculous to pray for an eclipse?, 102
Shakespeare, William
...Advance our half-face sun..., 102
Unknown
That is an important eclipse, 103
Wordsworth, William
...That darkening of his radiant face..., 103
electron
Bragg, Sir William
...an electron springs into existence, 104
Frankel, Felice
Electrons know two verbs..., 104
Lederman, Leon
The "naked" electron is..., 104
Sullivan, J.W.N.
The electron is not an enduring something..., 104
ellipse
Hardy, Thomas
His world was an ellipse..., 105
energy
Feynman, Richard
...there is a certain quantity, which we call energy..., 106
...we have no knowledge of what energy is, 106
Huxley, Julian
I am Energy, 106
Meyerson, Emile
Energy is only an integral..., 106
Moulton, Forest Ray
...their energies will in some way be integrated again, 107
Soddy, Frederick
Energy, someone may say, is a mere abstraction..., 107
Unknown
I'm busy conserving energy..., 108
equation
Hawking, Stephen
What is it that breathes fire into the equations..., 109
Holton, Gerald
...equations in physical science always have hidden limitations..., 109
London, Jack
...is called the Equation of Time, 109
Saaty, Thomas L.
Equations are the lifeblood of..., 109
error
van de Kamp, Peter
...a cosmic phemomenon trying to reveal itself in a sea of errors?, 110
Whitehead, Alfred North
...avoid adding other people's errors to our own, 110
Wright, Wilbur
If a man is in too big a hurry to give up an error..., 110
eternity
Harrison, Edward
If eternity is silliness..., 111
Paine, Thomas
...to conceive an eternal duration of what we call time..., 111
Vaughan, Henry
I saw eternity the other night..., 111
Young, Edward
Eternity is written in the skies, 111

events
Ferguson, Kitty
Events in the heavens happen in their own good time..., 113
Milne, Edward
...the events occurring in nature..., 113
experiment
Ehrlich, Paul
Much testing, 114
Gardner, Martin
When reputable scientists correct flaws in an experiment..., 310
Gore, G.
...relieved by preparing and making experiments, 114
Rutherford, E.
Experiment without imagination..., 114
Sagan, Carl
Experiment is the touchstone of science..., 114
exposition
Einstein, Albert
Your exposition is of matchless clarity..., 426
extraterrestrial
Dickinson, Terence
...wondering who else is out there..., 116
Eddington, Sir Arthur Stanley
...at the present time our race is supreme..., 117
Jones, Sir Harold Spencer
Can it be that...but on our own little Earth is life to be found?, 118
Milton, John
Dream not of other worlds..., 119
Pope, Alexander
What varied Being peoples every star..., 120
Sakharov, Andrei
In infinite space many civilizations are bound to exist..., 121
Shakespeare, William
There are more things in heaven and earth..., 122
extraterrestrial life
Butler, Samuel
...Th' Inhabitants of the Moon..., 115
Dick, Steven J.
Humanity 3,000 will know whether or not it is alone..., 116
Eiseley, Loren
...there must be life out there beyond the dark..., 117
Of men...there will be none forever, 117
...perhaps the only thinking animals in the entire sidereal universe..., 117
Fuller, R. Buckminster
Sometimes I think we're alone, 118
Giraudoux, Jean
...prowling around in space looking for a little company..., 118
Huygens, Christianus
...nay their Inhabitants too..., 118
Metrodorus of Chios
...if a single ear of corn grew..., 119
Oparin, A.I.
...the origin of life in the universe, 119
Pallister, William
Some other planets, peopled like our own..., 119

Sagan, Carl
...there are a million other civilizations..., 120
...Milky Way Galaxy is teeming with civilizations..., 121
...the subject of extraterrestrial life has finally come of age, 120
Shakespeare, William
I can call spirits from the vasty deep, 122
Shaw, George Bernard
...do you believe in life on other planets?, 122
Tsiolkovsky, Konstantin
Is it probable for Euroope to be inhabited..., 122
von Braun, Wernher
...to think that we are the only living things..., 122
Wells, H.G.
...this world was being watched keenly and closely by intelligences greater than man's..., 123

fact

Bridgman, P.W.
...the fact has always been for the physicist..., 124
Chesterton, G.K.
Facts as facts do not always creat a spirit of reality..., 124
Collins, Wilkie
"Facts?" he repeated..., 124
Faraday, Micahael
It is always safe...to distinguish...fact from theory..., 124
Huxley, Thomas
...proved to be contrary to any fact..., 125
Jacks, L.P.
Facts are popularly regarded as antidotes to mysteries, 125
Krough, A.
Facts are necessary..., 125
Mayer, J.R.
If a fact is known on all its sides..., 125
Michelson, A.A.
The more important fundamental laws and facts of physical science..., 125
Poincaré, Henri
...the physicist, must make a choice among facts..., 126
Shaw, George Bernard
...an Englishman was not daunted by facts, 126
Snow, C.P.
A fact is a fact is a fact, 126
Tyndall, John
...to blink at facts because..., 126
Whewell, William
When we inquire what Facts are to made of..., 126
Facts are the materials of science..., 151

force

Moleschott, Jakob
Force is not an impelling God..., 128
Weil, Simone
Two forces rule the universe..., 128
Whitman, Walt
You unseen force..., 128

formula

Emerson, Ralph Waldo
The formulas of science are like..., 129
Mitchell, Maria
Every formula which expresses a law of nature is..., 129

Planck, Max
...where possible, in a single formula, 129
Stoppard, Tom
...you could write the formula for all the future..., 129
fusion
Pauli, Wolfgang
Let no man join..., 130
future
Albran, Kehlog
I have seen the future..., 131
Einstein, Albert
The future...is every whit as necessary and determined as the past, 131
Hilbert, David
...the veil behind which the future lies hidden..., 131
Poincaré, Lucien
...to seek to predict the future which may be reserved for physics, 131
Tennyson, Alfred
When I dipt into the Future..., 131
galaxy
Eddington, Arthur
Let us first understand what a galaxy is, 132
Hoyle, Fred
...think of a galaxy as a collection of specks..., 132
Jeffers, Robinson
Galaxy on galaxy, innumerable swirls of innumerable stars..., 132
Wolf, Fred Alan
...tend to cluster...into galaxies, 133
geometry
Chern, Shiing Shen
Physics and geometry are one family, 134
de Fontenelle, Bernard Le Bovier
...the hand of geometry, 134
Manning, Henry Parker
The greatest advantage to be derived from the study of geometry of more than three dimensions..., 134
Suspended o'er geometry..., 134
Morgan, Frank
...how geometry rules the universe, 135
Russell, Bertrand
...Geometry throws no more light upon the nature of space than..., 135
God
Card, Orson Scott
All the universe is just a dream in God's mind..., 136
Dawkins, Richard
If God is a synonym for the deepest principles of physics..., 136
Feynman, Richard
God was invented to explain mystery, 136
Greenstein, George
Was it God who stepped in..., 137
Hawking, Stephen
...profound implications for the role of God in the affairs of the universe, 137
Herrick, Robert
Science in God..., 137
Jastrow, Robert
When an astronomer writes about God..., 137
Keillor, Garrison
We wondered if there is a God..., 137
Kepler, Johannes

How can we be the master of God's handiwork?, 138
Orgel, Irene
"Certainly," said God patiently, 138
Polyakov, Alexander
...because God created it, 138
Reade, Winwood
Man will then be...what the vulgar worship as God, 139
Thomson, J. Arthur
The heavens are telling the glory of God, 139
Twain, Mark
If I were going to construct a God..., 139
von Braun, Wernher
The more we learn about God's creation..., 139
Ziman, John
...God himself is waiting to see what will happen!, 139
grains
Heiles, Carl
...needle-like grains tend to spin..., 140
Seab, C.G.
Once the newly formed grains..., 140
gravitation
Arnott, Neil
Attraction, as gravitation..., 142
Einstein, Albert
...but gravitation cannot be held responsible for it, 143
gravitational lens
Drake, Frank
...the Sun, as a gravitational lens..., 141
gravity
Bierce, Ambrose
The tendency of all bodies to..., 142
Blake, William
God keep me...from supposing Up and Down to be the same thing..., 142
Feynman, Richard P.
...a mysterious attraction to a spinning ball..., 142
Lockyer, Joseph Norman
The force of gravity on their surface must be..., 143
Longfellow, Henry Wadsworth
Every arrow that flies feels the attraction of the earth, 143
Newton, Sir Isaac
...what hinders the fixed stars..., 143
heaven
Addison, Joseph
The ways of Heaven are dark and intricate..., 144
Alighieri, Dante
Heaven calls you..., 144
Browning, Robert
...what's a heaven for?, 144
Chesterton, G.K.
To see that glimpse of Heaven..., 145
Cicero
If I had ascended the very heaven..., 145
Dickinson, Emily
What once was "Heaven"/Is "zenith" now..., 145
Donne, John
And then that heaven, which spreads so farre..., 145
Frost, Robert
...for anything much/To happen in heaven..., 68
Shelley, Percy Bysshe
Heaven's ebon vault..., 146

heavens
Burnham, Robert Jr
...in the dark unknown immensity of the heavens..., 144
Donne, John
...this net throwne/Upon the Heavens..., 145
Grondal, Florence Armstrong
The silent song of the heavens..., 145
Herschel, William
The Heavens...are now seen to resemble..., 146
Kepler, Johannes
...the treasures hidden in the heavens so rich..., 400
Seneca
...the contemplation of the starry heavens..., 146
Simes, James
The prose of the heavens..., 147
hypothesis
Huxley, Thomas
Every hypothesis is bound to explain..., 148
Osiander, Andrew
There is no need for these hypotheses to be true..., 148
Steinbeck, John
When a hypothesis is deeply accepted..., 148
Sterne, Laurence
It is in the nature of a hypothesis..., 148
Whewell, William
The hypothesis which we accept..., 149
idea
Sagan, Carl
Someone has to propose ideas..., 150
Toulmin, Stephen
New ideas are the tools of science..., 150
Tucker, Abraham
...an idea on being displaced by another..., 150
Whitehead, Alfred North
In the study of ideas..., 151
ignorabimus
Hilbert, David
...in mathematics there is no ignorabimus, 152
ignorance
Hellman, C. Doris
...whose character we are still in ignorance of..., 152
Pratchett, Terry
...the universe was full of ignorance..., 152
imagination
Douglas, Vibert
Then it is that imagination, like a glorious greyhound..., 153
Einstein, Albert
Imagination is more important than knowledge, 153
Herschel, William
If we indulge in fanciful imagination..., 153
Jeans, Sir James
...prevent our imaginations from passing into a further space beyond, 153
Lowell, Percival
Imagination is the single source of..., 154
Imagination supplies the motive power..., 154
Mitchell, Maria
We especially need imagination in science, 154
Thoreau, Henry David

The imagination, give it the least license…, 155
Tombaugh, Clyde
You have to have the imagination to recognize…, 155
Velikovsky, Immanuel
Imagination coupled with skepticism…, 155
Wheeler, John Archibald
…an immense labor of imagination and theory, 155
Whitrow, G.J.
…remains a product of the imagination, 156
impossible
von Braun, Wernher
…to use the word 'impossible' with utmost caution, 157
infinite
Aristotle
It is impossible that the infinite should move…, 158
Descartes, Rene
We call infinite that thing whose limits we have not perceived…, 158
Emerson, Ralph Waldo
…the infinite lies stretched in smiling repose, 159
Green, Brian
…an infinite answer is nature's way of telling us…, 159
Kasner, Edward
The infinite in mathematics is…, 159
Whitehead, Alfred North
…the infinitely small and the infinitely vast…, 159
infinitesimal
Mitchell, Maria
Do not forget the infinite in the infinitesimal, 159
infinity
Blake, William
The nature of infinity is this…, 158
Harrison, Edward
Only a cosmic jester could perpetrate eternity and infinity, 159
von Haller, Albrecht
Infinity! What measures thee?, 159
Wolf, Fred Alan
…infinity is just another name for mother nature, 160
Zebrowski, George
Science, when it runs up against infinities…, 160
instrument
Bacon, Francis
Effects are produced by means of instruments…, 161
Bridgman, P.W.
Not only do we use instruments to give us fineness of detail…, 161
Egler, Frank E.
…the most important instrument in science must always be…, 161
Eisenhart, Churchill
For while science and gadgets are fine…, 161
Kuhn, Thomas
…scientists see new and different things when looking with familiar instruments…, 162
Lavoisier, A.
…we must be provided with good instruments, 162

...very expensive and complicated instruments..., 162
von Goethe, Johann Wolfgang
Ye instruments, ye surely jeer at me..., 162
Whitehead, Alfred North
...because we have better instruments, 163
interaction
Hugo, Victor
How do we know that the creations of worlds is not..., 164
Jupiter
Ackerman, Diane
Jupiter floods the night's black scullery..., 265
know
Carson, Rachel
...to pave the way for the child to want to know..., 172
knowledge
Addison, Joseph
The utmost extent of man's knowledge..., 165
Alighieri, Dante
...to have heard without retaining does not make knowledge, 165
Clerke, Agnes M.
...our knowledge will...appear in turn the merest ignorance..., 165
Confucius
...this is knowledge, 165
Gauss, Carl Friedrich
The knowledge whose content makes up astronomy..., 165
Gore, George
New knowledge is not like a cistern..., 166
Hinshelwood, C.N.
To some men knowledge of the universe has been..., 166
James, William
...the world of our present natural knowledge..., 166
Jeans, Sir James
Our knowledge of the external world must..., 166
Latham, Peter
...nothing so captivating as NEW knowledge, 166
Rabi, I.I.
...a beautiful structure of knowledge, 167
Russell, Bertrand
With equal passion I have sought knowledge, 167
Smith, Theobald
...and pruning the tree of knowledge..., 168
Tennyson, Alfred
To follow knowledge like a shining star..., 168
Thoreau, Henry David
Such is always the pursuit of knowledge, 168
Whewell, William
...these convey no knowledge to us..., 168
lady in the moon
Longfellow, Henry Wadsworth
'Tis her body that you see there, 204
laws
Holton, Gerald
...the law or set of laws is the net..., 169
Hoyle, Fred
...controlled by the laws of physics, 169
LaPlace, Pierre Simon
...do not seem to follow the great laws of nature..., 169

Mitchell, Maria
The laws of nature…, 170
The laws which regulate…, 170
The immense spaces of creation…, 170
Pagels, Heinz
The fact that the universe is governed by simple natural laws is…, 170
Rowland, Henry
…he who obscurely worked to find the laws…, 171
Russell, Bertrand
…mathematics follows inevitably from a small collection of fundamental laws…, 171
Schwarzschild, Martin
If simple perfect laws uniquely rule the universe…, 171
Unknown
The universe does not have laws…, 171
Weyl, Hermann
…the indestructible world of laws…, 171

learn
Dennett, Daniel
We often learn more from bold mistakes…, 172

Libra
Longfellow, Henry Wadsworth
I heat the Scales…, 72

life
Glaskov, Yuri
The winds scatter across the planet the seeds of life…, 173
Jeans, Sir James
Is this, then, all that life amount to?, 173
Pagels, Heinz
…the possibility of life in a universe…, 173
Wald, George
…the whole burden of life…, 174

light
Feynman, Richard
Things on a very small scale…, 175
Mullaney, James
The light we see coming from…, 175
Newton, Sir Isaac
…emit Light and shine…, 175
Pratchett, Terry
No matter how fast light travels…, 176
Thomas, Dylan
Light travels where no sun shines, 176
Unknown
I know the speed of light…, 176
Light travels faster than sound…, 176
Whitesides, George M.
We know light in its diluted form…, 175

loadstone
Regnault, Pére
Till you have discovered to me the Myseries of the Loadstone…, 424

logic
Joubert, Joseph
Logic operates, metaphysics contemplates, 177
Rexroth, Kenneth
Crossed only by the bright fences/Of logic, 177
Shakespeare, William
He dreweth out the thread of his verbosity…, 177
Unknown

logic is a systematic method of..., 177
Lost Pleiad
Hemans, Felicia
...no more art seen of mortal eye, 72
magnetic
Zirin, H.
If the Sun had no magnetic field..., 179
magnetic field
Parker, E.N.
...the continuing thread of cosmic unrest is the magnetic field, 179
Magnetic fields (and their inevitable offspring..., 179
magnetism
Hale, G.E.
...discovery of the effect of magnetism on radiation..., 179
man
Bradley, John Hodgdon
...man is the only animal who can..., 180
Landau, Lev
...has shown that man can tear himself away from..., 180
Lowell, Percival
...man is but a detail..., 180
Miller, Perry
...man is not at home in this universe..., 180
Steinbeck, John
Man is...related inextricably to..., 181
man of science
Oppenheimer, J. Robert
Both the man of science and the man of art..., 305
mankind
Shapley, Harlow
Mankind is made of star stuff..., 180
Mars
Bradbury, Ray
...if we are interested in Mars at all..., 264
Longfellow, Henry Wadsworth
...to the red planet Mars, 264
Lowell, Percival
...so profoundly impressive as the canals of Mars, 264
matter
Darling, David
You are roughly eighteen billion years old and made of matter that has been..., 182
Dyson, Freeman J.
...matter is weird enough..., 182
Huxley, Julian
I am Matter, 183
Reeves, Hurbert
...demands that matter abandon itself to games of chance, 183
Updike, John
There is infinitely more nothing in the universe than anything else, 183
Weyl, Hermann
...has swept away space, time and matter..., 183
measurement
Bell, J.A.
The concept of measurement becomes so fuzzy..., 184
measuring
Rankine, William John Macquom
A party of astronomers went measuring of the earth..., 184

mechanics
Gauss, Carl Friedrich
...shall appear simply as a special case of mechanics, 185
Oliver, David
Mechanics is the wellspring from which physics flows..., 185
Mechanics is the vehicle of all physical theory, 185
Mercury
Ackerman, Diane
The Sun never sets on the Mercurian empire..., 260
mere
Feynman, Richard P.
Nothing is "mere", 186
metaphors
Cole, K.C.
The words we use are merely metaphors, 187
Moore, James R.
Clever metaphors die hard, 187
meteor
Butler, Samuel
The hairy meteor did announce..., 188
Caithness, James Balharrie
Wonderful, shimmering trail of light..., 188
Darwin, Erasmus
You chase the shooting stars..., 188
Devaney, James
A trail of whitest fire you shone..., 188
Hoffman, Jeffrey
Suddenly I saw a meteor go by underneath me, 189
Plum, David
...countless as snowflakes are meteors blazing..., 189
Revelation 9:1–2
...I saw a star fall from heaven..., 189
Shakespeare, William
And certain stars shot madly from their spheres, 190
Smythe, Daniel
A curve of fire traces the dark..., 190
Tennyson, Alfred
Now slides the silent meteor..., 190
Teasdale, Sara
I saw a star slide down the sky..., 190
Unknown
...to circle endlessly a barren stone, 190
Virgil
...stars fall headlong from the skies..., 191
meteor shower
Revelation 12:3–4
...and did cast them to the earth, 189
meteorite
Unknown
A rock from space that falls to earth is called a meteorite, 190
method of science
Thomson, Sir George
[The method of science]..., 307
milky way
Alighieri, Dante
...the Galaxy so whitens..., 192
Donne, John
...the milkie way, there is not one starre of any..., 192
Hearn, Lafcadio
Then I no longer behold the Milky Way..., 192
Herschel, William

...where it is surrounded with a bright zone, the Milky Way..., 193
Kilmer, Joyce
God be thanked for the Milky Way..., 193
Milton, John
The galaxy, that milky way..., 193
Poincaré, Henri
...the Milky Way would seem only a bubble of gas, 193
Rich, Adrienne
Driving at night I feel the Milky Way..., 193
Thoreau, Henry David
...is not our planet in the Milky Way?, 194

mind

Blaise, Clarke
Our minds soar with instant connection..., 195
Dyson, Freeman J.
...the tendency of mind to infiltrate and control matter..., 195
Einstein, Albert
The human mind is not capable of grasping the Universe, 195
Gauss, Carl Friedrich
...the compass of my mind ever turns, 196
Jevons, W. Stanley
...the mind of the great discoverr must..., 196
Kepler, Johannes
A mind accustomed to mathematical deduction..., 196
Land, Edwin
Each stage of human civilization is defined by our mental structures..., 196

model

Born, Max
...men who made free use of models..., 198
Feynman, Richard
...to make a model that explains..., 198
Weisskopf, Victor F.
A model is like an Austrian timetable, 198

molecule

Frankel, Felice
Molecules... are happiest when surrounded by their own kind, 199
Maxwell, James Clerk
...built up of molecules of the same kinds..., 199
...the molecules out of which these systems are built..., 199

momentum

Unknown
A rolling stone gathers momentum, 200

moon

Atwood, Margaret
The moon is a stone and the sky is full of..., 201
Blake, William
The moon like a flower..., 201
Brontë, Charlotte
Where, indeed, does the moon not look well?, 201
Burton, Sir Richard
That gentle Moon, the lesser light..., 201
Carroll, Lewis
The moon was shining sulkily..., 202
...this land of the skies..., 202
Coleridge, Samuel Taylor

The moving moon went up the sky..., 202
Collins, Michael
It was a totally different moon than I had ever seen before, 202
Fry, Christopher
The moon is nothing/But a circumambulatory aphrodisiac..., 202
Homer
As when the moon..., 203
Huxley, Julian
By death the moon was gathered in..., 203
Jastrow, Robert
The moon is the Rosetta stone of the solar system..., 203
Lear, Edward
They danced by the light of the moon, 203
Lightner, Alice
Queen of Heaven, fair of face..., 204
Milton, John
To behold the wandering moon..., 204
Shelley, Percy Bysshe
Art thou pale for weariness..., 204
Whom mortals call the moon, 205
Tennyson, Alfred
And portals of pure silver, walks the moon, 205
Verne, Jules
...who has not seen the moon..., 205
moon landing
Armstrong, Neil
That's one small step for a man..., 206
Hoffer, Eric
Our passionate preoccupation with the sky..., 206
Koestler, Arthur
Prometheus is reaching out for the stars..., 206
Nabokov, Vladimir
Treading the soil of the moon..., 207
Plaque
...First Set Foot upon The Moon, 206
motion
Butterfield, Herbert
...relating to the problem of motion, 208
Galilei, Galileo
...the essential features of observed accelerated motions, 208
Gleick, J.
...trying to account for the motion of a billiard ball..., 208
Meredith, George
The music of their motion may be ours, 208
Regnault, Pére
Nothing seems more clear at first than the Idea of Motion..., 209
muon
Penman, Sheldon
...the muon itself qualifies as a..., 210
mystery
von Weiszaecker, Karl Friedrich
Physics begins by facing a mystery..., 255
nature
Agassiz, Louis
The study of Nature is..., 211
Bohm, David
In nature nothing remains constant, 211

Boyle, Robert
Nature always looks out for..., 211
Browne, Sir Thomas
...nature is the art of God, 211
Einstein, Albert
Nature is the realization of..., 211
Emerson, Ralph Waldo
Nature is an endless combination and repetition of a very few laws, 212
It is very odd that Nature should be so unscrupulous, 212
Feynman, Richard
There was a moment when I knew how nature worked, 212
Gould, Stephen Jay
I do not believe that nature frustrates us by design..., 212
Heraclitus
The real constitution of things is accustomed to hide itself, 213
Herschel, J.F.W.
...Nature builds up by her refined..., 213
Holton, Gerald
The study of nature is..., 213
Hooke, Robert
The footsteps of Nature to be..., 213
Lawrence, Louise de Kiriline
Nature is a deep reality..., 213
McLennan, Evan
There is a charm for man in the study of Nature, 213
Muir, John
How lavish is Nature..., 214
When we are with Nature we are awake..., 214
Musser, George
The basic rules of nature are simple..., 214
Petrarch
...fools who seek to understand the secrets of nature, 214
Sayers, Dorothy L.
Nature never worked by rule and compass, 214
Seneca
Nature does not turn out her work according to..., 214
Nature does not reveal all her secrets at once, 214
Tennyson, Alfred
A void was made in Nature..., 215
Thoreau, Henry David
If we knew all the laws of Nature..., 215
von Schelling, F.W.J.
...that secret bond which couples our mind to Nature..., 215
...he spontaneously operates upon Nature..., 215
Whitehead, Alfred North
You cannot vaguely about Nature in general, 216
Nature, even in the act of satisfying anticipation..., 216
Whitman, Walt
The fields of Nature long prepared and fallow..., 216
Wordsworth, William
...Nature hides/Her treasures less and less-..., 216
Neptune
Clerke, Agnes M.

...the sympathetic thrillings of Neptune..., 266

neutrino

Crane, H. Richard
...believes in the existence of the neutrino..., 217
Eddington, Arthur
I am not much impressed by the neutrino theory, 217
The neutrino is just barely a fact, 217
Pontecorvo, Bruno
...the case of the neutrino invention by Pauli, 217
Stenger, Victor J.
Neutrinos are neither rare nor anomalous..., 217

night

Ackerman, Diane
It is nighttime on the planet Earth, 218
Amaldi, Ginestra Ciovene
The night sky looks like..., 218
Atwood, Margaret
Night falls. Or has fallen, 218
Murdin, Paul
...need the interruption of the night, 219
Stevenson, Robert Louis
Night is a dead monotonous period..., 219

novae

Gaposchkin, Sergei
Of exploring the Novae..., 221

observation

Adams, Douglas
See first, think later, then test, 222
Ayres, C.E.
...Moses...announced that his laws were based on direct observation, 222
Bolles, Edmund Blair
...a difference between scientific and artistic observation, 222
Cohen, Morris Raphael
Observation unilluminated by theoretic reason is sterile, 222
Dampier, Sir William Cecil
As observed by, Yours Faithfully, God, 223
Eddington, Sir Arthur
...divorced from opportunity for observational test, 223
Faraday, Michael
...an obscure and distorted vision is better than none, 223
Gay-Lussac, Joseph Louis
...wait until more numerous and exact observations have provided a solid foundation..., 223
Holton, Gerald
All intelligent endeavor stands with one foot on observation..., 224
Jeans, Sir James
Every observation destroys a bit of the universe..., 224
Jonson, Ben
Let me alone to observe..., 224
Lee, Oliver Justin
...depends finally upon some observation or measurement, 224
Lewis, Gilbert N.
...my eye touches a star..., 225
Lubbock, Sir John
What we see depends mainly on what we look for, 225
Meredith, George
Observation is the most enduring of the pleasures of life, 225

Mitchell, Maria
...nothing comes out more clearly in astronomical observations than..., 225
Orwell, George
To see what is in front of one's nose..., 225
Osler, Sir William
Man can do a great deal by observation..., 225
Shakespeare, William
By my penny of observation, 226
Smith, Theobald
It is the care we bestow on apparently trifling...minutiae..., 226
Thompson, W.R.
In the last analysis observation..., 226
Unknown
...others just say I'm seeing things, 226
The obscure we see eventually..., 226
von Goethe, Johann Wolfgang
Every act of looking turns into observation..., 226

observatory
Lowell, Percival
...to put up observatories where they can see..., 228
Mitchell, Maria
There is no observatory in this land..., 228
Rosseland, S.
...an astronomical observatory of to-day looks more like a..., 228

observe
Heisenberg, Werner
What we observe is not nature itself..., 223

observer
de Chambaud, J.J. Menuret
The name observer has been given to the physicist who..., 229
Deuteronomy 4:19
And lest thou lift up thine eyes upto heaven..., 229
von Goethe, Johann Wolfgang
...and not to suit observers..., 226

order
Browne, Sir Thomas
All things begin in Order..., 230
Frankel, Felice
Order is repetition..., 230
Kline, Morris
Is there a law and order in this universe..., 230
Mann, Thomas
Order and simplification are the first steps toward..., 230
Moulton, Forest Ray
Now we find ourselves a part of a Universal Order..., 230
Yang, Chen Ning
Nature possesses an order that..., 230

Orion
Teasdale, Sara
I saw Orion in the east..., 72
Tennyson, Alfred
...those three stars of the Airy Giants' zone..., 72

other worlds
Hippolytus
...somewhere worlds are coming to be..., 232
King, Stephen
There are other worlds than these, 232

Magnus, Albertus
Do there exist many worlds…, 232
paradox
Gilbert, W.S.
A most ingenious paradox…, 233
particle
Gleick, James
Quantum mechanics taught that a particle was not a particle but a smudge…, 234
Heisenberg, Werner
We can no longer speak of the behavior of the particle…, 234
Regnault, Pére
…if you divide a Particle into the most inconceivably minute Parts…, 234
particle physicists
Johnson, George
…the particle physicists place themselves with the angels…, 234
past
Barrow, John
…because they were as they were, 236
Whitman, Walt
The past, the infinite greatness of the past!, 236
pattern
Burns, Marilyn
Seeking patterns is a way of thinking…, 237
Gardner, Martin
Simple geometrical patterns…, 237
Jeffers, Robinson
…organized on one pattern…, 237
Peterson, Ivars
In their search for patterns and logical connections…, 237
phenomenon
du Noüy, Pierre Lecomte
When we speak of a phenomenon…, 239
Jevons, W. Stanley
…every strange phenomenon may be a secret spring…, 239
LaPlace, Pierre Simon
The phenomena of nature are…, 239
Wilson, Edward O.
…all tangible phenomena, from the birth of stars to…, 239
von Goethe, Johann Wolfgang
Search nothing beyond the phenomena…, 375
philosopher
Faraday, Michael
The philosopher should be a man willing to…, 241
philosopher of science
Ziman, John M.
…appointed authorities the Philosophers of Science, 242
philosophy
Dennett, Daniel
There is no such thing as a philosophy-free science…, 241
Durant, Will
…only philosophy can give us wisdom, 241
Inge, William
Science and philosophy can not be kept in…, 241
MacLaurin, Colin
It is not the business of philosophy…, 241
Raether, H.

Than are dreamt of in your philosophy, 242
Updike, John
Discomfort our philosophy, 242
Whitehead, Alfred North
Philosophy asks the simple question..., 242
Philosophy begins in wonder, 242
Philosophy is the product of wonder, 242

photon
Einstein, Albert
Every physicist thinks he knows what a photon is, 244
Unknown
If photons have mass, who is their priest?, 244

physicist
Adams, Douglas
Very strange people, physicists..., 245
Adams, Henry
...lies in the hands of physicists..., 245
Einstein, Albert
...the theoretical physicist as he stands before Nature..., 245
Green, Celia
If you say to a theoretical physcist..., 245
Hanson, N.R.
Physicists do not start from hypotheses..., 245
Johnson, George
Trying to capture the physicists' precise mathematical description of..., 246
Kush, Polykarp
...the role of the experimental physicist..., 246
Michelson, Albert A.
If a poet could at the same time be a physicist..., 246
Newman, James R.
...philosophers have let physicists get away with murder, 246
Rorty, Richard
...the physicists are men looking for new interpretations of the Book of Nature, 246
Toulmin, Stephen
...physicists seek the form of given regularities, 247
Unknown
...a physicist whose existence is postulated..., 247

physics
Bohr, Niels
The new situation in physics is..., 248
Born, Max
Hope is a word one is unlikely to find in the literature of physics, 248
Bronowski, Jacob
One achievement of physics in the twentieth century has been..., 248
Calvin, Melvin
...physics impinges on astronomy..., 248
Condon, E.U.
...the object of physics to organize past experience..., 249
Duhem, Pierre
Physics makes progress because..., 249
Edelstein, Ludwig

Physics… in antiquity remained closely connected with…, 249
Ehrenfest, Paul
Physics is simple, but subtle, 249
Gardner, Martin
In physics… there is never a sharp line separating pseudo-scientific speculation from…, 249
Goeppert-Mayer, Maria
Physics is puzzle solving, too…, 250
Hasselberg, K.B.
As for physics…, 250
Heidegger, M.
Modern physics is not experimental physics because…, 250
Heisenberg, Werner
… Physics advances by two distinct roads, 250
… a good description of how one should proceed in theoretical physics, 251
Jeans, Sir James
The classical physics seemed to bolt and bar the door…, 251
Koyre, Alexander
Good Physics is made *a priori*, 251
Lindsay, R.B.
… viewing the purpose of physics as…, 252
Mach, Ernst
… to adopt in physics a point of view…, 252
Maritain, J.
… physics thus advancing towards its destiny…, 252
Morgan, Thomas H.
Physics has progressed because…, 252
Pines, David
The central task of theoretical physics…, 253
Rabi, Isidor Isaac
I think that physics should be…, 253
Rutherford, Ernest
All science is either physics or stamp collecting, 253
Standen, Anthony
Physics is not about the real world…, 253
Teilhard de Chardin, Pierre
The true physics is that which will…, 253
Truesdell, Clifford
… the aim of theoretical physics is to construct mathematical models…, 254
Unknown
Introductory physics courses are taught at…, 254
… the words "theoretical physics" came out as…, 254
von Goethe, Johann Wolfgang
Physics must be sharply distinguished from mathematics, 254
Weinberg, Steven
Physics is not a finished logical system, 254
Wigner, Eugene P.
We have ceased to expect from physics…, 255
Zukav, Gary
… when most people think of 'physics'…, 255
pion
Marshak, Robert E.

But the glue that holds the nucleus of the atom together is a mystery..., 256

planet

Aristotle
That which does not twinkle is near..., 257

Burroughs, William
After one look at this planet..., 257

Chapman, Clark R.
Planets are like living creatures, 257

Chaucer, Geoffrey
The seven bodies I'll describe..., 257

Emerson, Ralph Waldo
He who knows what sweets and virtues are in... the planets..., 258

Joyce, James
Gasballs spinning about..., 258

Kahn, Fritz
...like bits of dust lost in immensity..., 258

Marlowe, Christopher
and measure every wand'ring planet's course..., 258

Miller, Hugh
The planet which we inhabit is but one vessel in the midst of..., 259

Moliére (Jean-Baptiste Poquelin)
A neighbouring planet did pass us close by..., 259

Siegel, Eli
The planets show grandeur..., 259

Standage, Tom
A planet is, by definition, an unruly object, 259

Teilhard de Chardin, Pierre
...it is only on the very humble planets..., 259

Tombaugh, Clyde
Many false planets shall appear..., 260

planetary science

Hammond, Allen L.
...planetary science has revived to become..., 258

planetry

Shapley, Harlow
Millions of planetry systems must exist..., 259

Pleiads

Tabb, John Banister
Who are ye with clustered light..., 73

Tennyson, Alfred
Many a night I saw the Pleiads..., 73

Pluto

Hoyt, William Graves
The planet was names Pluto..., 267

popular science

Mitchell, Maria
The phrase 'popular science'..., 305

positron

Eddington, Sir Arthur
A positron is a hole..., 268

problem

Douglas, A. Vibert
All the problems of the physical universe are..., 269

Einstein, Albert
Therre are so many unsolved problems in physics, 269

Halmos, Paul R.
A teacher who is not always thinking about solving problems..., 269

Hawkins, David

There are many things you can do with problems. . . , 269
Hilbert, David
. . . offers an abundance of problems. . . , 270
Kiepenheuer, Karl
. . . the inexhaustible store of problems. . . , 270
Unknown
What is the best you can do for this problem?, 270
progress
Clerke, Agnes M.
Progress is the result, not so much of. . . , 271
Duhem, Pierre
Scientific progress has often been compared to a mounting tide. . . , 271
Feyerabend, Paul
. . . the only principle that does not inhibit progress is. . . , 272
Poincaré, Lucien
There are no limits to progress. . . , 272
progress of science
Lee, Tsung Dao
The progress of science has always been the result of. . . , 272
Whitehead, Alfred North
The progress of Science consists in. . . , 272
proof
Aristotle
To prove what is obvious. . . , 273
Davis, Philip
He rests his faith on rigorous proof. . . , 273
Gleason, Andrew
Proofs really aren't there to. . . , 273
prove
Cabell, James Branch
But I can prove it by mathematics. . . , 273
quantum
Bohr, Niels
There is no quantum world, 277
Carroll, Lewis
Curiouser and curioser. . . , 277
Einstein, Albert
The more one chases after quanta. . . , 277
Lawrence, D.H.
I like relativity and quantum theories. . . , 278
Zee, A.
Welcome to the strange world of the quantum. . . , 279
quantum electrodynamics
Feynman, Richard P.
The theory of quantum electrodynamics describes Nature as. . . , 277
quantum mechanics
Heisenberg, Werner
Evidently there exists another "quantum mechanics", 278
quantum theory
Gell-Mann, Murray
. . . and thoroughly confusing discipline called quantum mechanics. . . , 277
Heisenberg, Werner
. . . quantum theory reminds us. . . , 278
Polkinghorne, J.C.
Quantum theory is both stupendously successful as an account of the small-scale structure of the world and. . . , 278

quasar
Mundell, Carole
...observing quasars is like..., 280
question
Kundera, Milan
...the only truly serious questions are..., 281
Landau, Lev
Physicists have learned that certain questions cannot be asked..., 281
Payne-Gaposchkin, Cecilia H.
...they remain a great, intriguing question, 281
Popper, Karl
These are the questions you should take up, 281
radio
Kraus, John
The radio sky is no carbon copy of the visible sky..., 283
radio astronomers
Mitton, Simon
...radio astronomers have led a revolution in our knowledge of the Universe..., 283
radio astronomy
Christiansen, Chris
Radio astronomy was not born with a silver spoon in its mouth, 283
Gingerich, Owen
But even if radio astronomy has not so much destroyed..., 283
Unsold, Albrecht
The old dream of wireless communication through space..., 284
radio data
Hoyle, Fred
...radio data serves like a good dog on a hunt, 85
real
Reeve, F.D.
...we cannot see nine-tenths of what is real..., 287
Weinberg, Steven
When we say that a thing is real..., 287
reality
Dürrenmatt, Friedrich
...to surrender to reality, 285
Egler, Frank E.
Reality is not what is..., 285
Einstein, Albert
...conceptually to grasp reality..., 285
In our endeavor to understand reality..., 285
Frankel, Felice
Our reality is an illusion, 286
Frost, Robert
You're searching, Joe/For things that don't exist, 286
Goodwin, Brian
Many people think reality is prosaic, 286
Jeans, Sir James
...not yet in contact with ultimate reality, 286
Mead, George H.
The ultimate touchstone of reality is..., 286
Raymo, C.
Science is a map of reality, 287
Weyl, Hermann
A picture of reality drawn in a few sharp lines..., 287
Wilde, Oscar
To test Reality we must see it on the tight rope, 287
reason
Gore, George

That which is beyond reason..., 288
John of Salisbury
Reason... is a mirror..., 288
Shakespeare, William
The reason why the seven stars are no more..., 289
von Goethe, Johannes Wolfgang
...Human reason holds the balance between them..., 289
reasoning
Kline, Morris
What mathematics accomplishes with its reasoning..., 288
redshift
Gamow, George
The discovery of the redshift..., 290
Gray, George
...the shifting of starlight into the red indicates..., 290
Jeans, Sir James
The redshift we observe in the spectra..., 290
Stapledon, Olaf
I noticed that the sun and all the stars... were ruddy, 291
references
de Cervantes, Miguel
Now let us come to those references..., 428
relativity
Eddington, Sir Arthur Stanley
...the moral of the theory of relativity is..., 292
Einstein, Albert
We shall endeavor to do this for the Theory of Relativity..., 292
Page, Leigh
...the truth of the relativity theory..., 292
research
Bachrach, Arthur J.
...people don't usually do research the way people who write books..., 293
Bush, Vannevar
...workers sometimes proceed in erractic ways, 293
Veblen, Thorstein
The outcome of any serious research..., 293
rule of nature
Ford, Kenneth W.
One of the elementary rules of nature is..., 212
Sagittarius
Longfellow, Henry Wadsworth
The centaur, Sagittarius..., 73
Saint Augustine Era
Cardenal, Ernesto
Before the big explosion there wasn't even empty space..., 414
Saint Augustine
What did God before He made heaven and earth?, 415
Tzu, Lao
Born before the heaven and earth..., 415
Saturn
Huygens, Christiaan
It is surrounded by a thin flat ring..., 265
Melville, Herman
...among the moons of Saturn, 265
Pallister, William
The planet Saturn, to the naked eye..., 265
Thayer, John H.
...go with me to Saturn, 266

scattering
Rutherford, Ernst
...this scattering backward must be the result of a single collision..., 294
science
Astbury, W.T.
...science is truly one of the highest expressions of human culture..., 295
Black, Hugh
The limits of science are..., 295
Bloom, Allan
Science, in freeing men..., 295
Born, Max
...no philosophical highroad in science..., 295
Buffon, Georges
The only good science is the knowledge of facts..., 296
Bury, J.B.
Science has been advancing without interruption during..., 296
Buzzati-Traverso, Adriano
Science is a game, 296
Campbell, Norman
...science is a body of useful and practical knowledge..., 296
Camus, Albert
...science that was to teach me everything..., 296
Carrel, Alexis
...a strange disparity between the sciences of..., 297
Chargaff, Erwin
Science cannot be a mass occupation..., 297
Science is much better in explaining than in understanding..., 297
The sciences have started to swell, 297
To be a pioneer in science..., 298
In science, there is always one more Gordian knot..., 298
Davis, W.M.
Science is not final any more than it is infallible, 298
del Rio, A.M.
...he who has once imbibed a taste for science..., 298
Dobshansky, Theodosius
Science is cumulative knowledge, 298
Science does more than collect facts..., 299
Science has been called..., 299
Egler, Frank E.
...science...ever reflecting a faith in the intelligibility of nature, 299
Science is a product of man..., 299
Einstein, Albert
...the goal of science is to discover rules..., 299
Feyerabend, Paul
The idea that science can..., 299
Feynman, Richard
Science is a way to teach..., 300
Fulbright, J. Williams
Science has radically changed the conditions of human life..., 300
Gould, Stephen Jay
The net of science covers..., 300
Heisenberg, Werner
Almost every progress in science has..., 300
Science no longer confronts nature as..., 300
Hippocrates

There are in fact two things, science and opinion..., 300
Hodgson, Leonard
Science can only deal with what is..., 301
Holton, Gerald
Science is an ever-unfinished quest to discover facts..., 301
...science has grown almost more by what it has learned to ignore..., 301
Hoyle, Sir Fred
The fragmentation of science is a source of difficulty..., 301
Huxley, Aldous
Science is dangerous..., 301
Jeffers, Robinson
Science is not to serve but to know..., 301
Jourdain, Philip E.B.
Our ideal in natural science is to..., 302
Keyser, Cassius Jackson
Science is destined to appear as the child..., 302
Kuhn, Thomas
To understand why science develops..., 303
Lowell, Percival
In science there exists two classes of workers, 303
Mach, Ernst
Science acts and only acts in the domain of..., 304
Maffei, Paolo
...entering a world in which science and fantasy intertwine, 304
Margulis, Lynn
...Science has become a social method of..., 304
Mendeléeff, D.
...free and joyful is life in the realms of science..., 304
While science is pursuing..., 304
Millikan, Robert A.
...Science walks forward on two feet..., 305
Peacock, Thomas Love
...it is the ultimate destiny of science to exterminate the human race, 306
Popper, Karl
...science is most significant as one of the greatest spiritual adventures..., 306
Russell, Bertrand
A life devoted to science is therefore a happy life..., 306
Smolin, Lee
Science is above everything else..., 306
Soddy, Frederick
As science advances..., 306
Whitehead, Alfred North
Science can find no individual enjoyment in Nature..., 307
scientific discoverers
Verne, Jules
...no more envious race of men than scientific discoveres, 309
scientific honesty
Renan, Ernest
Orthodox people have as a rule very little scientific honesty, 308
scientific method
Bridgman, P.W.
The scientific method is...is nothing more than..., 308

scientific methodology
Medawar, Peter
If the purpose of scientific methodology is to..., 308
scientific spirit
Rabi, I.I.
The essence of scientific spirit is..., 308
scientific truth
Wells, H.G.
Scientific truth is the remotest of mistresses, 309
scientist
Calder, Alexander
Scientists leave their discoveries..., 310
Erasmus, Desiderius
The scientists, reverenced for their beards..., 310
Gray, George W.
The modern scientist is like a detective..., 311
Mitchell, Maria
The true scientist must be..., 311
...the highest joy of the true scientist..., 311
Ting, Samuel C.C.
Scientists must go beyond what is taught in the textbook..., 311
Unknown
...the scientist must ever seek in vain to..., 311
Young, J.Z.
One of the characteristics of scientists and their work..., 311
Scorpio
Longfellow, Henry Wadsworth
Though on the frigid scorpion I ride..., 73
seeing
Herschel, William
Seeing is in some respect an art which must be learnt, 224
sense organs
Born, Max
...observed with the help of our sense organs..., 312
senses
Einstein, Albert
...see the universe by the impression of our senses..., 312
shadow
Aesop
...grasping at the shadow, 313
Eddington, Sir Arthur Stanley
The shadow of my elbow rest on the shadow-tables..., 313
simple
Chandrasekhar, Subrahmanyan
The simple is the seal of the true..., 315
simplicity
Bailey, Janet
...hides an ultimate simplicity, 315
Einstein, Albert
...the fundamental laws are simplified more and more as experimental research advances, 315
Gore, George
Simplicity, whether truthful or not..., 315
Haldane, J.B.S.
The catch in this criterion lies in the word "simplest", 316
Heisenberg, Werner
...speaking of simplicity and beauty..., 316
Poincaré, Henri

It is because simplicity, because grandeur, is beautiful..., 316
Reid, Thomas
...led into errors by the love of simplicity..., 316
Wheeler, John
The glittering central mechanism of the world in all its...simplicity, 316
Sirius
Grondal, Florence Armstrong
...sparkling brilliance would pale beside Sirius..., 343
Tennyson, Alfred
...the firey Sirius alters hue..., 351
sky
Astronomy Survey Committee
...the starry sky on a clear, dark night, 317
Brandt, John C.
...to all who look to the sky..., 317
Browning, Robert
Sky-what a scowl of cloud..., 317
de Saint-Exupéry, Antoine
A sky as pure as water..., 317
Emerson, Ralph Waldo
The sky is the daily bread of the eyes, 318
Friedman, Herbert
...to look at a star-filled sky..., 318
Kremyborg, Alfred
The sky is that beautiful old parchment..., 318
Lockwood, Marion
...watched the sky passing through its changing gamut of color..., 318
Lowell, Amy
Watching the stars pass across the sky..., 318
Manilus
...spend my life touring the boundless skies..., 318
Maunder, E. Walter
The oldest picture book...is the Midnight Sky, 319
Moulton, Forest Ray
...awe-inspiring spectacle than that presented by the sky..., 319
Schaefer, Bradley E.
The sky is beautiful and vast..., 319
Shakespeare, William
My soul is in the sky, 319
Shore, Jane
Each night the sky splits open like a melon..., 319
Smoot, George
There is something about looking at the night sky..., 319
Swings, Pol
The sky belongs to everyone..., 319
Whitman, Walt
Over all the sky..., 320
Solar System
Carlyle, Thomas
...he had not two Planets to conquer, 321
Davies, Paul
Our solar system is a tiny island of activity..., 321
Horowitz, Norman H.
If the exploration of the solar system in our time bring home to us..., 321
Lambert, Johann Heinrich
Nothing is more simple than the Solar System..., 321
Lowell, Percival

...the centre of a solar system grander than our own..., 321
MacLennan, Hugh
We have just reached the outer fringes of the Solar System, 322
Southern Cross
Meredith, Owen
The great cross of the South, 74
space
Archytas of Tarentum
If I am at the extremety of the heaven..., 323
Arnold, James R.
Space is the empty place next to the full place..., 330
Bradley, John Hodgdon Jr
...such is the sea of space, 323
Carlyle, Thomas
...Space is but a mode of our human Sense..., 323
Deudney, Daniel
Space is only 80 miles from every person on earth..., 324
Eddington, Sir Arthur Stanley
...space is not a lot of points close togetner..., 325
Empson, William
Space is like earth..., 325
Gibran, Kahlil
Space is not space between the earth and the sun to one who..., 325
Glenn, John Jr
In space one has the inescapable impression that..., 325
Hale, George Edward
...the lure of the uncharted seas of space..., 325
Joubert, Joseph
There is something divine about the ideas of space and eternity..., 325
Murray, Bruce
Space...is a colorful thread..., 326
Ockels, Wubbo
Space is so close..., 326
Reade, Winwood
...mankind will migrate into space..., 326
Russell, Bertrand
...form an impassable barrier to space..., 326
Siegel, Eli
Space won't keep still..., 326
Smith, Logan Pearsall
...crashing blindly forever across the void of space, 326
Valéry, Paul
Space is an imaginary body..., 327
vas Dias, Robert
...outer space is as much a territory of the mind as it is a..., 327
von Braun, Wernher
Don't tell me that man doesn't belong out there, 327
Weyl, H.
...as in the problem of space, 327
space exploration
Murray, Bruce
I think space exploration is as important as..., 331
space flights
Jung, C.G.
Space flights are merely an escape..., 331
spacetime
Berlinski, David

...thus creating time and space..., 328
MacLeish, Archibald
Space–time has no beginning..., 328
Minkowski, Hermann
The views of space and time which I wish to lay before you..., 328
Thorne, Kip S.
...spacetime is like a piece of wood impregnated with water, 329
Wheeler, John A.
...life as a spacetime expert..., 328
Space and time, unified as spacetime..., 329
space travel
Blagonravov, Anatoly A.
The exploration of the cosmos..., 330
Burroughs, William
Man is an artifact designed for space travel, 330
Dyson, Freeman J.
...a million species spreading through the galaxy..., 330
Firsoff, V.A.
...if we go into space..., 331
Lucretius
He has ventured far beyond the flaming ramparts of the world..., 331
Thoreau, Henry David
...the navigation of space..., 332
Tsiolkovsky, Konstantin Eduardovich
Mankind will not stay on earth forever, 332
Unknown
...an annual free trip around the Sun, 332
Verne, Jules
...projectiles providing comfortable transportation between Earth and the Moon, 332
von Braun, Wernher
[Space travel] will free man from..., 332
Wells, H.G.
...and reach out their hands amidst the stars, 333
spectra
Huggins, William
I looked into the spectroscope, 334
spectral
Stedman, Edmund
Of spectral line..., 334
spectroscope
Maxwell, James Clerk
...full tale into the spectroscope..., 334
spectrum analysis
Twain, Mark
Spectrum analysis enabled the astronomer to tell..., 335
spin
Goudsmit, Samuel A.
...most young physicists do not know that spin had to be introduced..., 336
spiral arms
van de Hulst, H.C.
The discovery of spiral arms..., 337
star
Acton, Loren
When you look out the other way toward the stars..., 338
Aeschylus
I know yon midnight festival/Of swarming stars..., 338

Alighieri, Dante
...again to see the stars, 338
Aurelius, Marcus
Look round at the courses of the stars..., 338
Brown, Fredric
...the lights in the sky that are stars, 339
Bryant, William Cullen
The glorius host of light..., 339
Bunting, Basil
The star you steer by is gone, 339
Burke, Edmund
...the stars lie in such apparent confusion..., 339
Burnet, Thomas
They lie carelessly scatter'd..., 339
Campbell, Thomas
And the sentinel stars..., 340
In yonder pensile orb..., 340
Carlyle, Thomas
When I gaze into these stars..., 340
Clarke, Arthur C.
...it's full of stars, 340
Cole, Thomas
When the stars come out to watch the daylight die, 340
Coleridge, Samuel Taylor
...the stars hang bright above..., 341
Crane, Hart
Stars scribble on our eyes..., 341
Darwin, Erasmus
Roll on, ye stars..., 341
de Saint-Exupéry, Antoine
All men have stars..., 341
de Tabley, Lord
The star a million years..., 341
Eddington, Sir Arthur Stanley
...we shall be competent to understand so simple a thing as a star, 342
...the conditions in the stars not sufficiently extreme..., 342
Eliot, George
The stars are golden fruit..., 342
Emerson, Ralph Waldo
Hitch your wagon to a star, 343
The stars awaken a certain reverence..., 342
...let him look at the stars, 342
...Those delicately emerging stars..., 342
Flecker, James Elroy
...the fleet of stars is anchored..., 343
Goodenough, Ursula
I lie on my back under the stars..., 343
Grondal, Florence Armstrong
...all the wondrous phenomena of visible stars..., 343
Guiterman, Arthur
He touches his wand to the Evening Star, 343
...Stars are underdeveloped Fireflies, 344
Habington, William
The starres, bright sentinels of the skies, 344
...Celestial sphere,/So rich with jewels hung..., 344
Hardy, Thomas
The kingly brilliance of Sirius..., 344
...the twinkling of all the stars seemed to be..., 344
Hearn, Lafcadio
...whose foam is stars..., 344
Heine, Heinreich

...the stars in the sky only appear to us to be so beautiful and pure because..., 345
Herrick, Robert
The starres of the night..., 345
Hodgson, Ralph
My eyes were blind with stars..., 345
Homer
...the stars shine clear..., 345
Hoyle, Fred
The stars are best seen as a spectacle..., 345
Huxley, Julian
Dead stars innumerable lie..., 346
Isaiah 40:26
Lift up your eyes on high and see..., 346
Jacobson, Ethel
Crystal fish..., 346
Jeffers, Robinson
...and watched the stars pass, 346
Job 38:7
The morning stars sang together..., 346
Keats, John
Bright star, would I were steadfast as thou art..., 347
Krutch, Joseph Wood
The stars are twinkling rogues..., 347
Longfellow, Henry Wadsworth
The stars arise, and the night is holy, 347
Lowell, Percival
...cause one to see stars, 347
Mandino, Og
...I will endure the darkness for it shows me the stars, 347
Meredith, George
...the stars/Which are the brains of heaven..., 347
Milton, John
...and every star perhaps a world of destined habitation..., 348
And all the spangled hosts..., 348
Mitchell, Maria
...a look at the stars will show us the littleness of our own interests, 349
We call the stars garnet and sapphire..., 349
Pagels, Heinz R.
Stars are born, they live and they die, 349
Plato
...telling about all the figures of the stars..., 349
Poe, Edgar Allan
...the multitudinous vistas of the stars..., 349
Ptolemy
close-knit encompassing convolutions of the stars..., 349
Raymo, Chet
I put up summer stars like vegetables in jars..., 350
Service, Robert
But the stars sing an anthem of glory..., 350
Shakespeare, William
...these blessed candles of the night..., 350
Smythe, Daniel
...the depths of stars and space, 350
Spenser, Edmund
He that strives to touch the stars..., 351
Taylor, Bayard

Each separate star..., 351
Teasdale, Sara
Stars over snow..., 351
Thompson, Francis
Without troubling of a star, 351
Thoreau, Henry David
The stars are the apexes of..., 351
Truly the stars were given for..., 352
Travers, P.L.
...Mary Poppins sticking on the stars..., 352
Trevelyan, G.M.
The stars out there rule the sky..., 352
Twain, Mark
Look what billions and billions of stars there is, 352
There are too many stars in some places..., 352
Unknown
...a star is a pretty simple structure..., 353
...the rest of us will go to the stars, 352
Vaughan, Henry
The jewel of the just..., 353
Webster, John
We are merely the stars' tennis-balls..., 353
Whitman, Walt
...myriads of other globes, 353
Wordsworth, William
Look for the stars..., 353
star shower
Frost, Robert
To see the star shower known as Leonid..., 189
study
Rowland, Henry Augustus
The whole universe is before us to study, 354
Skinner, B.F.
...drop everything else and study it, 354
Tsiolkovsky, K.E.
...the era of more intensive study of the heavens, 354
stupidity
Einstein, Albert
...sacrifice at the altar of stupidity..., 355
sun
Bourdillon, Francis William
With the dying Sun, 356
Deutsch, Armin J.
The face of the sun is not without expression..., 356
Dryden, John
The glorious lamp of heaven, the radiant sun..., 356
Ecclesiastes 11:7
...it is pleasant for the eyes to behold the sun..., 356
Falconer, William
High in his chariot glow'd the lamp of day, 356
Gilbert, W.S.
The Sun, whose rays/Are all ablaze..., 357
Heraclitus
The sun...is new each day, 357
Kelvin, Lord
Now, if the sun is not created a miraculous body..., 357
Longfellow, Henry Wadsworth
Down sank the great red sun..., 357
Macpherson, James
Whence are thy beams, O sun!, 357
Muir, John
The sun...seems conscious of the presence of every living thing..., 357
Nietzsche, Friedrich
...the glowing Sun..., 358

Parker, E.N.
The riddles the sun presents are..., 358
Pascal, Blaise
Let him gaze on that brilliant light..., 358
Starr, Victor P.
It has always been easier to record and describe solar events than to..., 358
Swift, Jonathan
...the sun...will at last be wholly consumed and annihilated..., 358
Thoreau, Henry David
...I never assisted the sun materially in his rising..., 359
Wells, H.G.
The sun, red and very large..., 359
Xenophanes
The sun comes into being each day..., 359

sunspots
Galilei, Galileo
...nor are the sunspots stars, 360
Zirin, Harold
...the sunspots, the elegant prominences..., 360

supernova
Crowley, Abraham
A Star, so long unknown, appears..., 361
Herschel, J.F.W.
...subsequent extinction of a new and brilliant fixed star..., 361
Schaaf, Fred
...a star gone to seed..., 362
Woosley, Stan
During the supernova's first 10 seconds..., 362

Supreme Disposer
Lambert, Johann Heinrich
If we admit the existence of a Supreme Disposer..., 138

symmetry
Ferris, Timothy
...at symmetry's bubbling spring, 363
Weyl, Hermann
Symmetry is a vast subject..., 363
Wickham, Anna
God, Thou great symmetry,/Who put a biting lust in me..., 363
Yang, C.N.
...the simple mathematical representations of the symmetry laws, 363

teach
Gauss, Carl Friedrich
...two courses of lectures to three students..., 365
Regnault, Pére
Will you discover to me..., 365
Stoppard, Tom
If you don't teach me the true meaning of things..., 365

telescope
Bierce, Ambrose
A device having a relation to the eye..., 366
Emerson, Ralph Waldo
The sight of a planet through a telescope..., 367
Holmes, Oliver Wendell
But most of all I love the tube that spies..., 367
Hubble, Edwin
...the utmost limits of our telescope, 367
Kepler, Johannes
What now...shall we make of our telescope?, 367

Lovell, A.C.B.
Astronomy has marched forward with the growth in size of its telescopes, 368
Mitchell, Maria
The tube of Newton's first telescope..., 368
Mullaney, James
The telescope is not just another gadget..., 368
Panek, Richard
The relation between the telescope and..., 368
Peltier, Leslie C.
Old telescopes never die..., 368
Rowan-Robinson, Michael
Now it is the astronomer's telescope..., 368
Ryder-Smith, Roland
Through tempered glass, his window on the sky..., 369
Toogood, Hector B.
If held the wrong way... the telescope is of little or no use, 369
Tsiolkovsky, Konstantin
...limited power of our telescope..., 369
Vezzoli, Dante
Cyclopean eye that sweeps the sky..., 369
Wordsworth, William
A Telescope upon its frame..., 370

telescope shelter

Vehrenberg, Hans
...working peacefully in my telescope shelter..., 369

theorize

von Goethe, Johann Wolfgang
...we theorize every time we look carefully at the world, 375

theory

d'Abro, A.
...a theory of mathematical physics is..., 371
Bethe, Hans
Scientific theories are not overthrown..., 371
Duhem, Pierre
...which will organize them into a theory!, 371
Einstein, Albert
For the creation of a theory..., 371
Creating a new thoery is not like destroying a barn..., 372
Physical theories try to form a picture of reality..., 372
Goldhaber, Maurice
Theories in physics are like that, 372
Koestler, Arthur
The history of cosmic theories..., 372
Laszlo, E.
...theories, like window panes..., 372
Libes, Antoine
Let us add a word in favour of theories..., 373
Lowell, Percival
The test of a theory is..., 373
Popper, Karl
We have no reason to regard the new theory as better than the old theory..., 374
...the more a theory forbids..., 374
Rothman, Tony
...the expected lifetime of a theory..., 374

Slater, John C.
A theoretical physicist... asks just one thing of his theories..., 374
Synge, J.L.
A well built theory has three merits..., 375
Wheeler, John A.
So maybe I'll end up loving your theory, 375
Wisdom, J.O.
Sometimes it is used for a hypothesis..., 375
thermodynamics
Atkins, P.W.
Everything is driven by motiveless, purposeless decay, 376
Cardenal, Ernesto
The second law of thermodynamics! ..., 376
Dickerson, Richard E.
It is possible to know thermodynamics without understanding it, 376
Hoffmann, Roald
My second law, your second law, ordains..., 376
Pippard, A.B.
It may be objected by some..., 376
Reiss, H.
...the almost certain truth that nobody... understands thermodynamics..., 377
Almost all books on thermodynamics contain some errors which are..., 377
Truesdell, Clifford
Thermodynamics is the kingdom of deltas, 377
Every physicist knows exactly what the first and second law mean..., 377
Unknown If you think things are mixed up now..., 377
time
Barnett, Lincoln
Time itself will come to an end, 378
Carlyle, Thomas
...no hammer in the horologue of Time..., 378
Eddington, Sir Arthur
Whatever may be time de jure..., 378
Lightman, Alan
There is a place where time stands still..., 378
Lyell, Charles
...to find that in time also the confines of the universe lie beyond..., 378
Mann, Thomas
Time has no divison marks..., 379
McLuhan, Marshall
Our time is a time for crossing barriers..., 379
Poinsot, Louis
If anyone asked me to define time..., 379
St Augustine
What is time?, 379
Shakespeare, William
...events in the womb of time..., 379
...to inquire the nature of time, 380
Silesius, Angelus
...One step across/that line called Time..., 380
Swinburne, Richard

. . . each state of the universe at each instant of time. . . , 380
Unknown
. . . not defining time, but only with measuring it, 380
Weil, Simone
Time is an image of eternity. . . , 380
Wells, H.G.
There are really four dimensions, Three of which we call the three planes of Space and a fourth, Time, 380
time travel
Allen, Elizabeth Akers
Backward, turn backward, O Time. . . , 382
travel
Lowell, Percival
. . . travel and discovery have called with strange insistence to him who. . . , 331
true
Faraday, Michael
Nothing is too wonderful to be true, 383
truth
Balfour, Arthur James
. . . our beliefs grow less inadequate to the truths. . . , 383
Black, Max
Scientists can never hope to be in possession of the truth. . . , 383
Cole, William
. . . in our inquiries after truth. . . , 383
Feynman, Richard
You can recognize truth by its beauty and simplicity, 384
Gore, George
The deepest truths require. . . , 384
Halmos, Paul
. . . the joy of suddenly discovering a hitherto unknown truth. . . , 384
Hawkins, Michael
Scientific truths is simply another way of saying. . . , 384
Teilhard de Chardin, Pierre
. . . do we pursue truth. . . , 384
UFO
Sagan, Carl
Do you believe in UFOs?, 385
uncertain
Hoyle, Fred
If matters still seem very uncertain. . . , 386
uncertainty
Unknown
Heisenberg might have slept here, 386
understand
Galilei, Galileo
. . . an opinion on things that you do not understand?, 387
Pagels, Heinz, R.
The attempt to understand the origin of the universe. . . , 388
understanding
Atiyah, Michael
It is hard to communicate understanding. . . , 387
Ferris, Timothy
. . . bedrock understanding of cosmic structure, 387
Heisenberg, Werner
. . . the degree of understanding. . . , 388
Rabi, Isidor

Scientific understanding... is an essential step to..., 388
Walker, Kenneth
...all understanding of the universe comes from..., 388
universe
Adams, Douglas
The universe... is an unsettlingly big place..., 389
Amaldi, Ginestra Giovene
...roamed far and wide through distant reaches of the Universe, 389
Bacon, Francis
The Universe is not to be narrowed down..., 389
But the universe to the eye..., 389
Barrow, John D.
...we can never know the origins of the universe, 408
Barth, John
All the scientists hope to do is describe the universe mathematically..., 390
Blount, Sir Thomas Pope
Whoever surveys the curious fabric of the universe..., 390
Bruno, Giordano
The universe is then one..., 390
...the universe is all center..., 390
Camus, Albert
...the benign indifference of the universe, 390
Clark, Roger N.
The beauty of the universe defies description, 391
Nothing can compare to viewing the universe directly, 391
Clarke, Arthur C.
...the universes that drift like bubbles..., 391
...the universe has no purpose and no plan..., 391
...that the universe has the slightest interest in..., 391
Cook, Peter
I am very interested in the Universe..., 391
Copernicus, Nicolaus
...they try to conclude that the universe is finite, 391
Darling, David
...the universe has performed its most astonishing creative act, 392
Davy, Humphry
The more the phenomena of the universe are studied..., 392
Dawkins, Richard
The universe we observe has..., 392
Day, Clarence
A universe capable of giving birth to so many accidents is..., 392
de Fontenelle, Bernard Le Bovier
...methought the universe was too straight..., 393
de Sitter, W.
...the Universe bears all the marks of..., 393
de Vries, Peter
Anyone informed that the universe is expanding..., 393
The universe is like a safe..., 393
Deutsch, Karl W.

Any universe uneven enough to..., 393
Dyson, Freeman J.
I have found a universe growing without limit..., 393
The hypothesis is that the universe is constructed according to..., 394
Eddington, Sir Arthur Stanley
...the universe should accomplish some great scheme..., 394
Ehrmann, Max
...the universe is unfolding as it should, 394
Emerson, Ralph Waldo
...the universe is the property of every individual in it, 394
Engard, Charles J.
We accept the universe as far as we know it..., 395
Estling, Ralph
I do not know what...the Universe has in its mind..., 395
...there being design in the Universe, 395
Farmer, Phillip José
The universe is a big place..., 395
Flammarion, Camille
...the forces which rule the universe cannot remain inactive, 395
...the universe will move on as at present..., 395
France, Anatole
The universe which science reveals to us is..., 396
Frayn, Michael
The complexity of the universe is beyond expression..., 396
We look at the taciturn, inscrutable universe, and cry..., 220
Galilei, Galileo
Philosophy is written in this grand book, the universe..., 396
Giraudoux, Jean
...the whole universe is listening to us..., 397
Guth, Alan
The inflationary model of the universe provides..., 397
Halacy, D.S. Jr
Our universe operates not at the whims of..., 397
Harrison, Edward
...if the universe were structured in any other way..., 397
The universes are our models of the Universe, 397
When we doubt the Universe we doubt ourselves, 397
...we know of only one Universe, 398
Haught, James A.
The universe is a vast, amazing, seething dynamo..., 398
Hinshelwood, C.N.
...knowledge of the universe has been an end possessing in itself..., 398
Hogan, John
...attempts at understanding the universe..., 398
Hoyle, Fred
...our inquiry into the nature of the Universe..., 398

...the evaluation of this quite incredible Universe..., 398
There is a coherent plan in the universe..., 399
Hubble, Edwin Powell
...man explores the universe around him..., 399
Huygens, Chistrianus
...the magnificient Vastness of the Universe!, 399
Ionesco, Eugene
The universe seems to me infinitely strange..., 399
James, William
Whatever universe a professor believes in..., 399
Joad, C.E.M.
...and speculates about the universe as a whole..., 400
Kant, Immuanual
The world-edifice puts one into a quiet astonishment..., 400
Keillor, Garrison
...is the universe only one seed in one apple on a tree..., 400
Kirshner, Robert P.
Although the universe is under no obligation to make sense..., 401
Koestler, Arthur
...I regarded the universe as an open book..., 401
The universe has lost its core, 401
Mach, Ernst
The universe is like a machine..., 401
Melville, Herman
The universe is finished..., 401
Muir, John
...we find it hitched to everything else in the universe, 401
The clearest way into the Universe is..., 402
Ovenden, M.W.
...he knew that the Universe would turn out to be harmonious..., 402
Pascal, Blaise
By space the universe encompasses..., 402
Plato
...the accounts now given concerning the Universe..., 402
Poe, Edgar Alan
...of the Material and Spiritual Universe..., 402
...the perceptible universe exists as..., 402
...that the perceptible Universe exists..., 402
Polanyi, Michael
The universe is still dead..., 403
Ramon y Cajal, Santiago
...the universe will also be a mystery, 403
Reade, Winwood
The universe is anonymous..., 403
Reichenbach, Hans
Instead of asking for a cause of the universe..., 403
Renard, Maurice
Man, peering at the universe..., 403
Richards, Rheodore William
...mysteries of a universe..., 403
Rothman, Tony
...the order and beauty of the universe..., 404

Russell, Bertrand
The universe may have a purpose..., 404
...the universe has crawled by slow stages..., 404
Santanyana, George
The universe...is a wonderful and immense engine..., 404
Shelley, Percey Bysshe
Its easier to suppose that the universe has existed from..., 404
Siegel, Eli
The universe is Why, How, and What..., 405
The weight of the universe is..., 405
The universe, being clever..., 405
Spenser, Edmund
...of other worlds he happily should hear?, 405
Stern, S. Alan
The place we call our Universe..., 405
Swann, W.F.G.
There is one great work of art, it is the universe, 405
Swenson, May
What is it about the universe..., 406
Tennyson, Lord Alfred
...in a boundless universe..., 406
Thom, Rene
...the universe we see is a ceaseless creation..., 406
Thompson, Francis
The universe is his box of toys, 406
Thoreau, Henry David
The universe is wider than our views of it, 406
Toynbee, Arnold
...we gaze with frightened eyes into the dark universe, 406
Tyron, E.P.
...our universe is simply one of those things..., 407
Tzu, Lao
In the universe the difficult things are done as if..., 407
Unknown
...to look the universe in the face and not flinch, 407
Weinberg, Steven
The more the universe seems comprehensible..., 407
...the urge to trace the history of the universe..., 407
Wheeler, John Archibald
The Universe is a self-excited circuit, 407
...this is our Universe..., 407
We will first understand how simple the universe is..., 407
Whitman, Walt
...before a million universes, 408
Wiechert, Emil
The universe is infinite in all directions, 408
Young, Louise B.
The universe is unfinished..., 408
universe, cosmogenesis
Genesis
In the beginning..., 409
Singer, Isaac Bashevis
Who created the World?, 409
Spenser, Edmund
Through knowledge we behold the world's creation..., 409

Tagore, Rabindranath
...creation is not fettered by rules..., 410
universe, dying
Balfour, A.J.
The energies of our system will decay..., 410
Byron, George Gordon
...and the stars/Did wander darkling in the eternal space..., 411
Davies, Paul
...perpetual night will surely fall, 411
A universe that came from nothing..., 411
de Goncourt, Jules
Gentlemen it's closing time, 411
Dyson, Freeman J.
...the universe is on a one way slide..., 411
Eddington, Sir Arthur Stanley
...I may describe the end of the physical world as..., 411
Eliot, T.S.
This is the way the world ends..., 412
Frost, Robert
Some say the world will end in fire..., 412
Harrison, Edward
...a universe condemned to become a galactic graveyard, 412
Huxley, Julian
Not dead but doomed to die, 412
James, William
...the last expiring pulsation of the universe's life might be..., 413
Jeffers, Robinson
When the sun shall die..., 413
Nicholson, Norman
...if the universe/Reversed and showed/The colour of its money..., 413
Russell, Bertrand
In the vast death of the solar system..., 414
Wells, H.G.
...a steady twilight brooded over the Earth..., 414
Yeats, William Butler
When shall the stars be blown about the sky..., 414
unknown
Asimov, Isaac
If it is exciting to probe the unknown..., 416
France, Anatole
...the desire of the unknown makes beautiful the Universe, 396
Huxley, Thomas H.
The known is finite, the unknown infinite..., 416
Lindberg, Charles H.
...the farther we penetrate the unknown..., 416
Nicholson, Norman
The unknown is shown/Only by a bend in the known, 416
Nietzsche, Friedrich
To trace something unknown back to something known is..., 417
Rabi, I.I.
To science the unknown is a problem full of interest..., 417
unseen
Browning, Robert
Greet the unseen with a cheer!, 84
Reeve, F.D.

...we cannot see nine-tenths of what is real..., 84

Uranus

Herschel, William

...expressing my sense of gratitude, by giving the name Georgium Sidus..., 266

vacuum

Bacon, Roger

For vacuum rightly conceived..., 418

Huygens, Christiaan

...he has left an infinite Vacuum..., 418

Venus

Ball, Robert S.

...the splendid lustre of Venus..., 261

Tennyson, Alfred

And the planet of love is on high..., 261

vernal equinox

Cuppy, Will

Among things you might be thinking about today is the vernal equinox..., 419

vernier

Langley, Samuel Pierpoint

That little Vernier, on whose slender lines..., 420

Virgo

Longfellow, Henry Wadsworth

I am the Virgin..., 74

void

Pagels, H.R.

...the most complete void that we can imagine..., 415

wave mechanics

Gamow, George

In wave mechanics there are no impenetrable barriers..., 421

waves

Thomson, G.P.

...if the waves pass over an obstacle..., 421

wisdom

Descartes, Rene

Human wisdom...always remains one and the same..., 422

Milton, John

That which before us lies in daily life/Is the prime wisdom, 422

work

Littlewood, John E.

Most of the best work starts in hopeless muddle..., 423

Lowell, Percival

Gauge your work by its truth to nature..., 423

world

Bruno, Giordana

...an infinity of worlds, 424

Johnson, Lyndon B.

Think of the world as it looks from..., 251

Pascal, Blaise

The whole visible world is only an imperceptible atom..., 424

Regnault, Pére

...the World is round..., 424

Shelley, Percy Bysshe

Worlds on worlds are rolling over..., 425

Tennyson, Alfred

'Tis not too late to seek a newer world, 425

Wordsworth, William

...worlds unthought of till..., 425

write

Emerson, William

...how difficult a thing it is to write well..., 426
writing
Lowell, Percival
...all writing should be a collection of precious stones..., 426
Rossi, Hugo
It is extremely hard for mathematicians to do expository writing, 427
von Braun, Wernher
When a good scientific paper earns a student as much glory as..., 427

AUTHOR BY SUBJECT INDEX

-A-

Abbey, Edward
doctrines, 97
Ackerman, Diane
Earth, 261
Jupiter, 265
Mercury, 260
night, 218
Acton, Loren
stars, 338
Adams, Douglas
observation, 222
physicists, 245
universe, 389
Adams, Henry
physicist, 245
Addison, Joseph
heaven, 144
knowledge, 165
Aeschylus
stars, 338
Aesop
shadow, 313
Agassiz, Louis
nature, 211
Albran, Kehlog
Earth, 261
future, 131
Alighieri, Dante
heaven, 144
knowledge, 165
milky way, 192
stars, 338
Allen, Elizabeth Akers
time travel, 382
Amaldi, Ginestra Giovene
curiosity, 82
night, 218
universe, 389
Amos 8:9
eclipse, 100
Apollo 11
astronaut, 8
Aratus
Capricornus, 71
constellations, 67
Archilochus
eclipse, 100
Archytas of Tarentum
communication, 63
space, 323
Aristotle
infinite, 158
planet, 257
proof, 273
Armstrong, Neil
astronaut, 8
moon landing, 206
Arnold, James R.
space, 330
Arnott, Neil
gravitation, 142
Asimov, Isaac
black hole, 50
unknown, 416
Asphaung, Erik
asteroid, 5
Astbury, W.T.

science, 295
Astronomy Survey Committee
sky, 317
Atiyah, Michael
understanding, 387
Atkins, P.W.
thermodynamics, 376
Atwood, Margaret
moon, 201
night, 218
Aurelius, Marcus
stars, 338
Ayres, C.E.
observation, 222
Aytoun, William
aurora borealis, 39

-B-

Babylonian Inscription
comet, 58
Bachrach, Arthur J.
research, 293
Bacon, Francis
instruments, 161
universe, 389
Bacon, Roger
vacuum, 418
Bailey, Janet
simplicity, 315
Balfour, Arthur James
truths, 383
universe, dying, 410
Ball, Robert S.
Venus, 261
Barnett, Lincoln
time, 378
Barrow, John
past, 236
Barrow, John D.
universe, 408
Barth, John
universe, 390
Bell, J.A.
measurement, 184
Bennett, Arnold
astronomy, 22
Berlinski, David
spacetime, 328
Bernstein, Jeremy
analogy, 3
book, 52
Bethe, Hans
theories, 371
Bichot, Xavier
astronomy, 22
Bierce, Ambrose
gravity, 142
telescope, 366
Bigellow, David
astronomical song, 19
Bilaniuk, O.
compulsory, 64
Black, Hugh
science, 295
Black, Max
truth, 383
Blagonravov, Anatoly A.
space travel, 330
Blaise, Clarke
minds, 195
Blake, William
gravity, 142
infinity, 158
moon, 201
Bloom, Allan
science, 295
Blount, Sir Thomas Pope
universe, 390
Boethius
astronomical computation, 17
Bohm, David
nature, 211
Bohr, Niels
physics, 248
quantum, 277
Bolles, Edmund Blair
observation, 222
Bolyai, John

discovery, 92
Born, Max
models, 198
physics, 248
science, 295
sense organs, 312
Bourdillon, Francis William
sun, 356
Bowyer, Stuart
universe, cosmogenesis, 408
Boyle, Robert
nature, 211
Bradbury, Ray
Mars, 264
Bradley, John Hodgdon
man, 180
Bradley, John Hodgdon Jr
chaos, 57
space, 323
Bragg, Sir William
electron, 104
Brandt, John C.
sky, 317
Bridgman, P.W.
fact, 124
instruments, 161
scientific method, 308
Bridgman, Percy
cosmogony, 76
Bronowski, Jacob
astronomy, 22
physics, 248
Brontë, Charlotte
moon, 201
Brown, Fredric
stars, 339
Browne, Sir Thomas
nature, 211
order, 230
Browning, Robert
heaven, 144
sky, 317
unseen, 84
Bruner, Jerome
discovery, 92
Bruno, Giordano
universe, 390
worlds, 424
Bryant, William Cullen
stars, 339
Buffon, Georges
science, 296
Bunting, Basil
star, 339
Burke, Edmund
stars, 339
Burnet, Thomas
stars, 339
Burnham, Robert Jr
astronomy, 22
heavens, 144
Burns, Marilyn
patterns, 237
Burns, Robert
aurora borealis, 39
change, 56
constellation, 67
Burroughs, William
planet, 257
space travel, 330
Burton, Sir Richard
moon, 201
Bury, J.B.
science, 296
Bush, Vannevar
creation, 81
research, 293
Butler, Samuel
extraterrestrial life, 115
meteor, 188
Butterfield, Herbert
motion, 208
Buzzati-Traverso, Adriano
science, 296
Byron, George Gordon
astrology, 6
comet, 58
universe, dying, 411

-C-

Cabell, James Branch
prove, 273
Caithness, James Balharrie
eclipse, 100
meteor, 188
Calder, Alexander
scientists, 310
Calder, Nigel
astronomers, 10
Calvin, Melvin
physics, 248
Campbell, Norman
science, 296
Campbell, Norman Robert
analogies, 3
Campbell, Thomas
stars, 340
Camus, Albert
science, 296
universe, 390
Card, Orson Scott
God, 136
Cardenal, Ernesto
Saint Augustine Era, 414
thermodynamics, 376
Carlyle, Thomas
Solar System, 321
space, 323
stars, 340
time, 378
Carrel, Alexis
science, 297
Carroll, Lewis
moon, 202
quantum, 277
Carson, Rachel
know, 172
Cayley, Arthur
beauty, 44
Cernan, Gene
astronaut, 9
Chaisson, Eric
cosmological, 77
Chandrasekhar, Subrahmanyan
beauty, 44
simple, 315
Chapman, Clark R.
planets, 257
Chapman, Robert D.
sky, 317
Chargaff, Erwin
astronomy, 23
science, 297, 298
Chaucer, Geoffrey
planets, 257
Chern, Shiing Shen
geometry, 134
Chesterton, G.K.
facts, 124
heaven, 145
Christiansen, Chris
radio astronomy, 283
Church, Peggy Pond
atomic power, 37
Cicero
heaven, 145
Clark, Roger N.
universe, 391
Clarke, Arthur C.
stars, 340
universe, 391
Clerke, Agnes M.
astronomy, 23
discovery, 92
knowledge, 165
Neptune, 266
progress, 271
Cloos, Hans
Earth, 261
Coblentz, Stanton
distance, 96
Cohen, Morris Raphael
observation, 222
Cole, K.C.
metaphors, 187
Cole, Thomas
stars, 340

Cole, William
truth, 383
Coleridge, Samuel Taylor
Earth, 262
moon, 202
stars, 341
Collins, Michael
moon, 202
Collins, Wilkie
beauty, 44
facts, 124
Condon, E.U.
physics, 249
Confucius
knowledge, 165
Conrad, Joseph
astronomy, 23
Conrad, Pete
astronaut, 8
Cook, Peter
universe, 391
Cook, Joseph
astronomical rhyme, 17
Copernicus, Nicolaus
beauty, 45
universe, 391
Crane, H. Richard
neutrino, 217
Crane, Hart
stars, 341
Crowley, Abraham
supernova, 361
Cunningham, Clifford J.
astronomers, 10
Cuppy, Will
vernal equinox, 419
Curie, Marie
discovery, 93

-D-

d'Abro, A.
theory, 371
Dampier, Sir William Cecil
observation, 223
Darling, David
matter, 182
universe, 392
Darwin, Erasmus
Draco, 71
meteors, 188
stars, 341
Davies, Paul
atom, 34
design, 87
Solar System, 321
universe, dying, 411
Davis, Philip
proof, 273
Davis, W.M.
science, 298
Davy, Humphry
universe, 392
Dawkins, Richard
God, 136
universe, 392
Day, Clarence
universe, 392
de Cervantes, Miguel
constellation, 67
references, 428
de Chambaud, J.J. Ménuret
observer, 229
de Fontenelle, Bernard Le Bovier
comet, 58
geometry, 134
universe, 393
de Goncourt, Jules
universe, dying, 411
de Saint-Exupéry, Antoine
alien, 115
dark matter, 84
sky, 317
stars, 341
de Sitter, W.
universe, 393
de Tabley, Lord
star, 341

de Vries, Peter
universe, 393
del Rio, A.M.
science, 298
Democritus
atomism, 38
Dennett, Daniel
learn, 172
philosophy, 241
Descartes, Rene
infinite, 158
wisdom, 422
Deudney, Daniel
space, 324
Deuteronomy 4:19
observer, 229
Deutsch, Armin J.
sun, 356
Deutsch, Karl W.
universe, 393
Devaney, James
meteor, 188
Diamond, Jared
aliens, 115
Dick, Steven J.
extraterrestrial life, 116
Dick, Thomas
comet, 58, 59
Dickerson, Richard E.
thermodynamics, 376
Dickinson, Emily
Arcturus, 70
heaven, 145
Dickinson, Terence
extraterrestrial, 116
Dietrich, Marlene
alien, 117
Dirac, P.A.M.
age, 2
beauty, 45
Dobshansky, Theodosius
science, 298, 299
Donne, John
astronomers, 10
comet, 59
constellations, 68
heaven, 145
milky way, 192
Douglas, A. Vibert
astrophysicist, 32
imagination, 153
problems, 269
Doyle, Sir Arthur Conan
axiom, 43
Drake, Frank
gravitational lens, 141
Dryden, John
sun, 356
du Noüy, Pierre Lecomte
phenomenon, 239
Duhem, Pierre
beauty, 45
physics, 249
progress, 271
theory, 371
Durant, Will
astrology, 6
philosophy, 241
Dürrenmatt, Friedrich
reality, 285
Dyson, Freeman J.
matter, 182
mind, 195
space travel, 330
universe, 393, 394
universe, dying, 411

-E-

Ecclesiastes 11:7
sun, 356
Eco, Umberto
books, 53
Eddington, Sir Arthur Stanley
anti-chance, 55
atom, 34
differential equations, 89
discovery, 93
Earth, 262

extraterrestrial, 117
galaxy, 132
neutrino, 217
observation, 223
positron, 268
relativity, 292
shadow, 313
space, 325
star, 342
time, 378
universe, 394
universe, dying, 411
Edelstein, Ludwig
physics, 249
Egler, Frank E.
instrument, 161
reality, 285
science, 299
Ehrenfest, Paul
physics, 249
Ehrlich, Paul
experiment, 114
Ehrmann, Max
universe, 394
Einstein, Albert
concepts, 65
determined, 88
dimension, 91
exposition, 426
future, 131
gravitation, 143
imagination, 153
mind, 195
nature, 211
photon, 244
physicist, 245
problems, 269
quanta, 277
reality, 285
relativity, 292
science, 299
senses, 312
simplicity, 315
stupidity, 355
theory, 371, 372
Eiseley, Loren
extraterrestrial life, 117
Eisenhart, Churchill
instruments, 161
Eliot, George
age, 2
stars, 342
Eliot, T.S.
universe, dying, 412
Emerson, Ralph Waldo
astrology, 6
astronomy, 23
beauty, 46
formulas, 56, 129
infinite, 159
nature, 212
planets, 258
sky, 318
stars, 342, 343
telescope, 367
universe, 394
Emerson, William
astronomy, 23
write, 426
Empson, William
space, 325
Engard, Charles J.
universe, 395
Epictetus
creation, 81
Erasmus, Desiderius
scientists, 310
Esar, Evan
atom, 34
Estling, Ralph
universe, 395
Eustace, R.
astronomer, 14
nature, 214

-F-

Falconer, William
sun, 356

Faraday, Michael
fact, 124
observation, 223
philosopher, 241
true, 383
Farmer, Phillip José
universe, 395
Ferguson, Kitty
events, 113
Ferris, Timothy
cosmos, 79
symmetry, 363
understanding, 387
Feyerabend, Paul
progress, 272
science, 299
Feynman, Richard P.
atoms, 34
energy, 106
God, 136
gravity, 142
light, 174
mere, 186
model, 198
nature, 212
quantum electrodynamics, 277
science, 300
truth, 384
Fiore, Quentin
time, 379
Firsoff, V.A.
space travel, 331
Fitzgerald, Edward
creation, 81
Flammarion, Camille
eclipse, 100
universe, 395
Flecker, James Elroy
stars, 343
Ford, Kenneth W.
rule of nature, 212
France, Anatole
universe, 396
Frankel, Felice
electrons, 104
light, 175
molecules, 199
order, 230
reality, 286
Frayn, Michael
axioms, 43
notations, 220
universe, 396
Friedman, Herbert
astronomer, 10
sky, 318
Frost, Robert
Canis Major, 71
cosmochemistry, 75
heaven, 68
reality, 286
star shower, 189
universe, dying, 412
Fry, Christopher
moon, 202
Fulbright, J. William
science, 300
Fuller, R. Buckminster
chemistries, 75
extraterrestrial life, 118

-G-

Galilei, Galileo
motion, 208
sunspots, 360
understand, 387
universe, 396
Gamow, George
universe, cosmogenesis, 409
redshift, 290
wave mechanics, 421
Gaposchkin, Sergei
novae, 221
Gardner, Martin
black hole, 50
experiment, 310
patterns, 237
physics, 249

Gauss, Carl Friedrich
knowledge, 165
mechanics, 185
mind, 196
teach, 365
Gay-Lussac, Joseph Louis
observation, 223
physics, 249
Gell-Mann, Murray
quantum theory, 277
Genesis
universe, cosmogenesis, 409
Gibran, Kahlil
astronomer, 11
space, 325
Gilbert, William
paradox, 233
sun, 357
Gilman, Peter A.
sun, 358
Gingerich, Owen
radio astronomy, 283
Giraudoux, Jean
extraterrestrial life, 118
universe, 397
Glass, Bentley
discovery, 93
Glazkov, Yuri
life, 173
Gleason, Andrew
proofs, 273
Gleick, James
motion, 208
particle, 234
Glenn, John Jr
space, 325
Goeppert-Mayer, Maria
physics, 250
Goldhaber, Maurice
theories, 372
Goodenough, Ursula
stars, 343
Goodfield, June
idea, 150
Goodwin, Brian
reality, 286
Gore, George
experiments, 114
knowledge, 166
reason, 288
simplicity, 315
truths, 384
Görtniz, Thomas
cosmology, 77
Goudsmit, Samuel A.
spin, 336
Gould, Stephen Jay
nature, 212
science, 300
Gray, George W.
redshift, 290
scientist, 311
Green, Celia
physicist, 245
Greenberg, J. Mayo
dust, 98
Greene, Brian
infinite, 159
Greenstein, George
data, 85
God, 137
Greenstein, J.L.
astrophysics, 32
Grondal, Florence Armstrong
astronomer, 11
heavens, 145
Sirius, 343
Guiterman, Arthur
Earth, 262
star, 343, 344
Guth, Alan
big bang, 48
universe, 397

-H-

Haber, Heinz
curiosity, 83
Habington, William

stars, 344
Halacy, D.S. Jr
universe, 397
Haldane, J.B.S.
differential equation, 89
simplicity, 316
Hale, George Edward
magnetism, 179
space, 325
Haliburton, T.C.
aurora borealis, 39
Halley, Edmond
astronomers, 11
comet, 59
Halmos, Paul R.
problems, 269
truth, 384
Hammond, Allen L.
planetary science, 258
Hanson, N.R.
physicists, 245
Hardy, G. H.
mathematics, 2
Hardy, Thomas
eclipse, 101
ellipse, 105
star, 344
Harrison, Edward
eternity, 111
infinity, 159
universe, 397, 398
universe, dying, 412
Hasselberg, K.B.
physics, 250
Haught, James A.
universe, 398
Hawking, Stephen
cosmology, 77
equations, 109
God, 137
Hawkins, David
problems, 269
Hawkins, Michael
truth, 384
Hearn, Lafcadio
milky way, 192
stars, 344
Heidegger, M.
physics, 250
Heidmann, Jean
distance, 96
Heiles, Carl
grains, 140
Heine, Heinreich
stars, 345
Heinlein, Robert
analogy, 3
Heisenberg, Werner
observe, 223
particle, 234
physics, 250, 251
quantum mechanics, 278
science, 300
simplicity, 316
understanding, 388
Hellman, C.Doris
ignorance, 152
Hemans, Felicia
Lost Pleiad, 72
Heraclitus
nature, 213
sun, 357
Herbert, George
astronomer, 11
Herrick, Robert
God, 137
stars, 345
Herschel, J.F.W.
nature, 213
super nova, 361
Herschel, William
heavens, 146
imagination, 153
milky way, 193
seeing, 224
Uranus, 266
Hersh, Reuben
proof, 273

Hilbert, David
beauty, 46
future, 131
ignorabimus, 483
problems, 270
Hinshelwood, C.N.
knowledge, 166
universe, 398
Hippocrates
science, 300
Hippolytus
other worlds, 232
Hodgson, Leonard
science, 301
Hodgson, Ralph
star, 345
Hoffer, Eric
moon landing, 206
Hoffman, Jeffrey
meteor, 189
Hoffmann, Roald
thermodynamics, 376
Hogan, John
cosmologist, 77
universe, 398
Holmes, Oliver Wendell
comet, 59
telescope, 367
Holton, Gerald
discovery, 93
equations, 109
laws, 169
nature, 213
observation, 224
science, 301
Homer
constellation, 68
moon, 203
stars, 345
Hooke, Robert
nature, 213
Horowitz, Norman H.
Solar System, 321
Hoyle, Sir Fred
astronomer, 12
astronomical science, 17
big bang, 48
creation, 81
galaxy, 132
laws, 169
radio data, 85
science, 301
stars, 345
uncertain, 386
universe, 398, 399
Hoyt, William Graves
Pluto, 267
Hubble, Edwin Powell
telescope, 367
universe, 399
Huggins, William
spectra, 334
Hugo, Victor
interaction, 164
Huxley, Aldous
science, 301
Huxley, Julian
astronomy, 24
dogmatize, 97
energy, 106
matter, 183
moon, 203
stars, 346
universe, dying, 412
Huxley, Thomas H.
fact, 125
hypothesis, 148
unknown, 416
Huygens, Christiaan
Saturn, 265
vacuum, 418
Huygens, Christianus
extraterrestrial life, 118
universe, 399

-I-

Infeld, Leopold
concepts, 65

reality, 285
theory, 372
Inge, William
philosophy, 241
Ionesco, Eugene
universe, 399
Irwin, James
Earth, 262
Isaiah 40:26
stars, 346

-J-

Jacks, L.P.
facts, 125
Jacobson, Ethel
stars, 346
James, William
knowledge, 166
universe, 399
universe, dying, 413
Jastrow, Robert
God, 137
moon, 203
Jeans, Sir James
astronomer, 12
imaginations, 153
knowledge, 166
life, 173
observation, 224
physics, 251
reality, 286
redshift, 290
Jedicke, Peter
astronomical song, 19
Jeffers, Robinson
astronomer, 12
galaxy, 132
pattern, 237
science, 301
stars, 346
universe, dying, 413
Jenkins, E.B.
depletion, 86
Jevons, W. Stanley
chance, 55
mind, 196
phenomenon, 239
Joad, C.E.M.
universe, 400
Job 38:7
stars, 346
Joel 2:30, 31
eclipse, 101
John of Salisbury
chance, 55
reason, 288
Johnson, George
particle physicists, 234
physicist, 246
Johnson, Lyndon B.
world, 251
Johnson, Severance
astrology, 6
Jones, Sir Harold Spencer
astronomer, 12
extraterrestrial, 118
Jonson, Ben
observation, 224
Joubert, Joseph
logic, 177
space, 325
Jourdain, Philip E.B.
science, 302
Joyce, James
planet, 258
Jung, C.G.
space flights, 331

-K-

Kahn, Fritz
planet, 258
Kant, Immanual
universe, 400
Kapteyn, A.J.
distribution, 99
Kasner, Edward
infinite, 159
Keats, John

Andromeda, 70
astronomer, 13
star, 347
Keill, John
astronomy, 24
Keillor, Garrison
God, 137
universe, 400
Kelvin, Lord
sun, 357
Kepler, Johannes
discoveries, 93
God, 138
heavens, 400
mind, 196
telescope, 367
Keyser, Cassius Jackson
science, 302
Kiepenheuer, Karl
problems, 270
Kilmer, Joyce
milky way, 193
King, Stephen
other worlds, 232
Kingsley, Charles
Andromeda, 70
aurora borealis, 39
Kipling, Rudyard
universe, cosmogenesis, 409
Kirshner, Robert P.
universe, 401
Kline, Morris
order, 230
reasoning, 288
Koch, Howard
alien, 119
Koestler, Arthur
black hole, 50
moon landing, 206
theories, 372
universe, 401
Körner, T.W.
discovery, 94
Koyre, Alexander
physics, 251
Kraus, John
radio, 283
Kremyborg, Alfred
sky, 318
Krisciunas, Kevin
astronomical song, 19
Krough, A.
facts, 125
Krutch, Joseph Wood
stars, 347
Kuhn, Thomas
instruments, 162
science, 303
Kühnert, Franz
astronomers, 13
Kundera, Milan
questions, 281
Kush, Polykarp
physicist, 246

-L-

Lambert, Johann Heinrich
Solar System, 321
Supreme Disposer, 138
Lanczos, Cornelius
differential equations, 89
Land, Edwin
mind, 196
Landau, Lev
man, 180
questions, 281
Langley, Samuel Pierpoint
vernier, 420
LaPlace, Pierre Simon
astronomy, 24
laws, 169
phenomena, 239
Lasota, Jean-Pierre
black holes, 51
Laszlo, E.
theories, 372
Latham, Peter
knowledge, 166

Laurence, William L.
atomic power, 37
Lavoisier, Antoine
instruments, 162
Lawrence, D.H.
cosmos, 79
quantum,278
Lawrence, Louise de Kiriline
nature, 213
Lear, Edward
moon, 203
Lederman, Leon
electron, 104
Lee, Oliver Justin
comets, 60
observation, 224
Lee, Tsung Dao
progress of science, 272
Leibniz, Gottfried Wilhelm
beauty, 46
Leighton, R.B.
energy, 106
light, 175
mere, 186
Levy, David H.
comets, 60
Lewis, Gilbert N.
observation, 225
Libes, Antoine
theories, 373
Lightman, Alan
time, 378
Lightner, Alice
moon, 204
Lindberg, Charles H.
unknown, 416
Lindsay, R.B.
physics, 252
Littlewood, John E.
work, 423
Lockwood, Marion
sky, 318
Lockyer, Joseph Norman
gravity, 143
London, Jack
equation, 109
Long, Roger
astronomy, 24
Longfellow, Henry Wadsworth
Aries, 71
gravity, 143
lady in the moon, 204
Libra, 72
Mars, 264
Sagittarius, 73
Scorpio, 73
stars, 347
sun, 357
Virgo, 74
Lovell, A.C.B.
telescopes, 368
Lowell, Amy
sky, 318
Lowell, Percival
data, 85
discovery, 94
imagination, 154
man, 180
Mars, 264
observatory, 228
science, 303
Solar System, 321
stars, 347
theory, 373
travel, 331
work, 423
writing, 426
Lubbock, Sir John
observation, 225
Lucretius
atomism, 38
space travel, 331
Luminet, Jean-Pierre
astrophysicists, 32
Lyell, Charles
time, 378

-M-

Mach, Ernst
atomic theory, 35
communication, 63
physics, 252
science, 304
universe, 401
Mackay, Charles
astronomer, 13
MacLaurin, Colin
philosophy, 241
MacLeish, Archibald
Earth, 263
spacetime, 328
MacLennan, Hugh
Solar System, 322
Macpherson, James
sun, 357
Maddox, John
big bang, 48
Maffei, Paolo
science, 304
Magnus, Albertus
other worlds, 232
Mandino, Og
stars, 347
Manilus
skies, 318
Mann, Thomas
order, 230
time, 379
Manning, Henry Parker
geometry, 134
Marcet, Jane
cosmochemistry, 75
Margulis, Lynn
science, 304
Maritain, J.
physics, 252
Marlowe, Christopher
planet, 258
Marshak, Robert E.
pion, 256
Maunder, E. Walter
comets, 60
sky, 319
Maxwell, James Clerk
molecule, 199
spectroscope, 334
Mayer, J.R.
fact, 125
McLennan, Evan
nature, 213
McLuhan, Marshall
time, 379
Mead, George H.
reality, 286
Medawar, Peter
scientific methodology, 308
Melville, Herman
analogies, 3
Saturn, 265
universe, 401
Mencken, H.L.
cosmos, 79
Mendeléeff, D.
science, 304
Meredith, George
motion, 208
observation, 225
stars, 74, 347
Meredith, Owen
Southern Cross, 74
Metrodorus of Chios
extraterrestrial life, 119
Meyerson, Emile
energy, 106
Michelson, Albert A.
facts, 125
physicist, 246
Miller, Henry
chaos, 57
Miller, Hugh
planet, 259
Miller, Perry
man, 180
Millikan, Robert A.
science, 305

Milne, A.A.
concept, 51
discover, 94
Milne, Edward
events, 113
Milton, John
astronomer, 13
axial tilt, 42
celestial motion, 54
extraterrestrial, 119
milky way, 193
moon, 204
stars, 348
wisdom, 422
Minkowski, Hermann
spacetime, 328
Misner, Charles W.
beauty, 46
Mitchell, Maria
accurate, 1
astronomers, 13
astronomy, 25
book, 53
changes, 56
formula, 129
imagination, 154
infinitesimal, 159
laws, 170
observations, 225
observatory, 228
popular science, 305
scientist, 311
stars, 349
telescope, 368
Mitton, Simon
radio astronomers, 283
Moleschott, Jakob
force, 128
Moliére (Jean-Baptiste Poquelin)
planet, 259
Moore, James R.
metaphors, 187
Morgan, Frank
geometry, 135
Morgan, Thomas H.
physics, 252
Moulton, Forest Ray
energies, 107
order, 230
sky, 319
Muir, John
nature, 214
sun, 357
universe, 401, 402
Mullaney, James
light, 175
telescope, 368
Mundell, Carole
quasar, 280
Murdin, Paul
astronomy, 25
night, 219
Murray, Bruce
space, 326
space exploration, 331
Musser, George
nature, 214

-N-

Nabokov, Vladimir
atom, 35
moon landing, 207
Neal, Patricia
communication, 63
Newcomb, Simon
astronomy, 25
Newell, Homer E.
moon, 203
Newman, James R.
infinite, 159
physicists, 246
Newman, William I.
extraterrestrial life, 121
Newton, Sir Isaac
gravity, 143
light, 175
Nicholson, Norman

universe, dying, 413
unknown, 416
Nietzsche, Friedrich
sun, 358
unknown, 417
Noll, Walter
physics, 254
Noyes, Alfred
constellation, 68

-O-

Ockels, Wubbo
space, 326
Oliver, David
mechanics, 185
Oparin, A.I.
extraterrestial life, 119
Oppenehimer, J. Robert
man of science, 305
Orgel, Irene
God, 138
Orwell, George
observation, 225
Osiander, Andrew
astronomer, 14
hypotheses, 148
Osler, Sir William
observation, 225, 226
Ovenden, M.W.
universe, 402
Ovid
change, 56

-P-

Page, Leigh
relativity, 292
Pagels, Heinz R.
laws, 170
life, 173
stars, 349
understand, 388
void, 415
Paine, Thomas
eternity, 111
Pallister, William
extraterrestrial life, 119
Saturn, 265
Panek, Richard
telescope, 368
Paracelsus
astronomical, 17
Parker, E.N.
magnetic field, 179
sun, 358
Pascal, Blaise
sun, 358
universe, 402
world, 424
Pauli, Wolfgang
fusion, 130
Payne-Goposchkin, Cecila H.
question, 281
Peacock, Thomas Love
science, 306
Peltier, Leslie C.
comets, 60
telescopes, 368
Penman, Sheldon
muon, 210
Penrose, Roger
astronomy, 25
cosmology, 77
Penzias, Arno
astronomy, 25
Peterson, Ivars
patterns, 237
Petrarch
nature, 214
Pines, David
physics, 253
Pippard, A.B.
thermodynamics, 376
Pittendreigh, W. Maynard Jr
curiosity, 83
Planck, Max
axioms, 43
formula, 129

Plato
astronomy, 26
cosmos, 79
eclipse, 101
stars, 349
universe, 402
Plum, David
meteor, 189
Poe, Edgar Alan
big bang, 49
stars, 349
universe, 402
Poincaré, Henri
arbitrary, 4
beauty, 46
eclipse, 102
facts, 126
milky way, 193
simplicity, 316
Poincaré, Lucien
future, 131
progress, 272
Poinsot, Louis
time, 379
Polanyi, Michael
universe, 403
Polkinghorne, J.C.
quantum theory, 278
Polyakov, Alexander
God, 138
Pontecorvo, Bruno
neutrino, 217
Pope, Alexander
extraterrestrial, 120
Popper, Karl R.
cosmology, 77
determinism, 88
questions, 281
science, 306
theory, 374
Pratchett, Terry
ignorance, 152
light, 176
Ptolemy
stars, 349

-R-

Rabi, Isidor Isaac
knowledge, 167
physics, 253
scientific spirit, 308
understanding, 388
unknown, 417
Raether, H.
philosophy, 242
Ramón y Cajal, Santiago
universe, 403
Rankine, William John Macquorn
measuring, 184
Raymo, Chet
reality, 287
stars, 350
Reade, Winwood
God, 139
space, 326
universe, 403
Rees, Martin
astronomers, 14
Reeve, F.D.
real, 287
unseen, 84
Reeves, Hubert
cosmos, 80
matter, 183
universe, cosmogenesis, 409
Regnault, Pére
loadstone, 424
motion, 209
particle, 234
teach, 365
world, 424
Reichenbach, Hans
universe, 403
Reid, Thomas
simplicity, 316
Reiss, H.
thermodynamics, 377

Renan, Ernest
scientific honesty, 308
Renard, Maurice
universe, 403
Revelation 12:3–4
meteor, 189
Revelation 9:1–2
meteor, 189
Rexroth, Kenneth
logic, 177
Rich, Adrienne
milky way, 193
Richards, Rheodore William
universe, 403
Roller, Duane H.D.
laws, 109
observation, 224
science, 301
Rorty, Richard
physicists, 246
Rosseland, S.
observatory, 228
Rossi, Hugo
writing, 427
Rothman, Tony
theory, 374
universe, 404
Rowan-Robinson, Michael
telescope, 368
Rowland, Henry Augustus
atom, 35
laws, 171
study, 354
Rukeyser, Muriel
atoms, 35
Russell, Bertrand
data, 85
design, 87
geometry, 135
knowledge, 167
laws, 171
science, 306
space, 326
universe, 404
universe, dying, 414
Rutherford, Ernest
experiment, 114
physics, 253
scattering, 294
Ryder-Smith, Roland
telescope, 369

-S-

Saaty, Thomas L.
equations, 109
Sagan, Carl
astronomy, 26
constellation, 68
Earth, 263
experiment, 114
extraterrestrial life, 120, 121
ideas, 150
UFO, 385
Sagan, Dorian
science, 304
Saint Augustine
Saint Augustine Era, 415
time, 379
Sakharov, Andrei
extraterrestrial, 121
Sands, M.
energy, 106
light, 175
mere, 186
Santayana, George
cosmos, 80
universe, 404
Sayers, Dorothy L.
astronomer, 14
nature, 214
Schaaf, Fred
supernova, 362
Schaefer, Bradley E.
sky, 319
Schwarzschild, Martin
laws, 171
Scott, Dave

astronaut, 8
Scott, Robert F.
aurora borealis, 40
Scott, Sir Walter
aurora borealis, 40
Seab, C.G.
grains, 140
Seneca
comet, 60, 61
heavens, 146
nature, 214
Service, Robert W.
aurorea borealis, 40
stars, 350
Shakespeare, William
astrology, 7
astronomers, 15
comet, 61
eclipse, 102
extraterrestrial life, 122
logic, 177
meteor, 190
observers, 226
reason, 289
sky, 319
stars, 350
time, 379, 380
Shapley, Harlow
astronomy, 26
cosmos, 80
mankind, 180
planetry, 259
Shaw, George Bernard
exterterrestrial life, 122
facts, 126
Shelley, Percy Bysshe
heaven, 146
moon, 204, 205
universe, 404
worlds, 425
Shepherd, Alan
astronaut, 8
Sherrod, P. Clay
astronomy, 26
Shore, Jane
sky, 319
Siegel, Eli
planets, 259
space, 326
universe, 405
Silesius, Angelus
time, 380
Sillman, Benjamin
astronomy, 27
Simes, James
heavens, 147
Singer, Isaac Bashevis
universe, cosmogenesis, 409
Skinner, B.F.
study, 354
Slater, John C.
theory, 374
Smith, Logan Pearsall
space, 326
Smith, Theobald
knowledge, 168
observation, 226
Smolin, Lee
science, 306
Smoot, George
sky, 319
Smythe, Daniel
meteor, 190
stars, 350
Snow, C.P.
fact, 126
Sobel, Dava
gravitational lens, 141
Soddy, Frederick
energy, 107
science, 306
Spenser, Edmund
astrophysicist, 32
stars, 351
universe, 405
universe, cosmogenesis, 409
Standage, Tom
planet, 259

Standen, Anthony
physics, 253
Stapledon, Olaf
red shift, 291
Starr, Victor P.
sun, 358
Stedman, Edmund
spectral, 334
Steensen, Niels
beautiful, 47
Steinbeck, John
hypothesis, 148
man, 181
Steinhardt, Paul
universe, 397
Stenger, Victor J.
neutrino, 217
Stern, S. Alan
universe, 405
Sterne, Laurence
hypothesis, 148
Stevenson, Robert Louis
night, 219
Stoll, Clifford
astronomer, 15
Stoppard, Tom
formula, 129
teach, 365
Struve, Otto
astronomy, 27
Sturluson, Snorri
universe, cosmogenesis, 410
Sudarshan, E.C.
compulsory, 64
Sullivan, Arthur
paradox, 233
sun, 357
Sullivan, J.W.N.
electron, 104
Swann, W.F.G.
universe, 405
Swenson, May
universe, 406
Swift, Jonathan
sun, 358
Swigert, Jack
astronaut, 8
Swinburne, Richard
time, 380
Swings, Pol
sky, 319
Synge, J.L.
theory, 375

-T-

Tabb, John Banister
Pleiads, 73
Tagore, Rabindranath
universe, cosmogenesis, 410
Taylor, Bayard
aurora borealis, 41
star, 351
Taylor, Edwin F.
spacetime, 328
Teasdale, Sara
Arcturus, 70, 346
meteor, 190
Orion, 72
stars, 351
Teilhard de Chardin, Pierre
Earth, 263
physics, 253
planets, 259
truth, 384
Tennyson, Lord Alfred
future, 131
knowledge, 168
meteor, 190
moon, 205
nature, 215
Orion, 72
Pleiads, 73
Sirius, 351
universe, 406
Venus, 261
world, 425
Thayer, John H.
Saturn, 266

Thom, Rene
universe, 406
Thomas, Dylan
light, 176
Thomas, Lewis
Earth, 264
Thompson, Francis
astronomer, 15
star, 351
universe, 406
Thompson, W.R.
observation, 226
Thomson, G.P.
waves, 421
Thomson, J. Arthur
God, 139
Thomson, J.J.
discovery, 94
Thomson, James
comet, 61
Thomson, Sir George
method of science, 307
Thoreau, Henry David
discovery, 94
imagination, 155
knowledge, 168
milky way, 194
nature, 215
space travel, 332
stars, 351, 352
sun, 359
universe, 406
Thorne, Kip S.
beauty, 46
black hole, 51
spacetime, 329
Ting, Samuel C.C.
scientists, 311
Tolman, R.C.
cosmology, 78
Tolstoy, Leo
comet, 61
Tombaugh, Clyde
imagination, 155
planet, 260
Tomlinson, C.
cosmos, 80
Toogood, Hector B.
telescope, 369
Toulmin, Stephen
physicists, 247
tools, 150
Townes, Charles H.
universe, cosmogenesis, 410
Toynbee, Arnold
universe, 406
Travers, P.L.
stars, 352
Trevelyan, G.M.
stars, 352
Truesdell, Clifford
physics, 254
thermodynamics, 377
Tsiolkovsky, Konstantin Eduardovich
exterterrestrial life, 122
space travel, 332
study, 354
telescope, 369
Tucker, Abraham
idea, 150
Turok, Neil G.
cosmology, 78
Twain, Mark
astronomer, 15, 16
astronomy, 27
constellation, 69
God, 139
spectrum analysis, 335
stars, 352
Tyndall, John
facts, 126
Tyron, E.P.
universe, 407
Tzu, Lao First philosopher of Chinese Taoism
Saint Augustine Era, 415
universe, 407

-U-

Unknown
astronomers, 16
astronomer's drinking song, 20
astronomy, 27
atom, 35
discovery, 95
eclipse, 103
energy, 108
laws, 171
light, 176
logic, 177
meteorite, 190
momentum, 200
observation, 226
photons, 244
physicist, 247
physics, 254
problem, 270
proofs, 274
scientist, 311
space travel, 332
stars, 352, 353
thermodynamics, 377
time, 380
uncertainty, 386
universe, 407
universe, cosmogenesis, 410
Unsold, Albrecht
radio astronomy, 284
Updike, John
matter, 183
philosophy, 242

-V-

Valéry, Paul
space, 327
van de Hulst, H.C.
spiral arms, 337
van de Kamp, Peter
errors, 110
vas Dias, Robert
space, 327
Vaughan, Henry
eternity, 111
stars, 353
Veblen, Thorstein
research, 293
Vehrenberg, Hans
telescope shelter, 369
Velikovsky, Immanuel
imagination, 155
Verne, Jules
moon, 205
scientific discoverers, 309
space travel, 332
Vezzoli, Dante
telescope, 369
Virgil
astronomy, 28
meteors, 191
Vizinczey, Stephen
Earth, 264
von Braun, Wernher
extraterrestrial life, 122
God, 139
impossible, 157
space, 327
space travel, 332
writing, 427
von Goethe, Johann Wolfgang
instrument, 162
observers, 226
phenomena, 375
physics, 254
reason, 289
theorize, 375
von Haller, Albrecht
infinity, 159
von Humboldt, Alexander
comets, 61
von Lenard, Philipp E.A.
discovery, 95
von Schelling, F.W.J.
nature, 215
von Weiszaecker, Karl Friedrich
mystery, 255

-W-

Walcott, Derek
astronomer, 16
Wald, George
life, 174
Walker, Kenneth
understanding, 388
Weaver, Tom
supernova, 362
Weaver, Warren
concepts, 65
Webster, John
stars, 353
Weil, Simone
forces, 128
time, 380
Weinberg, Steven
physics, 254
real, 287
universe, 407
Weisskopf, Victor F.
curiosity, 83
model, 198
Wells, H.G.
extraterrestrial life, 121, 123
scientific truth, 309
space travel, 333
sun, 359
time, 380
universe, dying, 414
Weyl, Herman
laws, 171
matter, 183
reality, 287
space, 327
symmetry, 363
Wheeler, John Archibald
beauty, 46
imagination, 155
simplicity, 316
spacetime, 328, 329
theory, 375
universe, 407
Whewell, William
facts, 126, 151
hypothesis, 149
knowledge, 168
Whipple, Fred L.
Earth, 264
Whitehead, Alfred North
differential equations, 89
errors, 110
ideas, 151
infinite, 159
instruments, 163
nature, 216
philosophy, 242
progress of science, 272
science, 307
Whitehead, Hal
astronomy, 28
Whitesides, George M.
electrons, 104
light, 175
molecules, 199
order, 230
reality, 286
Whitman, Walt
constellations, 70
force, 128
nature, 216
past, 236
sky, 320
stars, 353
universes, 156, 408
Whitrow, G.J.
imagination, 156
Wickham, Anna
symmetry, 363
Wiechert, Emil
universe, 408
Wigner, Eugene P.
physics, 255
Wilde, Oscar
aurora borealis, 41
reality, 287
Wilson, Edward O.
phenomena, 239

Wisdom, J.O.
theory, 375
Wolf, Fred Alan
galaxies, 133
infinity, 160
Woosley, Stan
supernova, 362
Wordsworth, William
eclipse, 103
nature, 216
stars, 353
telescope, 370
worlds, 425
Wright, Helen
curiosity, 83
Wright, Wilbur
error, 110

-X-

Xenophanes
sun, 359

-Y-

Yang, Chen Ning
order, 230
symmetry, 363
Yeats, William Butler
universe, dying, 414
Young, Edward
eternity, 111
Young, J.Z.
scientists, 311
Young, John
astronaut, 8
Young, Louise B.
universe, 408
Yudowitch, K.L.
age, 2

-Z-

Zebrowski, George
infinities, 160
Zee, A.
quantum, 279
Ziman, John M.
God, 139
philosophers of science, 242
Zirin, Harold
magnetic, 179
sunspots, 360
Zukav, Gary
physics, 255